Principles of Animal Biometeorology

BIOMETEOROLOGY

Volume 2

For further volumes:
http://www.springer.com/series/8101

Roberto Gomes da Silva
Alex Sandro Campos Maia

Principles of Animal Biometeorology

Springer

Roberto Gomes da Silva
Department of Animal Sciences
Universidade Federal Rural do Semi-Árido
Mossoró, RN, Brazil

Alex Sandro Campos Maia
Department of Animal Science
College of Agricultural and Veterinary Sciences (FCAV)
Universidade Estadual Paulista
Jaboticabal, SP, Brazil

ISBN 978-94-007-5732-5 ISBN 978-94-007-5733-2 (eBook)
DOI 10.1007/978-94-007-5733-2
Springer Dordrecht Heidelberg New York London

Library of Congress Control Number: 2012951127

Printed on acid-free paper

Springer is part of Springer Science+Business Media (www.springer.com)

Foreword

Biometeorology is a relatively new scientific area, with three principal subdivisions (human, plant, and animal) being recognized since the International Society of Biometeorology (ISB) was established in 1956. However, the basis for recognition as an area worthy of investigation can be traced back for much longer periods of time. For animal biometeorology in particular, the lineage includes studies of animal calorimetry and bioenergetics as long ago as the eighteenth century (Lavoisier), with more intensive investigations in the early twentieth century by subsequent researchers, resulting in classic references by Samuel Brody (*Bioenergetics and Growth*, 1945) and Max Kleiber (*The Fire of Life*, 1961, revised 1975).

Later in the mid-twentieth century, the focus of animal biometeorology expanded from the basic studies of bioenergetics to the application of the results for improved animal performance, health, and well-being in production systems. Animal scientists, engineers, and veterinarians collaborated in this effort, with numerous research reports and several books published. Notable among the books are those by Esmay (*Principles of Animal Environment*, 1969), Moberg (*Animal Stress*, 1985), Sainsbury and Sainsbury (*Livestock Health and Housing*, 1979), Clark (*Environmental Aspects of Housing for Animal Production*, 1981), Yousef (*Stress Physiology in Livestock,* three volumes, 1985), and DeShazer (*Livestock Energetics and Thermal Environmental Management*, 2009). Additionally, starting in 1974, there have been a series of International Livestock Environment Symposia (organized and proceedings published by the American Society of Agricultural and Biological Engineers) which have provided a forum for discussion and integration of research information for improving animal management in challenging environments. The Congresses of the International Society of Biometeorology, held every 3 years, have furthered that effort.

The authors of the present book, *Principles of Animal Biometeorology*, have extensive research experiences in the field. The material included provides another step in integration of current knowledge about biometeorological principles in assessing the impact of environments on animals of various types. Emphasis is on the physical aspects of heat transfer and heat exchanges in thermal environments (temperature, humidity, wind, and radiation), but with a view toward thermoregulatory responses of the animals. The book should serve as a useful resource for those interested in animal biometeorology and application of such information for understanding and improving animal environments.

G. LeRoy Hahn (Agricultural Engineer and Biometeorologist, retired from the U.S. Meat Animal Research Service, Agricultural Research Service, U.S. Department of Agriculture, Clay Center, Nebraska)

Contents

List of Symbols

Greek Alphabet

α		Absorptance of surface for radiation
α_s		Absorptance of surface for solar radiation
β	degrees	Solar elevation
δ	degrees	Solar declination
ε		Emissivity of surface
ε_{ac}		Apparent emissivity of clear sky
$\mathbf{\Phi}$	$W\ m^{-2}$	Energy flux density
Φ_{cap}	$L\ s^{-1}$	Air flux through a capsule
γ	C^{-1}	Psychrometric constant
$\mathbf{\Gamma}$	$W\ m^{-2}$	Storage of thermal energy
η	degrees	Hour angle of sun
λ	$J\ g^{-1}$	Latent heat of vaporisation of water
λ	μm	Wavelength of electromagnetic radiation
μ	$g\ m^{-1}\ s^{-1}$	Dynamic viscosity of air
v	$m^2\ s^{-1}$	Kinematic viscosity of air
θ	degrees	Solar zenith angle
ρ	$g\ m^{-3}$	Density of air
ρ		Reflectance of surface
ρ_g		Reflectance of ground surface
σ	$W\ m^{-2}\ K^{-4}$	Stefan-Boltzmann constant $= 5.67 \times 10^{-8}\ W\ m^{-2}\ K^{-4}$
τ		Transmittance of the air or a surface
Ψ	$g\ m^{-3}$	Absolute humidity of air
ω	degrees	Solar azimuth angle
	degrees h^{-1}	Angular speed of Earth

Latin Alphabet

A	m^2	Surface area
$\mathbf{C}$	$W\,m^{-2}$	Convective heat flux density
$\mathbf{C}_S$	$W\,m^{-2}$	Convective heat flux density at body surface
$\mathbf{C}_R$	$W\,m^{-2}$	Convective heat flux density at respiratory surfaces
c_0	$m\,s^{-1}$	Light speed = 2.99792458×10^8
c		Cloud-type coefficient
c_p	$J\,g^{-1}\,K^{-1}$	Specific heat of air at constant pressure
c_v	$J\,g^{-1}\,K^{-1}$	Specific heat of air at constant volume
c_{pt}	$J\,g^{-1}\,K^{-1}$	Specific heat of body tissues
c_{pv}	$J\,g^{-1}\,K^{-1}$	Specific heat of water vapour
d	m	Characteristic dimension of surface
d		Ordinal number of the year day (1–365)
D	$m^2\,s^{-1}$	Thermal diffusivity coefficient
D_a	$m^2\,s^{-1}$	Thermal diffusivity coefficient of air
D_v	$m^2\,s^{-1}$	Thermal diffusivity coefficient of water vapour in air
e		Exponential number = 2.7182818284...
$\mathbf{E}$	$W\,m^{-2}$	Evaporative heat flux density
$\mathbf{E}_R$	$W\,m^{-2}$	Evaporative heat flux density at respiratory surfaces
$\mathbf{E}_S$	$W\,m^{-2}$	Evaporative heat flux density at body surface
F_c		Shape factor of surface for radiation
F_r	breaths min^{-1}	Respiratory rate
$\mathbf{G}$	$W\,m^{-2}$	Irradiance of surface
G_r		Grashof number
g	$m\,s^{-2}$	Gravitational constant
H	$kJ\,kg^{-1}$	Enthalpy of air
h		Hour
h	J s	Planck's constant = 6.626076×10^{-34}
h_c	$W\,m^{-2}\,K^{-1}$	Convection coefficient
I_c	$W^{-1}\,m^2\,°C$	Thermal insulation of hair coat
I_t	$W^{-1}\,m^2\,°C$	Thermal insulation of tissues
$\mathbf{J}$	$W\,m^{-2}$	Radiosity of surface
k	$W\,m^{-1}\,K^{-1}$	Thermal conductivity
k	$J\,K^{-1}$	Boltzmann constant = 1.380658×10^{-23}
K	$W\,m^{-2}\,K^{-1}$	Thermal conductance of body
$\mathbf{L}$	$W\,m^{-2}$	Long-wave irradiance of surface
LCT	°C	Lower critical temperature
$\mathbf{L}_D$	$W\,m^{-2}$	Long-wave irradiance, downwards
$\mathbf{L}_U$	$W\,m^{-2}$	Long-wave irradiance, upwards
L_t	degrees	Latitude

L_g	degrees	Longitude
M	W m^{-2}	Metabolic rate
M	g mol^{-1}	Molecular mass
M_w	g mol^{-1}	Molecular mass of water = 18,016 g mol^{-1}
m		Air mass number
$\dot{m}$	kg s^{-1}	Mass flux
N_u	nondimensional	Nusselt number
n		Proportion of sky cloudiness
P_r	nondimensional	Prandtl number
P_a	kPa	Atmospheric pressure
$P_s(T)$	kPa	Saturation vapour pressure at temperature T
P_v	kPa	Partial air vapour pressure
R	W m^{-2}	Thermal flux by radiation
		Emissive power of a surface
		Thermal energy exchange by radiation
R_e	nondimensional	Reynolds number
R	J mol^{-1} K^{-1}	Universal gas constant = 8.3143 J mol^{-1} K^{-1}
R_a	J mol^{-1} K^{-1}	Gas constant of air =287.04 J mol^{-1} K^{-1}
r	m	Geometric radius
r		Correlation coefficient
r	s m^{-1}	Thermal resistance
r_t	s m^{-1}	Thermal resistance of tissues
r_H	s m^{-1}	Thermal resistance for convection
r_K	s m^{-1}	Thermal resistance for conduction
r_R	s m^{-1}	Thermal resistance for radiation
r_V	s m^{-1}	Thermal resistance for evaporation
S	W m^{-2}	Short-wave irradiance of surface
S_c	nondimensional	Schmidt number
S_h	nondimensional	Sherwood number
t	s	Time
t		Atmospheric turbidity coefficient
T	°C, K	Temperature
T_a	°C, K	Air temperature, dry bulb temperature
T_b	°C, K	Mean body temperature
T_c	°C, K	Body core temperature
T_{dp}	°C, K	Dew point temperature
T_{exp}	°C	Temperature of expired air
T_g	°C, K	Black globe temperature
T_o	°C, K	Operative temperature
T_{oe}	°C, K	Equivalent operative temperature
T_r	°C	Rectal temperature

T_{rm}	°C, K	Mean radiant temperature
T_{rm}^{*}	°C, K	Effective mean radiant temperature
T_s	°C, K	Surface temperature
T_w	°C, K	Wet bulb temperature
U	m s^{-1}	Wind speed
UCT	°C	Upper critical temperature
U_R	%	Air relative humidity
V	m^3	Volume
V_T	m^3 $breath^{-1}$	Tidal volume
W	kg	Body weight
z	m	Altitude

Tables

Table A1 Système International (SI) units

Quantity	Name	Symbol	Definition c.g.s.
Concentration of substance	Mole	mol	
Electric current	Ampère	A	
Electric potential difference	Volt	V	
Linear dimension	Metre	m	
Energy	Joule	J	kg m^2 s^{-2}
Force	Newton	N	kg m s^{-2}
Frequency	Hertz	Hz	Oscillations s^{-1}
Mass	Kilogram	kg	
	Gram	g	
Power	Watt	W	J s^{-1}
Pressure	Pascal	Pa	kg m^{-1} s^{-2}
Temperature	Kelvin	K	
	Celsius	°C	
Time	Second	s	
Velocity			m s^{-1}
Volume	Litre	L	10^{-3} m^3

Table A2 Multiples and submultiples of SI units

Multiplier	Prefix	Symbol
10^{12} = 1,000,000,000,000	Tera	T
10^{9} = 1,000,000,000	Giga	G
10^{6} = 1,000,000	Mega	M
10^{3} = 1,000	Kilo	k
10^{2} = 100	Hecto	h
10^{1} = 10	Deca	da
10^{-1} = 0.1	Deci	d
10^{-2} = 0.01	Centi	c
10^{-3} = 0.001	Milli	m
10^{-6} = 0.000001	Micro	μ
10^{-9} = 0.000000001	Nano	n
10^{-12} = 0.000000000001	Pico	p

Table A3 Equivalence of some units

Quantity	Unit	Equivalence
Linear dimension	Inch (″, in)	0.0254 m
	Foot = 12 in. (ft)	0.3048 m
	Yard (yd)	0.9144 m
	Mile (mi)	1,609.34 m
	Mile, nautical (mi)	1,852 m
Area	Square inches (sq.in)	0.000645 m^2
	Square foot (sq.ft)	0.092903 m^2
	Acre	4,047 m^2
	Hectare	10,000 m^2
Mass	Pound (lb)	0.453592 kg
Volume	Litre (L)	1,000 cm^3
	Gallon, British (ga)	4.546 L
	Gallon, American (ga)	3.785 L
Temperature	Celsius (°C)	K − 273.15
	Kelvin (K)	°C + 273.15
	Fahrenheit (F)	(9/5)°C + 32
	Rankin (R)	F + 460
Pressure	Kilopascal (kPa)	1.000 Pa
	Millibar (mb, mbar)	0.1 kPa
	Millimetres of mercury (mmHg)	0.1333224 kPa
Thermal energy	British thermal unit (BTU)	1,055.1 J
	Calorie (cal)	4.186938 J
	Watt (W)	3.412806 BTU h^{-1}
	Watt (W)	0.238838 cal s^{-1}
	cal $(cm^2\ min)^{-1}$	697.823071 W m^{-2}
	BTU $(h\ ft^2)^{-1}$	3.154723 W m^{-2}
	W m^{-2}	0.316985 BTU $(h\ ft^2)^{-1}$
	Mcal day^{-1}	48.459931 W

List of Figures

List of Tables

Chapter 1
The Environment

Abstract In this chapter, the following are presented: environmental factors and definition of tropical climate, definitions and discussion about radiation environment (solar constant, solar radiation variation with latitude and season), equations for determination of sun angles (zenith, elevation, declination and hour angles of sun), aspects of the short-wave radiation at the ground level, atmospheric transmittance, diffuse and reflected radiation, influence of clouds, reflectance of ground surfaces and estimation of total short-wave radiation. The determination of downward and upward long-wave radiation is discussed, along with the respective formulas. The concepts of radiant heat load and the mean radiant temperature are discussed, and the respective methods of determination are presented. The theory of black globe is presented, and its practical use is discussed. The structure of the atmospheric layers, together with the gas components of the atmosphere, is described. The importance of CO_2, ozone and methane is discussed, together with that of aerosol pollutants; the evolution of atmospheric CO_2 is also discussed. The equations to determine the physical properties of the air are presented: atmospheric pressure, specific heat, density, thermal conductivity, viscosity, latent heat of vaporisation, psychrometric constant and diffusivity of vapour; the methods of air humidity specification are presented: saturation and partial vapour pressure, relative humidity, absolute humidity and dew point temperature.

Keywords Atmospheric layers • Atmospheric properties • Carbon dioxide • Long wave • Ozone • Radiant heat load • Short wave • Solar radiation • Tropical climate

1.1 Environmental Factors

1.1.1 Domains of the Environment

Physical environment is constituted by four domains or spheres which are not superimposed one over another but exchange energy among them: *atmosphere,*

R. Gomes da Silva and A.S. Campos Maia, *Principles of Animal Biometeorology*,
Biometeorology 2, DOI 10.1007/978-94-007-5733-2_1,

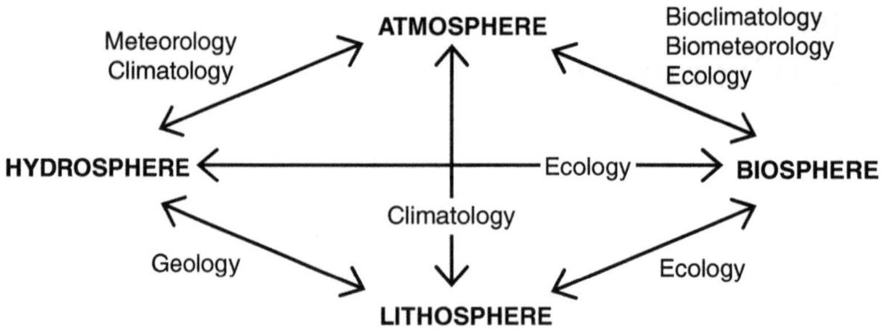

Fig. 1.1 Relationships among the four domains of the physical environment and the sciences which are associated to them

lithosphere, hydrosphere and biosphere. The first three include the air, the land and the waters, respectively, while the last is the universe of all live beings of the Earth. Between those four domains, there are close relationships and mutual interactions (illustrated in Fig. 1.1), which are of uttermost importance for sustaining life on the Earth.

The term environment includes all sources of energy, especially the radiant energy used for the plant photosynthesis that is stored as nutrients (proteins, carbohydrates, lipids). Such a material is the prime source of energy for other terrestrial and aquatic forms of life. All the live organisms in the biosphere are constituted by the water from hydrosphere; by the nitrogen, oxygen and other gases from the atmosphere; and by the minerals from the lithosphere.

Environmental factors as temperature and photoperiod determine growth rates of plants and affect reproduction of both animals and plants; besides, they control food availability and searching for animals. Gravity and light stimulate animals and plants and constitute references for them in terms of space and time, involving important processes as the sense of equilibrium and the biological clocks. Climatic and geological factors, among others, affect the spreading and viability of pathogenic agents and parasites which attack the organisms.

1.1.2 Weather and Climate

Among the environmental factors, variation of the atmospheric conditions is one of the most important effects upon the biosphere. In order to understand that variation, we must refer to two different terms: *weather* and *climate*.

Weather is the instantaneous state of the atmosphere and its study is the object of meteorology. The term "instantaneous" means here different periods of time, but in general, it is a 24-h period. Then, meteorological reports refer to the weather giving, as, for example, maximum and minimum air temperatures for a given location, or mentioning the occurrence of rain or winds in that period.

On the other hand, climate is the average weather conditions prevailing in a given location; it is established in the long range, after 10–30 years of continuous meteorological observations. Climatology is the science of the climate and has similar content to that of meteorology; however, they differ one from another with respect to the methodology. Climatologists use mainly statistical and cartographic techniques, while meteorologists use the laws of physics and mathematical techniques. Another difference is that the main concern of meteorology is the development and dynamics of the atmospheric phenomena, while that of climatology is the consequence of those phenomena.

Biometeorology is just a specialised field of meteorology that refers to the relationships between atmosphere and biosphere, which are aimed also by ecology and bioclimatology.

1.1.3 Definition of Tropical Climate

The word "tropical" refers to that zone of the Earth lying between the tropics of Cancer and Capricorn, which are, respectively, the 23°27′ north and south parallels. It is a belt that includes all the regions where the sun can be in some time at zenith. According to Ayoade (1983), the following definitions of a tropical region are valid:

(a) A region without winter or cold season
(b) A region where the annual average temperature is equal to or less than the average daily variation
(c) A region where the average temperature at sea level never is less than 18°C

It is generally considered that, despite the so-called tropical zone remains between the two 23°27′ parallels, these limits are not adequate because they are very rigid ones. In fact, some regions that are clearly nontropical can be found close to the equatorial line, such as parts of the Andes mountains in South America, the region of Mount Kilimanjaro in Kenya and some regions in Mexico, Central America and New Guinea (Nieuwolt 1978).

The use of temperature limits to demonstrate absence of winter in a tropical region seems also to be a misunderstanding; for example, considering an average temperature of 18°C for the coldest month can leave to the exclusion of the tablelands and highlands in tropical regions, where air temperature often falls beneath such a limit. Until some years ago, climates were described by meteorological standards, and several climatic classifications were used, as that of Köppen. However, such classifications were purely descriptive, leaving often too erroneous conceptions of the climatic variations in terms of biological significance, because of the used methodology – which was unable to explain the processes involved in the climatic variations.

In fact, interactions of the atmosphere with the other physical domains are dynamic and not static, involving continuous changes in the conditions of the ground surface, water distribution and alterations of the biosphere – especially the vegetation.

Therefore, climatic zoning and mapping could sometimes leave to a wrong idea about a given situation, because changing from a climate type to another is gradual. On the other hand, the atmospheric characteristics of a given region can be well understood only when they are considered within the environment as a whole.

Modern climatology concerns with long- or medium-term alterations of the climatic conditions, involving a number of factors – for example, solar radiation, environmental pollution and forest depletion. One of its basic instruments is the remote sensing by means of specialised satellites.

Then, in the present study, we do not consider any climatic classification, using rather a more practical division into *tropical*, *temperate* and *cold* climates. The term tropical is used to indicate a region located within the two tropics. The word "subtropical" has been extensively used to name a region with climatic conditions intermediary between tropical and temperate; it must be avoided, as showed by Nieuwolt (1978).

1.2 Radiation Environment

1.2.1 Short-Wave Solar Radiation

1.2.1.1 Solar Constant

All the energy at the surface of the Earth comes from the sun to be used for the biological and physical processes. Analyses of the solar spectrum show that the sun behaves almost as a perfect radiator, whose surface has an apparent temperature of $T_S = 5{,}755$ K.

Considering the Stefan-Boltzmann law, the radiation flux reaching the Earth outside atmosphere can be given as

$$\mathbf{S}_0 = \sigma T_s^4 r_s^2 r_t^{-2} \ \text{W m}^{-2} \tag{1.1}$$

where $r_s \cong 0.7 \times 10^6$ km is the radius of the sun, $r_t = 149 \times 10^6$ km the mean radius of Earth's orbit and σ is the Stefan-Boltzmann constant. Then,

$$\begin{aligned}\mathbf{S}_0 &= 5.67 \times 10^{-8}(5{,}755)^4 \left(0.7 \times 10^6\right)^2 \left(149 \times 10^6\right)^{-2} \\ &= 1{,}372.7 \ \text{W m}^{-2}\end{aligned}$$

which is the *solar constant*. However, it is not really a constant, since the flux of solar energy reaching the Earth varies according to the level of sun activity. The above given $\mathbf{S}_0$ figure is close to the average of observations (ranging from 1,369 to

Table 1.1 Spectral distribution of solar radiation at sea level

Band	Wavelength (μm)	% Irradiance
Ultraviolet		6.3
UVC	0.200–0.280	
UVB	0.280–0.315	
UVA	0.315–0.400	
Visible light		52.1
Violet	0.400–0.424	
Blue	0.424–0.492	
Green	0.492–0.535	
Yellow	0.535–0.586	
Orange	0.586–0.647	
Red	0.647–0.780	
Infrared	0.780–3.000	41.6

1,375 W m^{-2}) done by the Nimbus-7 satellite, according to Hickey et al. (1982). A greater variation was estimated by Willson (1993), between 1,346 and 1,389 W m^{-2}.

1.2.1.2 Radiation at the Sea Level

The solar beam passing through the atmosphere is altered in quantity, quality and direction as the radiation is absorbed and scattered by the atmospheric gases. Although the solar radiation includes a great part of the known electromagnetic spectrum, most measurements of the radiation at the ground are within the wavelengths from ultraviolet (0.2 μm) to near infrared (3 μm). Table 1.1 shows the bands measured at the sea level, assuming a perfect clean atmosphere.

As the solar radiation passes through the atmosphere, the quanta strike the gas molecules (whose sizes are much smaller than the wavelength of the radiation) and are scattered in all directions (*Rayleigh scattering*). Blue light is scattered nine times more than the red one, and this is the cause of the blue colour of the sky, while the red colour of the sky at sunset is due to the scattering of the red light by greater aerosol particles (e.g. dust).

Ultraviolet radiation with wavelength less than 0.3 μm is absorbed by the O_2 molecules in the ionosphere and stratosphere, leaving to about 3% reduction in the incoming radiation. It is a phenomenon associated to the formation of the ozone layer, and we will return latter to this point.

Depending on the geographic location and the circumstances, about 30% of the incoming energy is reflected back to space by the clouds. Other 15% were absorbed by water vapour, by CO_2 and by aerosol particles present in the atmosphere; more 15% of the radiation are dispersed. From the amount that reaches the ground surface, about 6% are reflected, and approximately 31% are absorbed.

One could assume the atmospheric reduction in the ultraviolet radiation as of little significance, but such a reduction is of great biological importance. In fact, the ultraviolet band is divided into three sub-bands: UVA (0.315–0.400 μm),

UVB (0.280–0.315 μm) and UVC (0.200–0.280 μm). This last C band has an intense biological effect that is associated to molecular changes in the tissues, which can leave to gene mutations and carcinogenic processes; however, this band is almost totally absent at the sea level because its absorption in the ionosphere and stratosphere.

The UVB is important for vitamin D synthesis, but excessive exposure to it induces many risks of cell damage. Finally, the UVA rays have weaker penetrating force than that of UVC and UVB; they are generally associated to the photosynthesis in plants and melanin production in animals.

Visible light band, as it is shown in Table 1.1, refer to human beings only; the colours that can be seen depend on the retinal cone cells, which are sensible to the different wavelengths of the visible band: the red, the blue and the green. Under any circumstance, depending on the radiation received, the various wavelengths detected by those cells are combined together, in order to permit the vision of a wide range of colours. There are other cell types – more sensible but unable to differentiate colours – that are responsible for the so-called night vision.

Vision ability in animals varies in a great extent. Many species do not have cone cells, and they are supposed as blind to colours. Other species have high sensibility to a range of colours that is much wider than that of the man. For example, some birds can see the ultraviolet, while night hunter species are able to see in the far infrared (>0.78 μm).

1.2.1.3 Variation of Solar Irradiance

The solar constant S_0 is a measure of the quantity of energy that strikes the Earth outside atmosphere, but the fraction of it that effectively reaches the ground surface – without any reference to the atmospheric condition – varies according to the season, the altitude and the time of day. The influence of the season is more important at high latitudes, particularly near and beyond the tropics of Cancer and Capricorn, but much less evident in the equatorial belt.

Figure 1.2 shows the variation of the solar radiation at sea level in the Southern Hemisphere; it has little variability in the equator during the year; in the 0° latitude, the solar radiation shows two peaks corresponding to the spring and autumn equinoxes, respectively. But it is lower in both the solstices. On the other hand, the radiation at a 66° latitude (Antarctic Polar Circle) is almost null between April and August, when the sun remains below the horizon; but even in the summer peak (December), the radiation is reduced (300 W m^{-2} or less).

The amount of solar radiation received at the surface of the Earth depends on the zenith angle of the sun, θ, which is determined by latitude, by Earth's rotation about its axis and by its elliptical orbit. The value of θ is obtained for a given location, in a given date and a given daytime by means of the following equation:

$$\cos\theta = (\cos L_t)(\cos\delta)(\cos\eta) + (\sin L_t)(\sin\delta) \tag{1.2}$$

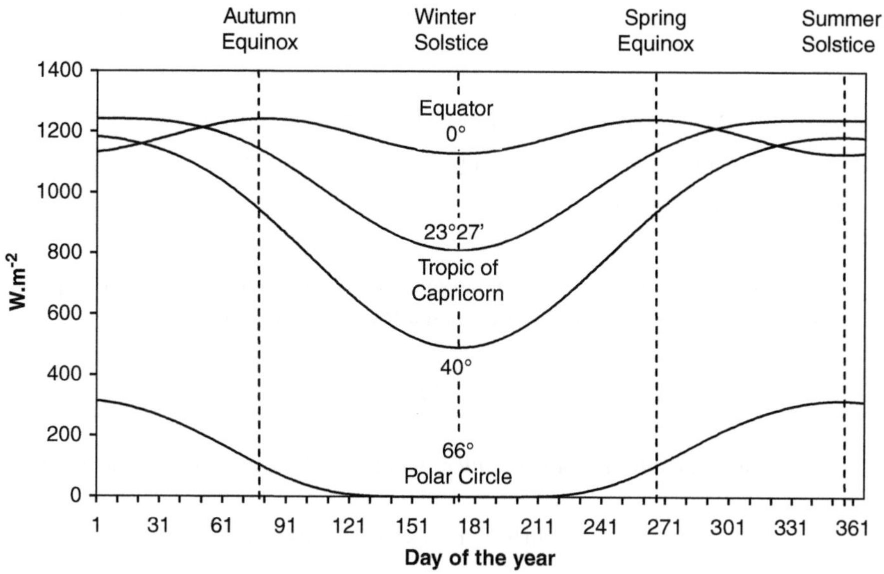

Fig. 1.2 Annual variation of the solar radiation at sea level in the Southern Hemisphere, considering four different latitude degrees and a standard clean atmosphere

where L_t is the latitude (negative value for the Southern Hemisphere); δ is the *solar declination*, the angle between the orbital plane and the Earth's equatorial plane; and η is the *hour angle of the sun*.

Solar declination is given by

$$\delta = 23.45 \sin\left[\left(\frac{360}{365.242}\right)(284 + \mathrm{d})\right] \text{ degrees} \tag{1.3}$$

after Meeus (1991), where d is the day of the year (1 for January 1, 365 for December 31 and so on). It must be remembered that the month of February has one additional day in leap years.

The hour angle of the sun is the fraction of 360° through which the Earth has turned in its rotation, each hour corresponding to 15°. Then,

$$\eta = 15(\mathrm{h} - \mathrm{h}_0) \text{ degrees} \tag{1.4}$$

where

$$h_0 = \text{solar noon} = 12 + \frac{M_\mathrm{h} - L_\mathrm{g}}{15} + e_\mathrm{t} \text{ hours} \tag{1.4a}$$

h is the local hour, M_h is the standard meridian, L_g is the longitude and e_t is the time correction. In general, this correction is used only when one wants more precision in the calculations; it is given by formula 1.5, adapted from Campbell and Norman (1998):

$$e_t = -0.05107 - 0.00655\,\mathrm{d} - 7.41244 \times 10^{-5}\mathrm{d}^2 4.72702 \times 10^{-6}\mathrm{d}^3 - 5.33999 \times 10^{-8}\mathrm{d}^4 + 2.55422 \times 10^{-10}\mathrm{d}^5 - 5.58785 \times 10^{-13}\mathrm{d}^6 + 4.59779 \times 10^{-16}\mathrm{d}^7 \quad (1.5)$$

The value of η is negative before noon and positive after noon. By applying Beer's law, the direct solar irradiance at sea level can be given as

$$\mathbf{S}_\mathrm{p} = \mathbf{S}_0 \cos\theta\tau^m = 1{,}372.7 \cos\theta\tau^m \;\; \mathrm{W\,m^{-2}} \quad (1.6)$$

where τ is the atmospheric transmittance, which can be estimated by the formula

$$\tau = exp\left\{\frac{\ln \mathbf{S} - \ln(\cos\theta) - 7.224681}{m}\right\} \quad (1.7)$$

where **S** is the short-wave solar radiation measured directly at the ground level by means of a pyranometer and m is the air mass number, given by

$$m = \frac{P_\mathrm{a}}{101.325 \;\cos\theta} \quad (1.7a)$$

where P_a is the actual atmospheric pressure (kPa). In general, $\tau = 0.912$ for very clear, cloudless days, as it is often observed in Natal, Brazil; $\tau = 0.157$ for sunny days with many clouds; and $\tau = 0.638$ on the average for sunny days with clear atmosphere and some clouds. Those values were obtained during field observations, far from urban areas with sources of atmospheric pollution.

Term τ in formulas 1.6 and 1.7 is equivalent to e^{-t}, where t is the turbidity coefficient, extinction coefficient or optical thickness of the atmosphere. Monteith and Unsworth (2008) refer to values as $t = 0.05$ for the Scottish Highlands and $t = 0.6$ for the polluted English Midlands; from these values, the respective transmittances can be calculated as 0.951 and 0.549, respectively.

Latitude and longitude values are generally presented in degrees (GG), minutes (MM) and seconds (SS) in the form GG°MM′SS″. However, for the calculations above described, they must be converted to decimal degrees, as follows:

$$Decimal\ degrees = \mathrm{GG} + \left[\mathrm{MM} + \left(\frac{\mathrm{SS}}{60}\right)\right]/60 \quad (1.8)$$

It must also be stressed that southern latitudes have always negative values. Equation 1.8 can be used also to convert time into decimal values, by substituting GG (degrees) for HH (hours).

Sometimes, it is needed to know the day length for a given location at a given date. The following equation is used to estimate the day length:

$$Day\ length = \left(\frac{2}{15}\right)\cos^{-1}\left[\frac{-(\sin L_t)(\sin\delta)}{(\cos L_t)(\cos\delta)}\right] \text{ hours} \quad (1.9)$$

For biological purposes, however, additional twilight time must be considered. In general, civil twilight can be defined as beginning and ending when the sun is 6° below the horizon. Thus, in this case, $\theta = -6°$ and 96° at sunrise and sunset, respectively, and Eq. 1.9 must be replaced by the following:

$$Day\ length = \left(\frac{2}{15}\right)\cos^{-1}\left[\frac{-0.104528 - (\sin L_t)(\sin\delta)}{(\cos L_t)(\cos\delta)}\right] \quad (1.9a)$$

The time of the first twilight at sunrise will be h_0−(half-day length), and the time of sunset will be h_0 + (half-day length).

1.2.1.4 Diffuse Short-Wave Radiation

Diffuse radiation is caused by deflection of direct solar radiation by particles and molecules in the atmosphere and highly affected by clouds, reaching the ground level from several angles. Such a dispersion results in qualitative, quantitative and directional changes of the radiation, which is reduced in two forms:

First, a quantum of energy strikes one atmospheric gas molecule and is dispersed in any direction (*Rayleigh scattering*). In that process, the dispersion of the blue light (wavelength $\lambda = 0.424$–0.492 μm) exceeds that of the red light, for example, ($\lambda = 0.647$–0.780 μm), by a factor of 9, and this explains why sky is blue. The direct cause of this phenomenon is the smaller diameter of gas molecules relative to the radiation wavelength.

Second, the radiation is absorbed by water vapour and CO_2 molecules. Finally, there has the process of ozone formation. We will return latter to these matters.

There has diffuse radiation even when the atmosphere is clear and cloudless, if the zenith angle of the sun is less than 50°. The amount of diffuse radiation can increase with high values of atmospheric turbidity induced by pollution gases.

Diffuse short-wave radiation can be estimated by the equation

$$S_{dif} = 411.81(1 - \tau^m)\cos\theta \ \ \mathrm{W\,m^{-2}} \quad (1.10)$$

after Liu and Jordan (1960) and where τ is the atmospheric transmittance. These authors obtained τ values from 0.75 to 0.45 for clear days at two sites; if $\tau < 0.45$, the sky is overcast. As the m factor compensates the cosine law in the above equation, then the $\mathbf{S}_{dif}$ value is relatively constant for the clear days (Peterson and Dirmhirn 1981).

Monteith and Unsworth (2008) showed a different formula:

$$\mathbf{S}_{dif} = \mathbf{S}_t[0.1 + 0.68(-\ln \tau)] \quad (1.11)$$

where $\mathbf{S}_t$ is the total short-wave solar irradiance on a horizontal surface (see Eq. 1.15). This formula is valid for solar zenith angles less than 60° only. According to these authors, when $\theta > 60°$ the value $\mathbf{S}_{dif}/\mathbf{S}_t$ becomes a function of θ, then $\mathbf{S}_{dif}$ can be greater than the value calculated by Eq. 1.11. On the other hand, McCullough and Porter (1971) suggested that the diffuse radiation would amount 8% only of the total short-wave radiation in a clear, cloudless day; about 45% of this amount would be within a solid angle cone that is close enough to the sun in order to be assumed as direct solar radiation. So, 4.4% only of the short-wave radiation would be diffuse radiation, while 95.6% would come from the direct radiation cone. However, such assumptions would not be valid when the sun is close to the horizon (Weber and Baker 1982).

Finally, Eqs. 1.10 and 1.11 give different results for the same data; however, Eq. 1.10 is probably the better for biological studies in general.

1.2.1.5 Cloudy Sky

As the cloud coverage increases, the relationship $\mathbf{S}_{dif}/\mathbf{S}_t$ increases also due to the reduction of direct solar radiation. When the sky is fully overcast, $\mathbf{S}_{dif}$ equals $\mathbf{S}_t$, and the irradiance varies as a function of the elevation of the sun ($\beta = 90 - \theta$). List (1971) calculated several curves for the solar irradiance according to the cloud type and the zenith angle of sun (θ).

Measurements done in Australia (Hounan 1963, 1969; Stafford-Smith et al. 1985) suggest that an amount of 25–30% of the short-wave solar radiation is transmitted through the clouds, if those clouds were not too dark. Based on this assumption, Stafford-Smith et al. (1985) proposed an "attenuation coefficient" for the irradiance, which could be given as

$$\mathbf{S}_{dir} = 1{,}372.7(1 - 0.75n)\tau^m \cos\theta \ \ \mathrm{W\,m^{-2}} \quad (1.12)$$

where n is the proportion of the sky covered by clouds.

However, such attenuation depends on the cloud type, as it was suggested by List (1971). Data from this last author show that the different cloud types transmit solar radiation in the proportions presented in Table 1.2. So, Eq. 1.12 must be changed to

$$\mathbf{S}_{dir} = 1{,}372.7[1 - n(1 - c)]\tau^m \cos\theta \ \ \mathrm{W\,m^{-2}} \quad (1.13)$$

where c is the coefficient relative to the prevailing cloud type (see Table 1.2).

Table 1.2 Coefficients of short-wave solar radiation transmitted through some types of clouds

Cloud type	*c*	Cloud type	*c*
Cloudless sky	1.00	Altostratus	0.42
Cirrus	0.86	Stratocumulus	0.35
Cirrostratus	0.85	Stratus	0.25
Altocumulus	0.53	Nimbostratus	0.13

Data from List (1971)

The presence of small amounts of clouds in an otherwise clear sky leaves to increased diffuse radiation, but direct solar radiation is not affected unless the sun is covered. With sparse cumulus type clouds, for example, total irradiance can exceeds by 5–10%, the flux observed under a cloudless sky (Monteith and Unsworth 2008).

1.2.1.6 Reflected Short-Wave Radiation

The amount of radiation shown by Eq. 1.13 corresponds to the solar short-wave radiation reaching the ground level, but part of it is reflected upwards by the ground surface and the objects thereon. This reflected short-wave radiation can be estimated by the equation

$$\mathbf{S}_{\text{ref}} = \rho_s(\mathbf{S}_{\text{dir}} + \mathbf{S}_{\text{dif}}) \;\; \text{W m}^{-2} \tag{1.14}$$

where ρ_s is the mean reflectance of the ground surface. Table 1.3 shows the reflectance values (sometimes referred as *albedo*) of some surfaces, with respect to the solar short-wave radiation.

1.2.1.7 Total Short-Wave Radiation

Total irradiance at ground level depends on the solar zenith angle, sky cloudiness and turbidity; then, it can be estimated by combining Eqs. 1.10, 1.13 and 1.14, as follows:

$$\begin{aligned}\mathbf{S}_{\text{t}} &= \mathbf{S}_{\text{dir}} + \mathbf{S}_{\text{dif}} + \mathbf{S}_{\text{ref}} \\ &= (1+\rho_s)(\mathbf{S}_{\text{dir}} + \mathbf{S}_{\text{dif}}) \\ &= \cos\theta(1+\rho_s)\{1{,}372.7[1-n(1-c)]\tau^m + 411.87(1-\tau^m)\} \;\; \text{W m}^{-2}\end{aligned} \tag{1.15}$$

1.2.2 Terrestrial Radiation

Terrestrial radiation is the long-wave radiation originated in the atmosphere and at the ground surface. It can be measured as coming from the sky (downward radiation) or from the Earth's surface (upward radiation), in any case under the form of long waves.

Table 1.3 Reflectance values for short-wave solar radiation and emissivities for long-wave radiation of some surface

Surface	Emissivity, ε	Reflectance, ρ_s
Ground		
Dark, humid	0.90	0.08
Dark, dry	0.87	0.13
Light, humid	0.84	0.10 a 0.32
Light, dry	0.68	0.18 a 0.22
White sand, dry	0.60	0.35 a 0.40
Arid, stony ground	0.92	0.24
Clean asphalt surface	0.93	0.05 a 0.13
Concrete	0.88 a 0.93	0.40
Vegetation		
Green grass, pasture	0.67	0.24 a 0.26
Tropical forest	0.99	0.13
Conifer grove	–	0.16
Orange plantation	–	0.16
Eucalyptus grove	–	0.19
Sugar cane plantation	0.95	0.15
Natural grassland	0.95	0.20 a 0.25
Savanna	–	0.15 a 0.19
High, dry grass	0.90	0.32
Other surfaces		
Water	0.96	0.05 a 0.08
Asbestos cement	0.90	0.41
Common bricks	0.93	0.45
Metals		
Aluminium, polished	0.05	0.74
Aluminium, commercial	0.10	0.68
Galvanised iron, new	0.13	0.35
Galvanised iron, oxidised	0.28	0.20
Copper, polished	0.04	0.82
Paints		
Epoxy or acrylic, white	0.90	0.75
Epoxy or acrylic, matt black	0.91	0.05 a 0.15
Lampblack	0.95	0.02

Data from Chapman (1987), Esmay (1969), Campbell and Norman (1998), Gates (1980) and Henderson-Sellers and Robinson (1987)

1.2.2.1 Downward Radiation

The ground surface and the atmospheric particles and molecules absorb the solar short-wave radiation and emit the absorbed energy as long-wave radiation. Aiming at the sky with an infrared radiometer, one can observe that the measured effective radiant temperature is higher at the horizon than at zenith. It is explained by the concentration of water vapour and CO_2, whose molecules are the main sources of long-wave radiation from the sky.

The flux density of downward long-wave radiation can be given by the equation

$$\mathbf{L}_\mathrm{D} = \varepsilon_{\mathrm{ac}} \sigma T_{\mathrm{rc}}^4 \ \mathrm{W\,m^{-2}} \tag{1.16}$$

where ε_{ac} is the apparent emissivity of the sky, T_{rc} is the effective radiant temperature of the sky (K) and σ is the Stefan-Boltzmann constant.

With respect to the apparent emissivity of the sky, it can be estimated as

$$\varepsilon_{\mathrm{ac}} = a + b\,(\ln w + 0.5) \tag{1.17}$$

(Monteith and Unsworth 2008) where w is the precipitable water content of the atmosphere and a and b are constants of the regression equation. However, there are many difficulties for *a priori* estimation of those constants. A more practical equation is that of Swinbank (1963):

$$\varepsilon_{\mathrm{ac}} = 9.2 \times 10^{-6}\, T_{\mathrm{a}}^2 \tag{1.18}$$

The best formula is probably that was proposed by Brutsaert (1975, 1984):

$$\varepsilon_{\mathrm{ac}} = 1.72 \left(\frac{P_{\mathrm{v}}}{T_{\mathrm{a}}} \right)^{1/7} \tag{1.19}$$

where P_v is the partial vapour pressure of the atmosphere (kPa) and T_a is the air temperature (K). In this equation, it is assumed that thermal radiation from the atmosphere is a function of the water vapour concentration. For a cloudy sky, the apparent emissivity of the atmosphere is higher, because clouds have emissivity close to one. Then,

$$\varepsilon_{\mathrm{ac}}^* = 0.84n + (1 - 0.84n)\varepsilon_{\mathrm{ac}} \tag{1.20}$$

after Monteith and Unsworth (2008), where ε_{ac} is given by Eq. 1.17.

An alternative for Eq. 1.20 was given by Monteith and Unsworth (2008) based on results from Prata (1996) and Niemelä et al. (2001):

$$\varepsilon_{\mathrm{ac}} = 1 - (1 + 0.1w)\exp\left\{ -(1.2 + 0.3w)^{0,5} \right\} \tag{1.21}$$

Approximate values of w (precipitable water content of the atmosphere) can be obtained from the equation

$$w = 4{,}650\, P_{\mathrm{v}}/T_{\mathrm{a}} \ \mathrm{kg.m^{-2}} \tag{1.22}$$

where P_v is the air partial vapour pressure near the ground (kPa) and T_a is the air temperature (K).

As for the effective radiant temperature of the sky, it can be measured directly by means of a directional radiometer provided with adjustable emissivity (e.g. Horiba IT-330). During a test made in Jaboticabal, SP, Brazil (latitude 21°15′ south), a device of this type was aimed at a clean sky area under air temperature of 27.5°C and 80% relative humidity; the measured radiant temperature of the sky was −14.5°C; when the radiometer was aimed at a cloud in the same sky area, the radiant temperature increased to 9°C. Thus, the radiant temperature of the sky is affected by cloudiness: as greater the sky area covered by clouds (n), the higher the apparent radiant temperature. In such a case, we can use the formula suggested by Stafford-Smith et al. (1985) to estimate the average of the radiant temperature of the sky:

$$T_{\mathrm{rc}} = T_{\mathrm{a}} - 20 + 10n \quad {}^{\circ}\mathrm{C} \tag{1.23}$$

1.2.2.2 Upward Radiation

When the solar short-wave radiation reaches the ground surface, part of it is reflected and part is absorbed. The reflected portion returns to exterior space or is intercepted by clouds, objects or animals. The amount of that reflected radiation can be estimated if the solar radiation intensity and the ground absorptance are known.

On the other hand, the ground is warmed because the energy absorbed by it, thus resulting in an emission of infrared radiation ($\lambda > 3$ μm long waves), is proportional to the ground temperature (Wien's law). Such a warming is important to keep a mild air temperature during the night. In sandy deserts as the Sahara, air temperature reaches very high levels during the day; however, solar absorptance of the sand (0.60–0.65) is lower than that of the normal dry ground (0.78–0.82); then there has no sufficient heat storage, and air temperature is very low at night. Such absorptance values can be determined from the data given by Table 1.3, as follows:

$$\alpha_{\mathrm{s}} = 1 - \rho_{\mathrm{s}}$$

A surface of concrete absorbs less solar energy than the normal soil and emits also less thermal radiation; however, it can affect greatly the environmental thermal load because of its high reflectance value. An asphalt surface has high absorptance (0.93) and emissivity values (0.90–0.98). On the other hand, plant cover protects the soil from the direct solar radiation and avoids excessive heating; in addition, it reduces the reflected radiation incident on animals and objects. Soil heating is an important factor for the maintenance of the microbial flora and fauna equilibrium, which is of fundamental importance for the plants.

Once heated the soil surface (or the surface of its plant cover), it emits thermal energy upwards, and the respective flux density is given by

$$\mathbf{L}_{\mathrm{U}} = \varepsilon_{\mathrm{s}} \sigma T_{\mathrm{s}}^{4} \quad \mathrm{W\ m^{-2}} \tag{1.24}$$

where T_{s} is the surface temperature (K) and ε_{s} the respective emissivity (see Table 1.3).

1.2.3 *Radiant Heat Load*

1.2.3.1 Definition

It is unquestionable that thermal exchange among animals and environments is of great importance in tropical climates. In many cases, the amount of that exchange assumes for the animal, the only difference between a bearable and an unbearable environment.

Several thermal radiation sources in the surroundings of an animal can be mentioned: the sun, sky, ground, shelter (walls and roof), buildings, etc., in short, any object or surface, real or virtual, whose temperature is above the absolute zero (0 K or −273.15°C). In order to quantify the thermal radiation exchanged by an animal with its environment, the great complexity of the later must be reduced to the conditions of a standard surface.

Suppose an animal placed in the centre of an infinitely great spherical container whose inner surface is a black body at a temperature T_{rm} (K), namely, the *mean radiant temperature*. This temperature is the average temperature of all the surfaces, real or virtual, that exist around the body of an animal in a given place. Such an animal exchanges with all those surfaces an amount of thermal energy known as *radiant heat load* or RHL, which can be defined as the total amount of thermal energy a body exchanges by radiation with its environment.

1.2.3.2 Evaluation

In meteorology and climatology solar radiation is evaluated by using the device named *pyranometer*, from which several types are known; the most widely used is probably that of Eppley. The Eppley pyranometer consists essentially of a disc with two concentric rings, one of them covered with magnesium oxide (white) and the other with lampblack. Each ring is a thermopile junction, and the whole assemblage is protected by a glass hemispherical cover.

Solar radiation is absorbed by the black ring and reflected by the white one in order to have a temperature differential between them that is measured. However, this device is not practical for the determination of RHL, because the main interest in this case is the evaluation of the global thermal radiation, not the solar radiation only.

The *black globe* – known also as *globe thermometer* – is probably the most practical device for the evaluation of RHL. Its use was first proposed by Vernon (1932), and since then, it has been extensively used. Figure 1.3 depicts a black globe in use to evaluate RHL in a corral. The globe is suspended by a thread made of thermo-insulating material at the place an animal would be, remaining there until a balance is reached by its internal temperature. Such a temperature gives an estimate of the combined effects of the thermal radiation incoming from every possible direction, together with the effects of the air temperature and wind velocity.

Fig. 1.3 *Black globe* used to evaluate radiant heat load. Over the globe, there is the display of the digital thermometer that was introduced for the determination of the globe temperature, T_g

Several globe types have been proposed. Vernon (1932) used models made of cardboard, glass and copper; Bedford and Warner (1934) tested metal globes with several diameters; Pereira et al. (1967) proposed black-painted *ping-pong* balls; Bond and Kelly (1955) used water valve buoys. Small diameter globes were studied by Hey (1968).

A globe is built of two half spheres of copper sheet which can be moulded in by a hydraulic press and then soldered together. The globe is then painted matt black and provided with a hole to receive a thermometer.

1.2.3.3 Black Globe Theory

The globe is suspended at a given place, avoiding any contact with other surfaces. The globe exchanges thermal energy with the several environmental surfaces by means of radiation and convection, until it attained a state of equilibrium in which the incoming heat equals the dissipated one. The amount of radiant heat exchanged with the environment can be given by

$$\mathbf{R} = \varepsilon_g \sigma \left(T_g^4 - T_{rm}^4 \right) \ \mathrm{W\,m^{-2}} \tag{1.25}$$

where ε_g and T_g are the emissivity and the temperature of the globe (K), respectively; T_{rm} is the *mean radiant temperature* (K) of the environment; and σ is the Stefan-Boltzmann constant. The emissivity value of the external surface of the globe is often assumed $\varepsilon_g = 0.95$, but it depends on the used paint.

Thermal exchange by conduction is negligible, since the globe is suspended by a thread made of insulating material and avoiding contacts with any other surface or object. It must be maintained dry in order to avoid thermal losses by evaporation. Exchange by convection is calculated by the equation

$$\mathbf{C} = h_c(T_g - T_a) \ \ \text{W m}^{-2} \tag{1.26}$$

where h_c is the convection coefficient of the globe. Bedford and Warner (1934) determined that coefficient as

$$h_c = b\,U^{0.5} \tag{1.27}$$

in which U is the wind velocity and b is a parameter associated to the globe size; for a 0.15-m diameter globe, this value would be $b = 14.1672$. That formula have been used in several studies, until Kuehn et al. (1970) proposed an alternative,

$$h_c = 0.05213\,d^{-0.42}\,U^{0.58} \tag{1.28}$$

where d is the globe diameter (m). The ISO-7726 (1985) norm gives the equation

$$h_c = 2.5 \times 10^8\,U^{0.6} \tag{1.29}$$

for a 0.15-m diameter globe.

The following formula (Thornsson et al. 2007) has been used by human bioclimatologists:

$$h_c = \frac{1.1 \times 10^8\,U^{0.6}}{d^{0.4}} \tag{1.30}$$

where the globe diameter d is given in millimetres (mm).

However, Eqs. 1.27, 1.28, 1.29, 1.30 consider thermal exchange by forced convection only, an unrealistic assumption. In fact, even in open outside environments, the wind velocity is often zero or very close to zero, but the globe continues to lose heat by free convection, depending on the temperature differential between it and the atmosphere. In such a case, if one uses any of the above formulas (1.27, 1.28, 1.29, 1.30), the result will be $h_c = 0$ or very close to zero; it happens as if there were no convection at all and that the total heat exchange is done by radiation only. But free convection is effectively present, notwithstanding.

In order to solve such a problem, both forced and free convection[1] can be taken into account (Silva 2002) by using separate equations for free and forced convection, respectively, and calculating the Nusselt number (N_u). Then,

$$h_c = kd^{-1}\,N_u \tag{1.31}$$

[1] Convection process details will be discussed later in Chap. 2.

where k is the thermal conductivity of the air at a given temperature and d is the globe diameter (m) (see later in this chapter [Sect. 1.3] for definitions of k and other properties of the atmosphere). Definition and formulas of N_u are given in Chap. 2 under the title "Convection" (Eqs. 2.40 and 2.46). The calculation of h_c by using formula 1.31 is somewhat laborious, but the results are more precise than those of Eqs. 1.27, 1.28, 1.29, 1.30, which underestimate T_{rm} as wind speed increases.

It must be remembered that the wind velocity and the air temperature, which are needed for calculation of h_c, must be evaluated close to the black globe.

When the globe is at thermal equilibrium (which is attained within about 15 min after it was installed in a place), then the radiative heat exchange equals the convective one, or **R** = **C**. By substituting into Eqs. 1.23 and 1.24, we have

$$\varepsilon_g \, \sigma \left(T_{rm}^4 - T_g^4 \right) = h_c \left(T_g - T_a \right)$$

and after some manipulation

$$\varepsilon_g \sigma T_{rm}^4 = h_c \left(T_g - T_a \right) + \varepsilon_g \sigma T_g^4$$

$$\sigma T_{rm}^4 = \frac{h_c}{\varepsilon_g} \left(T_g - T_a \right) + \sigma T_g^4 \ \ \text{W m}^{-2} \tag{1.32}$$

But the radiant heat load is defined by the mean radiant temperature as

$$\text{RHL} = \sigma T_{rm}^4 \tag{1.33}$$

by which

$$T_{rm} = \left(\frac{\text{RHL}}{\sigma} \right)^{1/4} \ \text{K} \tag{1.34}$$

The calculated RHL value can be eventually negative. In such a circumstance, the globe is losing thermal energy, and this is observed, for example, into a barn with a wet pavement whose surface is colder than the atmosphere.

Results obtained by Silva et al. (2010) in open field under the conditions of a semiarid equatorial region showed a high correlation coefficient ($r = 0.77$, $P < 0.001$) of RHL measurements (average 742.5 ± 4.6 W m^{-2}) with the total long-wave radiation estimates $\mathbf{L_D} + \mathbf{L_U}$ (average 700.5 ± 2.6 W m^{-2}). On the other hand, the correlation of RHL with the direct solar short-wave radiation ($\mathbf{S_{dir}} + \mathbf{S_{dif}}$) was very low, $r = 0.04$. It was concluded that the long-wave radiation contribution is of greater importance as for RHL determination than that of the short wave. As in the above-mentioned study there were observed small differences between RHL and the total long-wave irradiance, about 5.6% of RHL, it was suggested that the mean radiant temperature can be estimated without significant error by

$$T_{rm} = \left(\frac{\mathbf{L}_D + \mathbf{L}_U}{0.944\ \sigma}\right)^{1/4} \text{K} \tag{1.35}$$

This equation can be particularly useful for open-field conditions when a black globe is not available.

1.2.3.4 Effective Radiant Heat Load

The use of black globes to measure RHL in tropical regions can often leave to underestimates of the radiant energy received by animals under open-field conditions. As a matter of fact, the intense short-wave radiation intercepted by an animal in those regions (depending on the relative position of its body with respect to sun) is absorbed in part, thus heating body surface. In order to take into account that short-wave contribution, Silva et al. (2010) proposed a new measurement for the case of animals exposed to sun, the *effective radiant heat load:*

$$\text{ERHL} = 0.5\,\mathbf{S}_t + \text{RHL} \tag{1.36}$$

where $\mathbf{S}_t$ is the total short-wave irradiance (Eq. 1.15). The 0.5 multiplier is the proportion of the body area receiving the short-wave radiation, direct and diffuse. The estimated ERHL value can be used to calculate the effective mean radiant temperature as

$$T^*_{rm} = \left(\frac{\text{ERHL}}{\sigma}\right)^{1/4} \text{K} \tag{1.37}$$

1.2.3.5 Determination of Black Globe Emissivity

The emissivity coefficient of the black globe surface is not the constant value of $\varepsilon_g = 0.95$, as it is generally assumed to be. Emissivity is associated with the nature of the surface and its cover, and then it can vary according to the paint used. Therefore, it is advisable the measurement of ε_g for each globe.

A practical determination procedure can be done by using (a) a radiometer or infrared thermometer, in which the emissivity can be adjusted, and (b) a contact thermometer.

First, the surface temperature of the globe is taken by means of the contact thermometer, whose proof tip must be protected from the external environment by means of a small thermo-insulating seal. Second, the radiometer is aimed at the globe surface near to the same place the contact thermometer is. Third, the emissivity control of the radiometer is adjusted until it shows the same temperature measured by the contact thermometer. The observed ε value will be the emissivity of the globe surface.

1.3 The Atmosphere

1.3.1 Structure

The atmospheric covering of the Earth has several major zones or layers, which present different thermal characteristics. For practical purposes, we will consider here four of them, in the ascending order: troposphere, stratosphere, mesosphere and ionosphere. They are separated with variable degrees of sharpness and occur at variable elevations.

1.3.1.1 Troposphere

It is the lowest layer and grows up to an altitude of about 8 km in the poles and 18 km in the equatorial line. Air temperature falls gradually with the altitude, until reaching about −55°C at the layer's top; at this point, the temperature stabilises to an isothermal condition, and there has a sub-layer named tropopause, in which the atmosphere is extremely clean.

1.3.1.2 Stratosphere

This layer extends from the higher limit of the troposphere to about 45-km altitude. The isothermal conditions of the tropopause end at about 26 km, and then the temperature increases until 0°C at 50 km. This temperature change is associated to the photochemical reactions during the ozone formation.

1.3.1.3 Mesosphere

It is the layer placed between about 45- and 80-km altitude. After an isothermal pause at about 50 km, there has a temperature inversion until a temperature of −85°C is reached at 80 km, and then it is stabilised at the altitude of 90 km (*mesopause*).

1.3.1.4 Ionosphere

This layer extends from about 80- to 160-km altitude. Starting from 90 km, there has again a temperature increase, and it reaches −20°C as a result of solar

Table 1.4 Average composition of the atmosphere, in terms of dry air

Component	Symbol	Molecular weight (g)	% Volume
Nitrogen	N_2	28.010	78.08
Oxygen	O_2	31.999	20.95
Argon	Ar	38.980	0.93
Carbon dioxide	CO_2	44.010	0.035
Methane	CH_4	16.043	1.7×10^{-4}
Nitrous oxide	N_2O	44.0095	3×10^{-5}
Ozone	O_3	47.9985	4×10^{-6}
Total		**28.9501**	**99.9952**

From Ahrens (1991)

short-wave absorption. There are formed layers of ionised particles that reflect the Hertzian short waves, allowing long-distance radiocommunications.

The ionosphere is followed by *exosphere* (160–550 km), in which O_2 concentration is very low, and then there is the outer space. The outer limit of the atmosphere has been estimated as 32,000 km; however, about 90% of the atmospheric mass is below the 35-km altitude.

1.3.2 Gas Components

Atmosphere is a mixture of gases and water vapour, each component having an independent action. However, as the water vapour content is much variable according to the circumstances, the atmospheric gas content is always described in terms of dry air as it can be seen in Table 1.4.

It is easy to see that three gases only amount to 99.96% of the atmosphere: nitrogen, oxygen and argon; the rest is constituted by inert gases (neon, helium, krypton, xenon) and hydrogen in various proportions. Other gases can occur in various proportions also: ozone, carbon dioxide, methane and others that are generally considered as "pollutants".

Near the ground surface, there exists a state of equilibrium among the main gases, especially O_2 and N_2. The latter is removed from atmosphere by biological (soil bacteria) and physical processes (electric discharges) and then absorbed by the plants, which are eaten by animals. Through the process of organic decomposition, the nitrogen comes back to the atmosphere. As for the O_2, it is removed from the atmosphere through respiration in organisms and oxidation processes in general, returning by means of the plant photosynthesis.

1.3.2.1 Carbon Dioxide

CO_2 is generated by combustion and by respiratory processes in animals and plants; in this last case, the contribution of the oceanic algae is of great importance. The absorption of CO_2 by the terrestrial plants and the marine algae maintain the atmospheric gas balance.

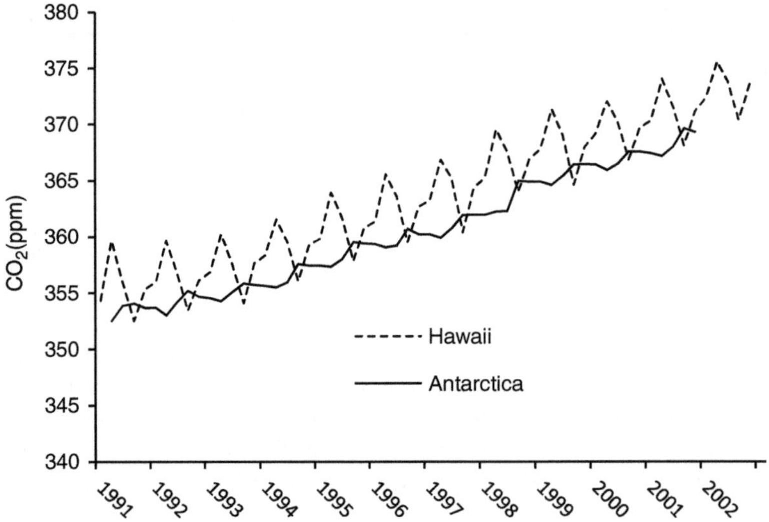

Fig. 1.4 Atmospheric CO_2 concentration in Hawaii (Mount Mauna Loa, 9°32′ north, 155°35′ west, 3,397-m altitude) and in Antarctica (89°59′ south, 24°48′ west, 2,810-m altitude) (Data from Keeling and Whorf 2003; Steele et al. 2002)

It must be noted that CO_2 plays a dramatic role in the functioning of the greenhouse effect, which is essential for life on Earth. This gas is permeable to the solar short-wave radiation but has a very low transmittance coefficient for the long-wave radiation. As a result, the ground surface is heated under the action of the solar direct short-wave radiation and then emits thermal long-wave radiation upwards; those long waves are absorbed by the CO_2 molecules and other particles in the atmosphere, thus preventing the heat to be dissipated in the outer space; part of this thermal energy is reradiated downwards to the ground surface. When under a state of equilibrium, CO_2 is essential for life, for without the greenhouse effect, the air temperature would be extremely variable, much cooler in the night and much hotter in the day, as it happens in the lifeless Moon.

However, because of the increasing use of fossil fuels (petroleum, coal) since the nineteenth century, the CO_2 content of the atmosphere has increased dramatically from 1900 onwards. Figure 1.4 depicts the results of atmospheric CO_2 evaluations done in locations free from anthropogenic contaminations (Mount Mauna Loa, Hawaii and Antarctica). The data show that CO_2 concentration increased from 352 ppm in 1991 to about 370 ppm in 2003. However, it can be noted a constant and regular annual variation in the two locations, with a maximum in the spring and a minimum in the fall. Those seasonal oscillations would be related probably to variations in the solar radiation.

As it is known, vast amounts of CO_2 were withdrawn from the atmosphere many million years ago, when the great geological events buried enormous quantities of biological material that was later converted into coal and petroleum. The stored CO_2 is now being released into the atmosphere by the burning of those fossil fuels.

Besides, another source of great importance is the occurrence of great forest fires and those caused by wars in the last 100 years. Increasing pollution of oceans can also reduce its ability to absorb CO_2 from the atmosphere.

1.3.2.2 Ozone

Most of the O_3 is present in the atmospheric layers from 16 up to 25-km altitude, where it is formed from the O_2 molecules under the ionising action of the ultraviolet radiation (especially in its shortest wavelengths, UVC). This radiation is absorbed when a photon strikes the O_2 molecule and transfer energy to it; the molecule becomes unstable because the excess energy divides itself into two oxygen atoms in a photochemical reaction like that:

$$O_2 + \mathbf{e} \rightarrow O + O$$

where **e** is the energy absorbed from the photon. It is known that the minimum amount of energy needed for such a reaction is associated to a wavelength $\lambda = 0.2422$ μm, which corresponds to the UVC radiation. The O_2 dissociation rate is proportional to the concentration of these molecules and to the radiation amount they are able to absorb in all the wavelengths shorter than 0.2422 μm.

Once formed, atomic oxygen attaches to O_2 to form O_3, which is later broken down into O_2 and O:

$$O_3 + O \rightarrow 2O_2$$
$$O_3 + \mathbf{e} \rightarrow O_2 + O$$

Thus, ozone is always being formed and dissociated in the stratosphere, in such a way that it attained a state of equilibrium. As a consequence of those processes, most solar radiation with $\lambda < 0.240$ μm is reduced in a great extent, while that with $\lambda < 0.175$ μm is totally extinguished above the stratosphere. Because of the decreased radiation of those wavelengths, O_3 is not formed in significant amounts at lower altitudes.

However, what about the danger of the "ozone holes"? Most of the ozone is produced above the equator, where the amount of UV radiation is maximal. On the other hand, this radiation is much reduced above the poles, especially in the winter. Thus, the "ozone holes" exist above the poles only. As the protection against short-wave UV is provided by the oxygen – not by the ozone, which is a by-product of this process – we will be at danger rather in the case of a failure in the atmospheric O_2 supply!

With respect to the problem of chlorofluorocarbons (CFCs), used in liquid propellants and foam-blowing agents, it is a very polemical matter. They are very stable into the troposphere, but some studies showed that they may escape into the stratosphere and may be decomposed under the influence of UV radiation with production of chlorine and chlorine oxide. Boeker and van Grondelle (1999) and Smith (2001), for example, described the following atmospheric reactions:

$$O_3 + Cl \rightarrow O_2 + ClO$$
$$ClO + O \rightarrow O_2 + Cl$$

Most atmospheric chlorine of natural origin is produced from methyl chloride originated from the burning plant material in forest and grassland fires, but only 10% of this chlorine reaches the stratosphere.

Sensors carried by planes flying over the Arctic region in 2004 showed that ClO molecules absorb the solar UV energy, breaking down into O_2 and Cl. These two compounds then react with the ozone, which was dissociated. In other words, those reactions involve absorption of the short-wave UV, thus preventing them to reach the ground surface.

1.3.2.3 Aerosols

In addition to the water vapour and the various gases mentioned above, the atmosphere often contains considerable amounts of particulate matter (*aerosols*) that are held in suspension, mainly dust, salt and sulphates from a wide range of sources. It is assumed that such aerosols are produced at an average rate of 10^9 kg per year (Barry and Chorley 1988):

Natural origin:

Marine salt
Soil dust
Volcanic dust
Smoke from forest and grassland fires
Sulphates from H_2S
Nitrates from NO_2
Plant hydrocarbons

Anthropogenic origin:

Stationary combustion
Industrial processes
Sulphates from SO_2

According to Thompson (1998), the most important sources of aerosols are the natural ones, especially salt, dust and sulphates (from H_2S), which amount up to 89% of the total aerosol content. As for the anthropogenic sources, the most important are the sulphates originated from the burning of fossil fuels, which are associated to the problem of acid rain. As a result of the acid rain, many populations

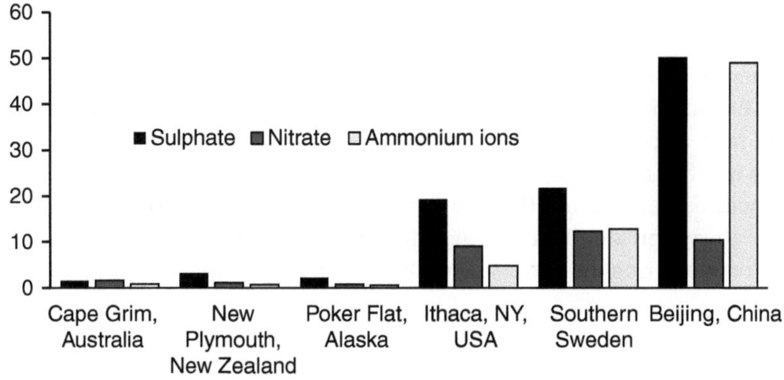

Fig. 1.5 Concentrations of sulphates (SO_4^{2-}), nitrates (NO_3^-) and ammonium ions (NH_4^-) in the rainwater of six regions, values in μEq-gL^{-1} (Data from Bridgman 1997)

of aquatic organisms are decimated, and the soil composition is greatly modified, for example, releasing toxic aluminium ions; as a result, forest life is severely affected, and biodiversity is threatened. Figure 1.5 illustrates the importance of the aerosols as pollutants, by comparing six regions with respect to their levels of industrial pollution.

1.3.3 *Physical Properties*

1.3.3.1 Atmospheric Pressure

The several atmospheric layers represent a mass of air laying over the terrestrial surface, whose weight at the sea level and a temperature of 0°C results in a 101.325-kPa pressure. However, such a pressure varies according to the altitude, the geographic position and the temperature, thus resulting in many different high- and low-pressure zones. The following equation can be used to estimate the normal atmospheric pressure of a given place owing to biometeorological studies:

$$P_a = 101.325\, exp\left\{-\frac{zg}{287.04\,T_a}\right\} \text{ kPa} \tag{1.38}$$

where z is the altitude (m), T_a the air temperature (K), g the acceleration of gravity,

$$g = 9.78013 + 8.18 \times 10^{-5} L_t + 1.168 \times 10^{-5} L_t^2 - 3.1 \times 10^{-6} z \text{ m s}^{-2} \tag{1.39}$$

and L_t the latitude (in decimal degrees).

1.3.3.2 Specific Heat (c_p)

According to the first law of thermodynamics, if one unit mass of air is heated without expansion, the increase in its heat content per degree of temperature is the specific heat at constant volume, symbolised by c_V. On the other end, if the air expands itself in a way that its pressure remains constant, an additional amount of energy will be needed, and then we have the specific heat at constant pressure, c_p, which is always greater than c_V. Once the difference between both values evaluates the work done during the expansion of the air per temperature degree, it follows that

$$c_p - c_v = \frac{R}{M}$$

in which R is the universal gas constant (8.3143 J mol^{-1} K^{-1}) and M is the molecular weight of the dry air (28.9501 mol g). The theoretical value of the quotient c_p/c_V is 7/5, in accord to experimental values (Monteith and Unsworth 2008). Then,

$$c_p = \left(\frac{7}{2}\right)\frac{R}{M} = \left(\frac{7}{2}\right)\frac{8.3143}{28.9501} = 1.0052\ \mathrm{J\ g^{-1}\ {}^{\circ}C^{-1}}$$

However, as R varies with temperature, in practice, it would be better to use the equation

$$c_p = 1.0052 + 4.577 \times 10^{-4}\ exp\left\{\frac{T_a}{32.07733}\right\}\ \mathrm{J\ g^{-1}\ {}^{\circ}C^{-1}} \tag{1.40}$$

in which the air temperature (T_a) is given in degrees °C.

1.3.3.3 Density (ρ)

When the atmosphere is heated, the air molecules absorb more energy and are displaced with greater intensity, in such a way that there has an expansion of the air mass in proportion to the temperature. The greater this expansion, the smaller the number of molecules into a unit volume; in other words, the smaller will be the atmospheric mass per unit volume. Then, as the density is defined by the relation of mass to volume, we will have

$$\rho = \frac{P_a M}{R T_a} = \frac{3481.965 P_a}{T_a}\ \mathrm{g\ m^{-3}} \tag{1.41}$$

where P_a is the atmospheric pressure (kPa) and T_a is the air temperature (K).

1.3.3.4 Thermal Conductivity (*k*)

As it is observed in solids, thermal energy transfer through the air is proportional to the temperature difference between two points of the space. Assuming a constant average density of the air, its thermal conductivity can be defined by the equation

$$k = \rho\, c_{\mathrm{p}} D_{\mathrm{a}} \quad \mathrm{W\,m^{-1}\,{}^{\circ}C^{-1}} \tag{1.42}$$

where D_a is the thermal diffusivity of the air, which depends on the temperature T_a (°C) as it is shown by the formula

$$D_{\mathrm{a}} = 1.888 \times 10^{-5} + 1.324 \times 10^{-7} T_{\mathrm{a}} \quad \mathrm{m^2\,s^{-1}} \tag{1.43}$$

1.3.3.5 Kinematic Viscosity (ν)

The coefficient of kinematic viscosity of the air depends on temperature and can be given by the equation

$$\nu = 1.32743 \times 10^{-5} + 9.22286 \times 10^{-8}\, T_{\mathrm{a}} \quad \mathrm{m^2\,s^{-1}} \tag{1.44}$$

where T_a is the air temperature (°C). Equation 1.44, as others related to the air properties here presented, is valid for air temperatures between −5 and 45°C only.

1.3.3.6 Dynamic Viscosity (μ)

It depends on the density (ρ) and the kinematic viscosity (ν) of the air at a given temperature (°C):

$$\mu = \rho \nu \quad \mathrm{g\,m^{-1}\,s^{-1}} \tag{1.45}$$

1.3.3.7 Latent Heat of Vaporisation (λ)

It is the amount of thermal energy needed to change 1 g of water into vapour and depends on the temperature. In the following formula,

$$\lambda = 2{,}500.788 - 2.37374\, T_{\mathrm{a}} \quad \mathrm{J\,g^{-1}} \tag{1.46}$$

where T_a is the air temperature (°C).

1.3.3.8 Psychrometric Constant (γ)

Even so the term *constant* is unfit because there has temperature dependence, it is yet used by tradition. It can be estimated for air temperatures between −5 and 45°C as

$$\gamma = \frac{c_p}{0.6223\,\lambda}\ {}^{\circ}\mathrm{C}^{-1} \tag{1.47}$$

1.3.3.9 Diffusivity of Water Vapour in the Air (D_v)

In order to evaluate water evaporation from any surface, it is needed to know the diffusivity coefficient of water vapour in the atmosphere, which depends on the air temperature (°C). The respective formula is

$$D_v = 2.12138 \times 10^{-5} + 1.4955 \times 10^{-7} T_a \ \ \mathrm{m}^2\,\mathrm{s}^{-1} \tag{1.48}$$

1.3.3.10 Specific Heat of the Water Vapour (c_{pv})

In the same way as the air, the water vapour can also be expanded at constant pressure, which requires an additional amount of energy. Such energy is the specific heat of the water vapour, which is estimated by

$$c_{pv} = 1.86641 - 2.61305 \times 10^{-5} T_a + 1.67249 \times 10^{-5} T_a^2 \ \ \mathrm{J\,g^{-1}\,{}^{\circ}C^{-1}} \tag{1.49}$$

This equation has a high determination coefficient, $r^2 = 0.998$.

1.3.4 *Atmospheric Humidity*

1.3.4.1 Definition of Air Saturation

Suppose a closed container having water and air thereover. The water molecules are displaced within the liquid mass in a speed that is proportional to its temperature; some of those molecules escape from the water surface, passing into the air. At the same time, some of the water molecules in the air enter the water surface again. The process of passing H_2O molecules into the atmosphere is the *evaporation*, while the inverse process is the *condensation.*

When there has a state of equilibrium between both processes of evaporation and condensation, that is, when the number of water molecules passing through the

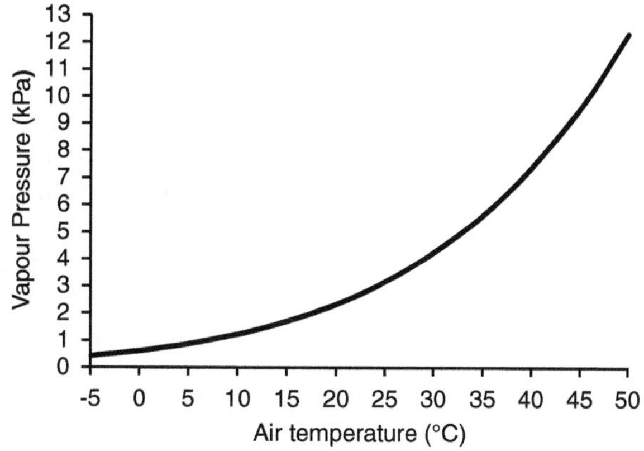

Fig. 1.6 Variation of the saturation vapour pressure of the air according to the temperature

surface into the atmosphere equals the number of those coming back to the water surface, then the air is said to be saturated. On the other hand, by increasing the temperature of the air leaves to an expansion of it; in the present case, such an expansion permits an increased number of water vapour molecules in a given volume of air. As a result, the higher the temperature, the greater the water vapour content of the atmosphere and higher its saturation pressure. It follows that the saturation pressure is a function of the temperature, given by Tetens' equation:

$$P_s(T) = 0.61078 \times 10^{7.5T/(T+237.5)} \text{ kPa} \tag{1.50}$$

Figure 1.6 shows the variation of the saturation pressure as a function of the air temperature.

If the temperature of the liquid surface is higher than that of the atmosphere, evaporation will occur even in the case that the atmosphere was saturated at the given temperature. This is a consequence of the fact that evaporation depends mainly on temperature of the liquid surface, not on that of the air. It is a very important fact as for the thermoregulation of animals.

1.3.4.2 Partial Vapour Pressure

It is the pressure carried off by the air content of water vapour in a non-saturated atmosphere. As it was previously discussed, there has a relationship of air temperature with its ability to absorb vapour; any change in the heat content of the air that was due to a temperature change must be equal to the latent heat of vaporisation, in order to give the partial vapour pressure of the atmosphere. The equation is

$$P_v = P_s(T_w) - P_a\gamma(T_a - T_w) \text{ kPa} \tag{1.51}$$

where T_w is the wet bulb temperature (°C), measured together with the dry bulb temperature (T_a, °C); P_a is the atmospheric pressure (kPa); γ is the psychrometric constant (°C^{-1}); and $P_s(T_w)$ is the saturation vapour pressure (kPa) of the air at temperature T_u.

Temperatures T_a and T_w are measured by means of a *psychrometer*, which is the standard instrument for air humidity evaluation. This device has two mercury or alcohol thermometers, one of which having a cloth cover that is maintained moist with water (the wet bulb thermometer) and the other is the dry bulb one, used for the dry bulb temperature or air temperature. The psychrometer must be exposed to the free air circulation under the shade and be protected against direct radiation. When the atmosphere is saturated, there has no evaporation from the wet bulb, and then both thermometers give the same temperature. If the atmosphere is not saturated, there has evaporation, and the temperature of the wet bulb thermometer is lower than that of the dry bulb thermometer; as greater the difference as drier the atmosphere.

1.3.4.3 Relative Humidity (U_R)

It is one of several atmospheric humidity units of measure, which gives the proportion of the observed humidity to the humidity of a saturated atmosphere, at the same temperature. The respective equation is

$$U_R = \frac{100P_v}{P_s(T_a)} \ \% \tag{1.52}$$

where P_v is the calculated partial vapour pressure (kPa) and $P_s(T_a)$ is the saturation pressure at the dry bulb temperature, T_a.

1.3.4.4 Absolute Humidity (Ψ)

It is the mass of water vapour in suspension per unit volume of air at a given temperature, calculated as

$$\Psi = \frac{10^3 M_W P_v}{RT_a} = \frac{2166.869 P_v}{T_a} \ \text{g m}^{-3} \tag{1.53}$$

where M_w is the molecular weight of water (18.016 mol g), R is the universal gas constant (8.3143 J mol^{-1} K^{-1}), P_V is the partial vapour pressure (kPa) and T_a is the air temperature (K).

1.3.4.5 Dew Point Temperature (T_{dp})

Dew point is the temperature in which the water vapour begins its condensation when cooled. In other words, it is the temperature of the air in which the saturation pressure equals the partial vapour pressure and can be estimated by the following equation (Campbell and Norman 1998):

$$T_{dp} = \frac{240.97 \ln(1.637251 \, P_v)}{17.502 - \ln(1.637251 \, P_v)} \ ^\circ\mathrm{C} \tag{1.54}$$

In general, as closer T_{dp} is to the air temperature T_a, the greater the probability of precipitation to occur.

1.3.4.6 Variation of the Air Humidity

When one measures the air relative humidity alongside the day, it can be observed that those measures are always higher by the morning and lower by the noon, following a sine curve inverse to that of the temperature. However, if it used the partial vapour pressure instead of the relative humidity, the curve will follow directly that of the temperature. During the day, the vapour pressure is higher at the ground surface and reduces with the distance from it; during the night, the variation is inverse.

The variation of the air humidity during the day is a result of temperature variations, but in general, there are no alterations in the water vapour content of the air. Because of this fact, some authors such as Campbell and Norman (1998) have questioned the utility of the relative humidity to evaluate atmospheric humidity, especially as long-time averages; in such cases, the average relative humidity would be meaningless. As an alternative, the above-mentioned authors recommended the conversion of the original psychrometric measures into partial vapour pressure or dew point temperature, then calculating the respective averages; these averages would be later converted without problems into relative humidity values.

1.3.5 Wind

1.3.5.1 Introduction

Wind is the movement of the air in relation to a given surface; if this surface is displaced in the same direction at the same velocity as the air, then there has no wind as for that surface, while it can be sensed by other motionless surfaces. On the other hand, if the atmosphere remains motionless, there has a wind over a displacing object; in such a case, the wind velocity is equal to the object's displacement speed. In other words, the wind can be caused by a movement of the atmosphere *per se* or by a displacement of the reference surface.

In meteorological terms, winds blow due to the conversion of solar energy into four other forms: internal, geopotential, latent and kinetic. Internal energy is the result from the molecular movement (heat); geopotential energy increases as the heated air expands itself and climbs to the sky while decreases when the air is cooled and goes down to the ground surface; latent energy is the energy stored in the atmosphere as water vapour; finally, the total atmospheric energy is the sum of the former three types, but part of it is converted into the kinetic energy for air displacement. There has a constant conversion from a type of atmospheric energy into another.

The kinetic energy is generated mainly by convective processes and represents a small fraction only of the total atmospheric energy, about 0.95% of the total solar radiation upon Earth (Thompson 1998); such a fraction is enough to sustain the atmospheric movements and balance the global energy.

In order to determine the relationship between organisms and atmospheric movements, it is necessary to know the wind velocity and the behaviour of the atmosphere within the so-called *terrestrial boundary layer*. Finally, it must be remembered that there has a conceptual difference between the wind and other environmental variables as air temperature and humidity. While these last variables are scalar quantities (i.e. they have one dimension), the wind is a vector quantity with two dimensions (magnitude and direction).

Several aspects such as the wind patterns over the Earth's surface and details about the respective mechanisms are beyond the purposes of this book. The interested readers will find more information about meteorological processes in the excellent books of Thompson (1998) and Houghton (2002), for example.

1.4 Problems

Problem 1.1. Find the zenith angle for Rio de Janeiro (latitude 22°54′10″ south, longitude 43°12′27″ west, 35-m altitude) at 14 h 30 min local time on March 25. Standard meridian is 45°.

Data:

L_t = latitude = −22°54′10″ = −22.902778°
L_g = longitude = 43°12′27″ = 43.2075°
d = day if the year = 84
h = day time = 14 h 30 min = 14.5

$$\delta = 23.45 \text{ sen}\left[\frac{360}{365.242}(284 + \text{d})\right] = 23.45 \text{ sen}\left[\frac{360}{365.242}(284 + 84)\right]$$

$$= 1.112176^\circ$$

$$e_t = -0.05107 - 0.00655\,d - 7.41244 \times 10^{-5}d^2 + 4.72702 \times 10^{-6}d^3 - 5.33999 \times 10^{-8}d^4 + 2.55422 \times 10^{-10}d^5 - 5.58785 \times 10^{-13}d^6 + 4.59779 \times 10^{-16}d^7 = -0.095723$$

$$\begin{aligned} h_0 &= 12 - \frac{M_h - L_g}{15} - e_t \\ &= 12 - \frac{45 - 43.2075}{15} - (-0.095723) = 11.976223 \text{ hours} \end{aligned}$$

$$\eta = 15(h - h_0) = 15(14.5 - 11.976223) = 37.856655°$$

$$\begin{aligned} \cos\theta &= (\cos L_t)(\cos\delta)(\cos\eta) + (\operatorname{sen} L_t)(\operatorname{sen}\delta) \\ &= \cos(-22.902778)\cos(1.112176)\cos(37.856655) \\ &\quad + \sin(-22.902778)\sin(1.112176) \\ &= 0.719615 \end{aligned}$$

$$\theta = \cos^{-1}(0.719615) = 43.977297° = 43°5838''$$

Problem 1.2. Consider the data of Problem 1.1 and calculate (a) duration of the photoperiod and (b) sunrise and sunset hours.

Data:

L_t = latitude = $-22°54'10''$ = $-22.902778°$
δ = declination of the sun = $1.112176°$
h_0 = noon = 11.976223 hr

$$\begin{aligned} \textit{Day length} &= \left(\frac{2}{15}\right)\cos^{-1}\left[\frac{-(\sin L_t)(\sin\delta)}{(\cos L_t)(\cos\delta)}\right] \\ &= \left(\frac{2}{15}\right)\cos^{-1}\left[\frac{-\sin(-22.902778)\sin(1.112176)}{\cos(-22.902778)\cos(1.112176)}\right] \\ &= 11.937341 = 11\,h\ 56\,\text{m}\ 14\,\text{s} \end{aligned}$$

The time at which sunrise begins is given by

$$\textit{Sunrise} = 11.976223 - 0.5(11.937341) = 6.007553 - 06:00:27\ \textit{hr}$$

and the time at the sunset is

$$\textit{Sunset} = 11.976223 + 0.5(11.937341) = 17.944894 = 17:56:42\ \textit{hr}$$

Problem 1.3. Suppose a pasture at latitude of 15°10′ south, longitude 48°35′ west and 225-m altitude. Determine the direct solar irradiance on the surface of the pasture at 14:30 h on July 7, 2011, assuming a sunny day covered with 10% of clouds. Air temperature is 26°C.

Data:

L_t = latitude $= -15°10' = -15.166667°$
L_g = longitude $= 48°35' = 48.583333°$
z = altitude = 225 m
h = day time = 14:30 h = 14.5 hr
d = day of the year = 188
T_a = air temperature = 26°C = 299.15 K
τ = atmospheric transmittance = 0.638 (see text)
n = cloudiness = 0.10
c = cloud-type coefficient = 0.35 (Table 1.2)

$$g = 9.78013 + 8.18 \times 10^{-5}(15.166667) + 1.168 \times 10^{-5}(15.166667)^2 - 3.1 \times 10^{-6}z = 9.783360\ \mathrm{m\,s}^{(-2)}$$

$$P_a = 101.325\ \exp\left\{\frac{-zg}{287.04T_a}\right\} = 101.325\ \exp\left\{\frac{-225(9.783360)}{287.04(299.15)}\right\} = 98.7605\ kPa$$

$$\delta = 23.45\ \mathrm{sen}\left[\frac{360}{365.242}(284 + \mathrm{d})\right] = 23.45\ \mathrm{sen}\left[\frac{360}{365.242}(284 + 188)\right] = 22.626866°$$

$$e_t = -0.05107 - 0.00655\,\mathrm{d} - 7.41244 \times 10^{-5}\mathrm{d}^2 + 4.72702 \times 10^{-6}\mathrm{d}^3 - 5.33999 \times 10^{-8}\mathrm{d}^4 + 2.55422 \times 10^{-10}\mathrm{d}^5 - 5.58785 \times 10^{-13}\mathrm{d}^6 + 4.59779 \times 10^{-16}\mathrm{d}^7 = -0.0691$$

$$\eta = 15\left(h - 12 + \frac{M_h - L_g}{15} + e_t\right) = 15\left(14.5 - 12 + \frac{45 - 48.583333}{15} - 0.0691\right) = 2.669789°$$

$$\begin{aligned}\cos\theta &= (\cos L_t)(\cos\delta)(\cos\eta) + (\mathrm{sen}\,L_t)(\mathrm{sen}\,\delta)\\ &= \cos(-15.166667)\ \cos(22.626866)\ \cos(2.669789) + \mathrm{sen}(-15.166667)\ \mathrm{sen}(22.626866)\\ &= 0.789257\end{aligned}$$

$$m = \frac{P_{\mathrm{a}}}{101.325 \cos \theta} = \frac{98.7605}{101.325(0.789257)} = 1.234947$$

$$\begin{aligned}\mathbf{S}_{\mathrm{dir}} &= 1,372.7[1 - n(1 - c)]\tau^{m} \cos \theta \\ &= 1,372.7[1 - 0.10(1 - 0.35)]0.638^{1.234947}(0.789257) \\ &= 581.5\ \mathrm{W\,m^{-2}}\end{aligned}$$

Problem 1.4. Consider the data of Problem 1.3 and include an air partial vapour pressure value of 2.13 kPa and a ground surface temperature of 39.2°C for the grassland. Calculate the following: (a) diffuse short-wave radiation and (b) long-wave radiation (downwards and upwards).

Data:

P_{v} = air partial vapour pressure = 2.13 kPa
T_{s} = grassland surface temperature = 39.2°C = 312.35 K
ρ_{s} = grassland surface reflectance = 0.25 (Table 1.3)
ε_{s} = grassland surface emissivity = 0.67 (Table 1.3)
$m = 1.234947$

Diffuse short-wave radiation:

$$\begin{aligned}\mathbf{S}_{\mathrm{dif}} &= 411.81(1 - \tau^{m}) \cos \theta = 411.81\left(1 - 0.638^{1.234947}\right)0.789257 \\ &= 138.4\ \mathrm{W\,m^{-2}}\end{aligned}$$

Reflected short-wave radiation:

$$\mathbf{S}_{\mathrm{ref}} = \rho_{\mathrm{s}}(\mathbf{S}_{\mathrm{dir}} + \mathbf{S}_{\mathrm{dif}}) = 0.25(581.5 + 138.4) = 180.0\ \mathrm{W\,m^{-2}}$$

Long-wave upward radiation:

$$\mathbf{L}_{\mathrm{C}} = \varepsilon_{s}\sigma T_{\mathrm{s}}^{4} = 0.67\left(5.67051 \times 10^{-8}\right)(312.35)^{4} = 361.6\ \mathrm{W\,m^{-2}}$$

Long-wave downward radiation:

$$T_{\mathrm{rc}} = T_{\mathrm{a}} - 20 + 10n = 299.15 - 20 + 10(0.10) = 280.15$$

$$\begin{aligned}\varepsilon_{\mathrm{c}} &= 0.84n + (1 - 0.84n)1.72\left(\frac{P_{\mathrm{v}}}{T_{\mathrm{a}}}\right)^{\frac{1}{7}} \\ &= 0.84(0.1) + [1 - 0.84(0.1)]1.72\left(\frac{2.13}{299.15}\right)^{1/7} = 0.86\end{aligned}$$

$$\mathbf{L}_{\mathrm{B}} = \varepsilon_{\mathrm{c}}\sigma {T_{\mathrm{rc}}}^{4} = 0.86\left(5.67 \times 10^{-8}\right)(280.15)^{4} = 300.4\ \mathrm{W\,m^{-2}}$$

Problem 1.5. Calculate the radiant heat load received by an animal in the pasture described above (Problems 1.3 and 1.4). A 0.15-m diameter black globe whose temperature was 40.1°C was used. The wind speed was 0.9 m s^{-1}.

Data:

T_a = air temperature = 26°C = 299.15 K
T_g = globe temperature = 40.1°C = 313.25 K
d = globe diameter = 0.15 m
ε_g = globe emissivity = 0.95
$\mathbf{L_B}$ = long-wave downward radiation = 300.4 Wm^{-2}
$\mathbf{L_C}$ = long-wave upward radiation = 361.6 Wm^{-2}
P_a = atmospheric pressure = 98.7605 kPa

Properties of the atmosphere at temperature T_a = 26°C:

$$\begin{aligned} \nu &= 1.32743 \times 10^{-5} + 9.22286 \times 10^{-8} T_a \\ &= 1.32743 \times 10^{-5} + 9.22286 \times 10^{-8}(26) = 1.567 \times 10^{-5}\ \text{m}^2\,\text{s}^{-1} \end{aligned}$$

$$\begin{aligned} k &= 0.02412 + 7.45143 \times 10^{-5} T_a \\ &= 0.02412 + 7.45143 \times 10^{-5}(26) = 0.026057\ \text{W}\,\text{m}^{-1}\,{}^\circ\text{C}^{-1} \end{aligned}$$

$$\rho = \frac{3{,}481.965\,P_a}{T_a} = \frac{3{,}481.965(98.7605)}{299.15} = 1{,}149.53\ \text{g}\,\text{m}^{-3}$$

$$\begin{aligned} c_p &= 1.00522 + 0.0004577\ \exp\left\{\frac{T_a}{32.07733}\right\} \\ &= 1.00522 + 0.0004577\ \exp\left\{\frac{26}{32.07733}\right\} = 1.006249\ \text{J}\ \text{g}^{-1}\,{}^\circ\text{C}^{-1} \end{aligned}$$

Non-dimensional numbers:

$$R_e = \frac{U\,d}{\nu} = \frac{(0.9)(0.15)}{1.567 \times 10^{-5}} = 8{,}615.188258$$

$$P_r = \frac{\rho\ c_p\ \nu}{k} = \frac{(1{,}149.53)(1.006249)\left(1.567 \times 10^{-5}\right)}{0.026057} = 0.695615$$

As the wind velocity is actually $U > 0.08$, there was forced convection at the globe surface (see Chap. 2 for details on the convection process). Then, the Grashof number G_r is not to be calculated.

$$h_c = 0.38\, k d^{-1}\, R_e^{\,0.6}\, P_r^{1/3}$$
$$= 0.38\,(0.026057)(0.15)^{-1}(8615.188258)^{0.6}(0.695615)^{0.333}$$
$$= 13.436446$$

$$\mathrm{RHL} = \frac{h_c}{0.95}(T_g - T_a) + \sigma T_g^4$$
$$= \frac{13.436446}{0.95}(313.25 - 299.15) + (5.67 \times 10^{-8})(313.25)^4$$
$$= 745.4\ \mathrm{W\ m^{-2}}$$

$$T_{rm} = \left(\frac{\mathrm{RHL}}{\sigma}\right)^{1/4} = \left(\frac{745.4}{5.67051 \times 10^{-8}}\right)^{1/4} = 338.6\ \mathrm{K}$$

If a black globe was not available, the following alternative method of Silva et al. (2010) can be used:

$$T_{rm} = \left(\frac{L_B + L_C}{0.944\ \sigma}\right)^{1/4} = \left(\frac{300.4 + 361.6}{0.944(5.67051 \times 10^{-8})}\right)^{1/4} = 333.5\ \ \mathrm{K}$$

$$\mathrm{RHL} = \sigma T_{rm}^4 = (5.67051 \times 10^{-8})(333.5)^4 = 701.5\ \mathrm{W\,m^{-2}}$$

The results of the two method presented small difference in T_{rm} values.

References

Ahrens, C.D.: Meteorology Today. West Publishing, St. Paul (1991)

Ayoade, J.O.: Introduction to Climatology for the Tropics. Wiley, New York (1983)

Barry, R.G., Chorley, R.J.: Atmosphere, Weather and Climate. Routledge, London (1988)

Bedford, T., Warner, C.: The globe thermometer in studies of heating and ventilation. J. Hygiene **34**, 458–473 (1934)

Boeker, E., van Grondelle, R.: Environmental Physics, 2nd edn. Wiley, Chichester (1999)

Bond, T.E., Kelly, C.F.: The globe thermometer in agricultural research. Agric. Eng. **36**, 251–260 (1955)

Bridgman, H.A.: Air pollution. In: Thompson, R.D., Perry, A. (eds.) Applied Climatology: Principles and Practice. Routledge, London (1997)

Brutsaert, W.: Evaporation into the Atmosphere: Theory, History and Applications. Reidel, Boston (1984)

Brutsaert, W.: On a derivable formula for long-wave radiation from clear skies. Water Resour. Res. **11**, 742–744 (1975)

Campbell, G.S., Norman, J.M.: An Introduction to Environmental Biophysics, 2nd edn. Springer, New York (1998)

Chapman, A.J.: Fundamentals of Heat Transfer. Macmillan, New York (1987)

Esmay, M.L.: Principles of Animal Environment. AVI Publishing, Westport (1969)

Gates, D.M.: Biophysical Ecology. Springer, New York (1980)

Henderson-Sellers, A., Robinson, P.J.: Contemporary Climatology. Longman, Harlow (1987)

Hey, E.N.: Small globe thermometers. J. Sci. Instrum. **1**, 955–957 (1968)
Hickey, J.R., Alton, G.M., Griffen, F.J., et al.: Extraterrestrial solar irradiance: two and one half years of measurement from Nimbus-7. Solar Energy **29**, 127–135 (1982)
Houghton, J.: The Physics of Atmospheres, 3rd edn. Cambridge University Press, Cambridge (2002)
Hounan, C.E.: Estimates of solar radiation over Australia. Aust. Meteorol. Mag. **11**, 1–14 (1963)
Hounan, C.E.: Revised regression equations for estimation of solar radiation over Australia. Aust. Meteorol. Mag. **17**, 91–94 (1969)
Keeling, C.D., Whorf, T.P.: Atmospheric CO_2 records from sites in the SIO air sampling network. In: Trends: A Compendium of Data on Global Change. Oak Ridge National Laboratory, Oak Ridge (2003)
Kuehn, L.A., Stubbs, R.A., Weaver, R.S.: Theory of the globe thermometer. J. Appl. Physiol. **29**, 750–757 (1970)
List, R.J.: Smithsonian Meteorological Tables, 6th edn. Smithsonian Institution Press, Washington (1971)
Liu, B.Y., Jordan, R.C.: The interrelationship and characteristic distribution of direct, diffuse, and total solar radiation. Solar Energy **4**, 1–19 (1960)
McCullough, E.C., Porter, W.P.: Computing clear day solar radiation spectra for the terrestrial ecological environment. Ecology **52**, 1008–1015 (1971)
Meeus, J.: Astronomical Algorithms. Willman-Bell, Richmond (1991)
Monteith, J.L., Unsworth, M.H.: Principles of Environmental Physics, 3rd edn. Academic, New York (2008)
Niemelä, S., Räisänen, P., Savijärvi, H.: Comparison of surface radiative flux parameterizations. Part 1: longwave radiation. Atmos. Res. **58**, 1–18 (2001)
Nieuwolt, S.: Tropical Climatology. Wiley, London (1978)
Pereira, N., Bond, T.E., Morrison, S.R.: "Ping-pong" ball into black-globe thermometer. Agric. Eng. **48**, 341–345 (1967)
Peterson, W.A., Dirmhirn, I.: The ratio of diffuse to direct solar irradiance (perpendicular do the sun's rays) with clear skies, a conserved quantity throughout the day. J. Appl. Meteorol. **20**, 826–828 (1981)
Prata, A.J.: A new longwave formula for estimating downward clear-sky radiation at the surface. Q. J. R. Meteorol. Soc. **122**, 1127–1151 (1996)
Silva, R.G.: Carga térmica radiante: revisando a técnica de avaliação (Radiant heat load: revisiting its evaluation technique). Revista de Engenharia Rural (Piracicaba) **13**, 77–84 (2002)
Silva, R.G., Guilhermino, M.M., Morais, D.A.E.F.: Thermal radiation absorbed by dairy cows in pasture. Int. J. Biometeorol. **54**, 5–11 (2010)
Smith, C.: Environmental Physics. Routledge, London (2001)
Stafford-Smith, D.M., Noble, I.R., Jones, G.K.: A heat balance model for sheep and its use to predict shade-seeking behavior in hot conditions. J. Appl. Ecol. **22**, 753–774 (1985)
Steele, L.P., Krummel, P.B., Langenfels, R.L.: Atmospheric CO_2 concentration from sites in the CSIRO Atmospheric Research GASLAB air sampling network. In: Trends: A Compendium of Data on Global Change. Oak Ridge National Laboratory, Oak Ridge (2002)
Swinbank, W.C.: Long-wave radiation from clear skies. Q. J. R. Meteorol. Soc. **89**, 339–348 (1963)
Thompson, R.D.: Atmospheric Processes and Systems. Routledge, London (1998)
Thornsson, S., Lindberg, F., Eliasson, I., Holmer, B.: Different methods for estimating the mean radiant temperature in an outdoor urban setting. Int. J. Climatol. **27**, 1983–1993 (2007)
Vernon, H.: The measurement of radiant heat in relation to human comfort. J. Ind. Hygiene **14**, 95–111 (1932)
Weber, M.R., Baker, C.B.: Comments on "The ratio of diffuse to direct solar irradiance (perpendicular do the sun's rays) with clear skies, a conserved quantity throughout the day". J. Appl. Meteorol. **21**, 883–885 (1982)
Willson, R.C.: Solar irradiance. In: Gurney, R.J., Foster, J.L., Parkinson, C.L. (eds.) Atlas of Satellite Observations Related to Global Change. Cambridge University Press, Cambridge (1993)

Chapter 2
Basic Physical Mechanisms

Abstract There were described the basic mechanisms of thermal exchange in animals: radiation, convection, conduction and evaporation. Radiation laws are discussed together with the radiative properties of the surfaces (absorptance, reflectance, transmittance and emissivity). Thermal exchange by radiation is discussed, and shape factors are described for different geometric structures taken as animal body models. Thermal exchange by conduction is described. There are presented and discussed equations for thermal exchange by free, forced and mixed convection in several cases; the respective equations for nondimensional numbers (Prandtl, Reynolds, Grashof and Nusselt) are presented together with some relevant information. The loss of heat by evaporation is discussed, and equations for nondimensional numbers (Schmidt, Sherwood) are presented. Several examples of application to animals are presented in detail.

Keywords Application to livestock • Conduction • Convection • Evaporation • Heat exchange • Radiation

2.1 Radiation

2.1.1 Definitions and Properties

Thermal radiation is the transfer of heat from a body to another through the displacement of electromagnetic energy by photons, which behave both as particles and waves. As for those electromagnetic waves, they constitute a fraction of the known electromagnetic spectrum (Table 2.1); this band is also known as infrared, from the end of the visible light to the beginning of the microwave band.

The waves are irradiated by atoms and molecules as a result of changes in their energetic content. The amount of radiation emitted depends on the nature of the material, the physical constitution and the absolute temperature of the emitting surface. Every surface whose temperature is above −273.15°C (or 0 K) emits

R. Gomes da Silva and A.S. Campos Maia, *Principles of Animal Biometeorology*, Biometeorology 2, DOI 10.1007/978-94-007-5733-2_2,

Table 2.1 Approximate division of the known electromagnetic spectrum

Radiation band	Wavelength (μm)	Frequency (GHz)
Cosmic rays	$\leq 4 \times 10^{-7}$	$\geq 7.5 \times 10^{11}$
Gamma rays	4×10^{-7} a 1.4×10^{-4}	2.1×10^{9} a 7.5×10^{11}
X-rays	10^{-4} a 2×10^{-2}	1.5×10^{7} a 3×10^{10}
Ultraviolet	5×10^{-3} a 0.39	7.7×10^{5} a 6×10^{7}
Visible light	0.39 a 0.78	3.8×10^{5} a 7.7×10^{5}
Infrared	0.78 a 1,000	3×10^{3} a 3.8×10^{5}
Hertzian waves	3×10^{7} a 10^{11}	3×10^{-6} a 3×10^{3}

The limits of some bands were not yet definitively established (GHz = 10^9Hz)

thermal radiation at wavelengths that depend on that temperature. The waves travel at the speed of light or 2.997925×10^8 m s^{-1} (in the vacuum), and when they attain a surface, a portion of the incoming energy is reflected, a portion is transmitted through the surface, and the rest is absorbed by it. Some important definitions are given as follows:

Radiant flux density. The total flux of radiant energy emitted by a given surface is the *radiant flux* density, which is given in W m^{-2}. The radiant flux density emitted by a surface at a given wavelength per unit area per unit time is its *emissive power* (**R**). In order to know the total amount of emitted radiation, we must know the amounts of radiation effectively emitted at each wavelength, $\mathbf{R}_\lambda$:

$$\mathbf{R} = \int_0^\infty \mathbf{R}_\lambda \, \mathrm{d}\lambda \tag{2.1}$$

Radiosity. However, some of the incoming radiation is reflected and this reflected, radiation leaves the surface together with that is emitted by the surface. Then, we can define *radiosity* (**J**) as the total flux of radiant energy leaving the surface by emission plus reflection per unit area per unit time:

$$\mathbf{J} = \int_0^\infty \mathbf{J}_\lambda \, \mathrm{d}\lambda \tag{2.2}$$

Irradiance. Finally, there is the total radiant flux density that attains the surface coming from the external environment, the *irradiance* (**G**):

$$\mathbf{G} = \int_0^\infty \mathbf{G}_\lambda \, \mathrm{d}\lambda \tag{2.3}$$

The three properties above described are related among themselves, in a way that each one can be deduced from the knowledge of the other two. Thus

$$\mathbf{J} = \mathbf{R} + \rho\mathbf{G} = \mathbf{R} + (1 - \alpha)\mathbf{G} \tag{2.4}$$

where ρ is the reflectance and α the absorptance of the surface, which will be defined as follows. A real surface can behave in four ways with respect to the thermal radiation: (a) reflecting the incident energy, (b) absorbing the energy, (c) transmitting the energy and (d) emitting the energy.

2.1.1.1 Reflectance (ρ)

It is the fraction of the radiant flux striking on a surface at a given wavelength and which is reflected depending on the direction of the incoming radiation and that of the reflected one.

2.1.1.2 Absorptance (α)

Fraction of the radiant flux reaching the surface at a given wavelength, which is absorbed by the surface. It depends on the direction and on the wave length of the incoming radiation.

2.1.1.3 Transmittance (τ)

Fraction of the radiant flux reaching the surface at a given wave length, that is transmitted through the surface.

In general, $\rho + \alpha + \tau = 1$. However, most gases present high τ but low α and ρ values. For example, the air under normal pressure conditions is virtually transparent to the thermal radiation, in such a way that for the atmosphere $\tau = 1$ and $\alpha \cong \rho \cong 0$. Other gases, as CO_2, present high absorptance values for thermal radiation.

On the other hand, solid surfaces in general (except glass) are opaque to the thermal radiation, and for them, $\tau = 0$ and $\rho + \alpha = 1$.

2.1.1.4 Emissivity (ε)

It is the ability of a surface to emit thermal radiation, as compared to a standard surface (*black body*). A black body would be any surface which can emit as radiation the entire energy amount contained in it, independently of the wavelength. By definition, $\varepsilon = 1$ for such a surface.

There has no perfect black bodies in the nature, except perhaps the astronomical phenomena known as *black holes*. An approximate model would be a great container whose inner walls were at a uniform, constant temperature. If it has a small hole in the wall, any radiation entering through the hole into the container will reflect successively on the inner surfaces; at each time, a fraction of this energy is absorbed and other fraction is reflected; at the end, it will be almost null the probability that any radiation amount escapes through the hole to the external environment. Thus, the virtual "surface" of the open hole is a black body, because it absorbs every incoming radiation and reflects none.

However, most real surfaces – named as *grey bodies* – behave in a different way, because α is always <1 for them; in addition, there has a dependence on the incoming radiation wavelength. As $\rho > 0$ in real surfaces, they always reflect some portion of the incoming radiation. It is interesting to note that $\alpha = \varepsilon$ for the grey bodies, a fact that can be explained as follows.

Turning back to the hole in the container, suppose that within the cavity the irradiance is equal to the emissive power of the inner walls, $\mathbf{G} = \mathbf{R}_b$. If in the cavity is placed a small object whose surface is a grey body with emissivity ε and absorptance α, being maintained at the same temperature T as the cavity, then the thermal balance will require that the emitted energy be equal to that absorbed. Thus

$$\alpha\mathbf{G} = \mathbf{R} = \varepsilon\mathbf{R}_b \tag{2.5}$$

where $\mathbf{R}_b$ is the emissive power of a black body. But as $\mathbf{G} = \mathbf{R}_b$, then we have $\alpha\mathbf{R}_b = \varepsilon\mathbf{R}_b$ and finally $\alpha = \varepsilon$. As a consequence, *every real surface with a high absorptance has an equally high emissivity with respect to a given wavelength and a given temperature*. On the other hand, for the most natural surfaces, $\rho = 1 - \alpha$ and $\rho = 1 - \varepsilon$.

2.1.2 Radiation Laws

2.1.2.1 Stefan-Boltzmann's Law

The emissive power of the radiation emitted by a black body is proportional to the fourth power of its absolute temperature:

$$\mathbf{R}_b = \sigma T^4 \ \text{W m}^{-2} \tag{2.6}$$

where $\sigma = 5.67 \times 10^{-8}$ W m^{-2} K^{-4} is the *Stefan-Boltzmann constant*. By considering this law, it is also possible to define the emissive power of a grey body as

$$\mathbf{R}_\lambda = \varepsilon_\lambda \sigma T^4 \ \text{W m}^{-2} \tag{2.7}$$

in which ε_λ is the emissivity of the surface to given wavelength λ. Table 1.3 shows the emissivity values for some surface types.

2.1.2.2 Planck's Law

The emissive power is a function of the wavelength and the surface temperature. Max Planck applied to this case his quantum theory to explain radiation by the model:

$$\mathbf{R}_{\lambda,T} = \frac{a}{\lambda^5 \left[e^{b/(\lambda T)} - 1\right]} \ \mathrm{m}^{-2}\,\mu\mathrm{m}^{-4} \tag{2.8}$$

Where:

$a = 2\pi h c_0 = 3.741775 \times 10^8$ W m^{-2} μm^4
$b = hc_0/k = 14{,}387.8$ μm K
h = Planck's constant = 6.626075×10^{-34} J s
k = Boltzmann's constant = 1.380658×10^{-23} J K^{-1}
c_0 = light speed in the vacuum = 2.99792458×10^8 m s^{-1}

The Stefan-Boltzmann constant is related to the above given constant values in a way that

$$\sigma = \left(\frac{\pi}{b}\right)^4 \left(\frac{a}{15}\right) = 5.67 \times 10^{-8}\ \mathrm{W\,m^{-2}\,K^{-4}}$$

By using Planck's law it is possible to determine a curve for the variation of emissive power of a surface, according to the wavelength. In animals, the skin behaves as a black body for the thermal radiation (infra-red), and Eq. 2.8 can be easily applied. Suppose a swine whose hairless skin is at a temperature of 35°C or 380.15 k. The emissive power is calculated for the several wavelengths from 0.5 to 100 μm by means of formula 2.8, thus obtaining the curve shown in Fig. 2.1.

In this example there was no power emission at wavelengths $\lambda < 2$ and >75 μm. In order to estimate the total radiation energy emitted, one would calculate by integration the area under the curve between the specified limits (2 and 75 μm).

2.1.2.3 Wien's Law

Deduced from Eq. 2.8, this law states that the emissive power of a surface varies inversely to the wavelength and proportionally to the surface temperature. Its maximum value is found for a wave length given by the relation

$$\lambda_{max} = 2897.8\,T^{-1}\ \mu\mathrm{m} \tag{2.9}$$

where T is the surface temperature (K).

2.1.2.4 Cosine Law

When a radiant beam reaches a surface with an area so small that the incoming rays can be considered as parallel ones, then the surface irradiance will depend on its angle with respect to the direction of the rays.

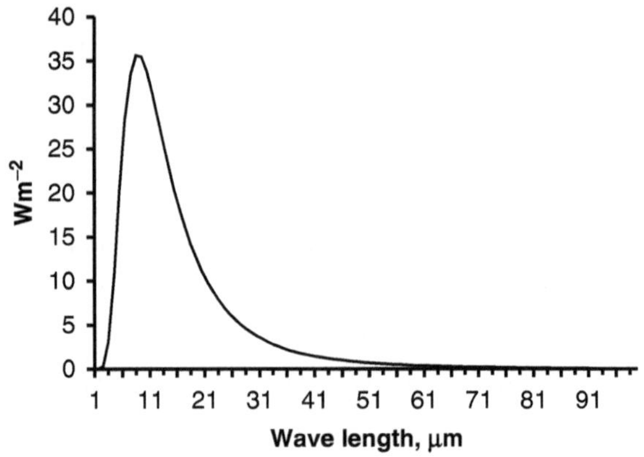

Fig. 2.1 Variation of the emissive power at the skin surface of an animal for a cutaneous temperature of 35°C

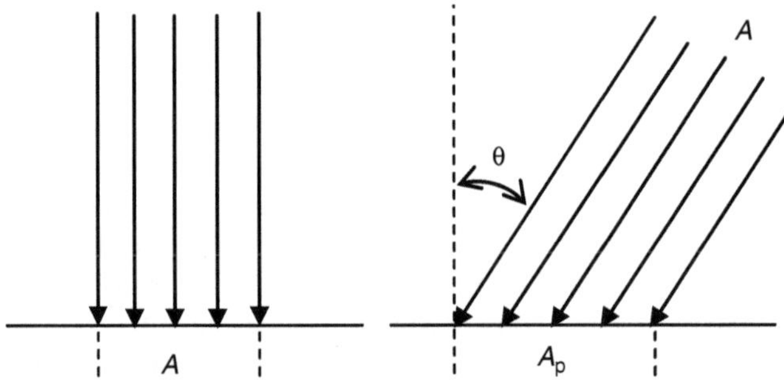

Fig. 2.2 The area of a horizontal surface illuminated by a radiation beam increases in proportion to its zenith angle to the surface, θ

This fact can be illustrated as follows. Suppose that a light beam is perpendicular (i.e. at a zenith angle of 90°) to a plane horizontal surface and results in a bright focus with area A_p. Then, as the zenith angle of the beam is reduced, the surface area of the beam focus increases proportionally to the beam angle, as it is shown in Fig. 2.2.

If A is the area of the focus relative to the angle $\theta = 0°$, then it follows that

$$A_p/A = \cos\,\theta \tag{2.10}$$

and the radiant flux density attaining effectively the surface will be given as

$$\mathbf{R} = \mathbf{R}_0 \cos\theta \text{ W m}^{-2} \tag{2.11}$$

where $\mathbf{R}_0$ is the radiant flux density normal to the surface.

2.1.2.5 Beer's Law

If the radiation beam passes through a given material before reaching a surface, then its intensity will be decreased at an exponential rate, depending on the thickness of the crossed layer and on its absorptance relative to the wavelength. Considering that the radiant flux reaches a given x point during its path across the layer, it will have there a flux density $\mathbf{R}_x$, and then, the absorption of radiation by a layer with thickness dx will be estimated as

$$\mathrm{d}\mathbf{R} = -\kappa\,\mathbf{R}\,\mathrm{d}x \tag{2.12}$$

where κ is the *attenuation* or *extinction coefficient*, defined as the probability a ray be intercepted along a distance dx through the layer. By integration, we can obtain the radiation flux density at any point x:

$$\mathbf{R}_x = \mathbf{R}_0\,\mathrm{e}^{-\kappa x} \tag{2.13}$$

by considering that $\mathbf{R}_0$ is the flux density at $x = 0$.

This law is applied strictly to the case in which the wave length band is sufficiently narrow so that the κ value is constant for the entire band. However, it is often used for the solar radiation crossing the atmosphere (in this case, absence of dispersion must be assumed); then, the distance dx is evaluated in terms of *atmospheric mass number*, which is given by:

$$m = \frac{P_\mathrm{a}}{101.325\cos\theta}$$

as defined previously in Chap. 1 (Eq. 1.7a).

2.1.3 *Radiation Geometry*

2.1.3.1 Effective Irradiation

It is often desired to evaluate the amount of thermal radiation received on a body's surface. In order to do this, the relationship between surface geometry and directional properties of the radiation must be taken into account.

Suppose a sphere exposed to a radiant flux that comes into it at an angle β with respect to the horizontal plane. The surface area of the sphere is given by $4\pi r^2$, where r is its radius; however, the radiant source "sees" the apparent area of the sphere, which is a circle whose area is $A_\mathrm{p} = \pi r^2$. This apparent area is projected over a horizontal surface below, producing a shadow whose form and area (A_s) depend on the angle of the radiation to the horizontal surface. Then

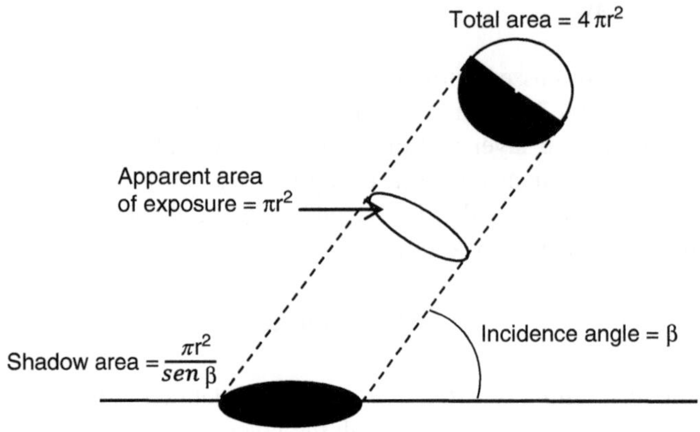

Fig. 2.3 Geometry of the shadow from a sphere exposed to radiation

$$A_s = \frac{\pi r^2}{\text{sen}\,\beta}\ \text{m}^2 \tag{2.14}$$

as it is shown in Fig. 2.3.

Irregular bodies as those of animals offer many difficulties for the evaluation of the area exposed to radiation; in such cases there were used equivalent bodies, as spheres or cylinders. These models are related to the real bodies by an analogy with the general body conformation. For example, a quadruped as a cow or a swine can be represented by a horizontal cylinder with the appropriate measures.

The proportion of the body area that receives radiation is determined by a *shape factor*, which depends on the geometry and on the radiation directional properties. For practical purposes, the shape factor is determined as the relation of the shadow area to the total area of the body or object:

$$F_c = \frac{A_s}{A} \tag{2.15}$$

Shape Factor for a Sphere

Let us have a sphere of radius r, as that in Fig. 2.3. The shadow projected by it has an area $A_s = \pi r^2/\sin\beta$, while the total area of the sphere is $A = 4\pi r^2$. The shape factor is then calculated as:

$$F_c = \frac{A_s}{A} = \frac{\pi r^2/\sin\beta}{4\pi r^2} = \frac{0.25}{\sin\beta} \tag{2.16}$$

Shape Factor for a Vertical Cylinder

Let us have a cylinder of radius r and length z, placed vertically on a horizontal surface. It receives radiation at an angle β with respect to the surface and casts a shadow on it. Total area of the cylinder and that of its shadow are given respectively, by

$$A = 2\pi r(\mathrm{r} + z)\ \mathrm{m}^2$$
$$A_\mathrm{s} = \pi r^2 + 2rz(\tan\beta)^{-1}\ \mathrm{m}^2$$

Then, the shape factor will be

$$F_\mathrm{c} = \frac{\pi r^2 + 2rz\,(\tan\beta)^{-1}}{2\pi r(r+z)} \tag{2.17}$$

Shape Factor for a Horizontal Cylinder

The shade area cast by a horizontal cylinder over a horizontal surface depends on the elevation angle of the radiation source (β) and the azimuth angle of the cylinder axis with respect to the radiation source (ω). See Fig. 2.4. Then the total area of the shade will be:

$$A_\mathrm{s} = 2rz(\sin\beta)^{-1}\left[1 - (\cos\beta)^2(\cos\omega)^2\right]^{1/2}\ \mathrm{m}^2 \tag{2.18}$$

As the total area of the cylinder is the same above given, the shape factor can be defined as

$$F_\mathrm{c} = \frac{2rz(\sin\beta)^{-1}\left[1 - (\cos\beta)^2(\cos\omega)^2\right]^{1/2} + \pi r^2(\tan\beta)^{-1}(\cos\omega)}{2\pi r(r+z)} \tag{2.19}$$

It must be stressed that, according to Monteith and Unsworth (2008), the shape factor of horizontal cylinders with relation $z/r = 4$ is independent from the azimuth angle, ω, if the elevation of the radiation source is high as $\beta > 40°$.

As it was suggested by Campbell and Norman (1998), a cylinder with hemispherical ends would be better than a simple cylinder as a model for quadruped animals. However, the respective equation given by those authors does not take into consideration the azimuth angle; this is an important aspect, as for the practical use in animals on range in the open field. Equation 2.19 can be modified in order to obtain F_c for a cylinder with hemispherical ends

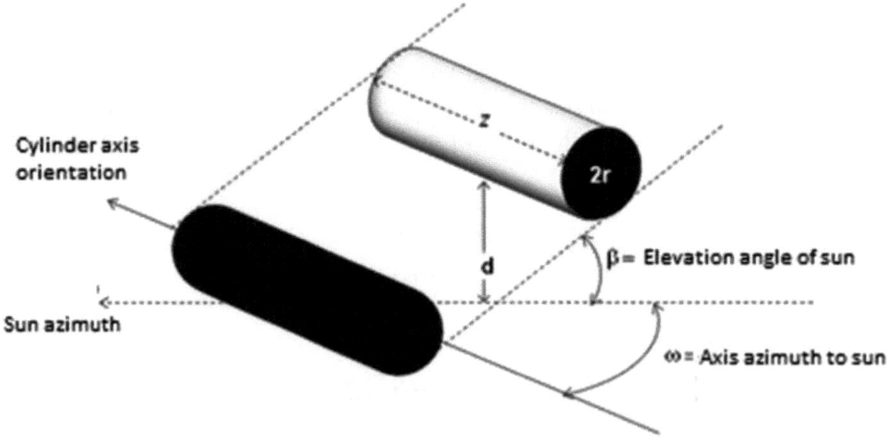

Fig. 2.4 Radiation geometry of a horizontal cylinder of radius r and length z, placed at a distance d from a horizontal surface. There are a solar elevation angle β and an azimuth angle ω

$$F_{\mathrm{c}} = \frac{(\sin\beta)^{-1}\left\{2rz\left[1-(\cos\beta)^{2}(\cos\omega)^{2}\right]^{1/2}+\pi r^{2}\right\}}{2\pi r(2r+z)} \tag{2.20}$$

Finally, as for the validity of the cylinder model for animals, Underwood and Ward (1966) measured 25 men and 25 women in swimsuits and photographed them at 19 different angles; in each case the shaded area was evaluated by planimetry methods. The average projected body area was close to that of a cylinder 1.65 m high and 0.12 m radius. A similar study was done by Clapperton et al. (1965) with sheep; it was found that the bodies of those animals (with a normal wool fleece) were equivalent to cylinders 0.91 m long with radius $r = 0.23$ m.

Shape Factor for a Horizontal Prolate Ellipsoid

As it is shown in Fig. 2.5, a prolate ellipsoid is a body with a circular transverse section and an elliptical longitudinal one which is often used as model for bird bodies. In most of those cases, there has a small difference between the greatest (R) and the smallest (r) radius, given in metres, in a way that the azimuth angle ω of the body axis can be neglected without much error. The surface area can be approximated by using Knud Thomsen's formula:

$$A = 4\pi\left(\frac{2(Rr)^{x}+r^{2x}}{3}\right)^{1/x}\ \mathrm{m}^{2} \tag{2.21}$$

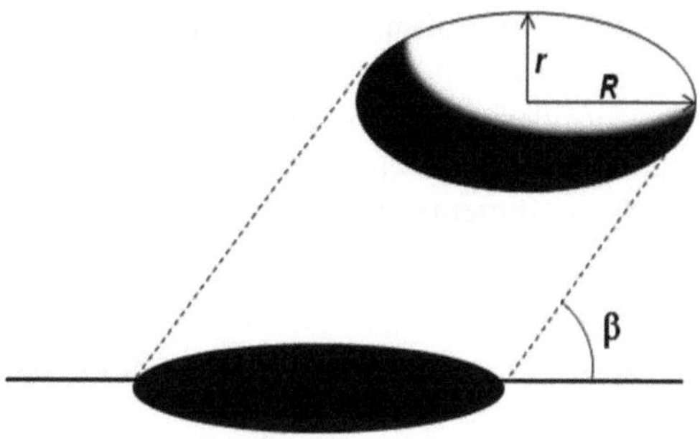

Fig. 2.5 Radiation geometry of a horizontal ellipsoid with greatest radius *R* and smallest radius *r*, exposed to a radiation source at an elevation angle *β*

where $x = 1.6075$.

The shaded area will be

$$A_s = \pi R^2 \left[1 + \frac{r^2}{R^2 tan^2 \beta}\right]^{1/2} \text{ m}^2$$

according to Monteith and Unsworth (2008). Finally:

$$F_c = \frac{A_s}{A} \tag{2.22}$$

2.1.3.2 Shape Factor and Irradiance

By means of the shape factor, it is possible to estimate the amount of thermal radiant energy reaching the surface of an animal or object. The flux of energy leaving the radiation source (radiosity) as evaluated at a horizontal plane is symbolised by $\mathbf{J}_h$. If it is assumed that this energy attains its target surface without losses, the irradiance of that surface can be considered as equal to the radiosity of the radiant source. Thus, the radiant flux that effectively reaches the target surface is defined by

$$\mathbf{R} = F_c \, \mathbf{J}_h \text{ W m}^{-2} \tag{2.23}$$

In other words, the shadow projected by a body over a horizontal surface allows the determination of the average irradiance of the body with respect to a radiation source, for example, the sun, since the values of the direct solar radiation, the surface area of the body, the solar elevation angle and the azimuth angle of the object with respect to sun.

2.1.3.3 Radiation Exchange Between Surfaces

By considering the simplest case, let us have two parallel planes, S_1 and S_2, each one at a known surface temperature T_1 and T_2, respectively. The respective emissivity coefficients are ε_1 and ε_2. Then

$$\mathbf{R}_1 = \varepsilon_1 \sigma T_1^4$$
$$\mathbf{R}_2 = \varepsilon_2 \sigma T_2^4$$

The amount of radiant energy leaving S_1 in direction to S_2 is given by

$$\mathbf{R}_1 = \mathbf{J}_1 - \mathbf{G}_1$$

where $\mathbf{J}_1$ is the radiosity and $\mathbf{G}_1$ the irradiance of S_1. However, we know that $\mathbf{J} = \mathbf{R} + (1 - \varepsilon)\mathbf{G}$ and also that $\mathbf{R} = \varepsilon\sigma T^4$. Thus, we have for the two surfaces

$$\mathbf{J}_1 = \varepsilon_1 \sigma T_1^4 + (1 - \varepsilon_1)\mathbf{G}_1$$
$$\mathbf{J}_2 = \varepsilon_2 \sigma T_2^4 + (1 - \varepsilon_2)\mathbf{G}_2$$

Now, suppose that all the radiant energy leaving S_1 attains S_2 without losses, in a way that any fraction reflected by a surface reaches another, it follows that

$$\mathbf{R}_1 = -\mathbf{R}_2$$
$$\mathbf{G}_1 = \mathbf{J}_2$$
$$\mathbf{G}_2 = \mathbf{J}_1$$

By rearranging those equations and expressing them in terms of matrices, we have

$$\begin{bmatrix} \varepsilon_1 \sigma T_1^4 \\ \varepsilon_2 \sigma T_2^4 \end{bmatrix} = \begin{bmatrix} 1 & -(1-\varepsilon_1) \\ -(1-\varepsilon_2) & 1 \end{bmatrix} \cdot \begin{bmatrix} \mathbf{J}_1 \\ \mathbf{G}_1 \end{bmatrix}$$

whose solution is

$$\begin{bmatrix} \mathbf{J}_1 \\ \mathbf{G}_1 \end{bmatrix} = \begin{bmatrix} 1 & -(1-\varepsilon_1) \\ -(1-\varepsilon_2) & 1 \end{bmatrix}^{-1} \begin{bmatrix} \varepsilon_1 \sigma T_1^4 \\ \varepsilon_2 \sigma T_2^4 \end{bmatrix}$$
$$= [1 - (1-\varepsilon_1)(1-\varepsilon_2)]^{-1} \begin{bmatrix} \varepsilon_1 \sigma T_1^4 + (1-\varepsilon_1)\varepsilon_2 \sigma T_2^4 \\ (1-\varepsilon_2)\varepsilon_1 \sigma T_1^4 + \varepsilon_2 \sigma T_2^4 \end{bmatrix}$$

As the thermal radiant flux is given by $\mathbf{R}_1 = \mathbf{J}_1 - \mathbf{G}_1 = -\mathbf{R}_2$, then

$$\begin{aligned}\mathbf{R} &= \frac{\varepsilon_1 \sigma T_1^4 + (1-\varepsilon_1)\varepsilon_2 \sigma T_2^4 - (1-\varepsilon_2)\varepsilon_1 \sigma T_1^4 - \varepsilon_2 \sigma T_2^4}{1-(1-\varepsilon_1)(1-\varepsilon_2)} \\ &= \frac{\varepsilon_1 \varepsilon_2 \sigma \left(T_1^4 - T_2^4\right)}{1-(1-\varepsilon_1)(1-\varepsilon_2)}\ \mathrm{W\,m^{-2}} \end{aligned} \tag{2.24}$$

2.2 Conduction

2.2.1 Definitions

Thermal conduction is the transfer of heat among parts of a body by means of the kinetic energy displacement of the molecules, or by the transportation of free electrons as in the case of metals. Such a flux passes from the highly energised molecules to those less energised ones, in other words, from a zone of high temperature to another at lower temperature.

A fundamental aspect of the conduction is the need of direct contact among the molecules of the bodies or surfaces involved. Thus, the transfer of thermal energy by such a process can occur only within the mass of a body, or between two bodies in mutual direct contact.

Thermal diffusivity is the physical property of the substance from which a body is constituted; it expresses the ability of the body to transfer thermal energy, in relation to its ability to store energy (*specific* heat). Then

$$D = \frac{k}{\rho c_{\mathrm{p}}}\ \mathrm{m^2\,s^{-1}} \tag{2.25}$$

where k is the *thermal conductivity*, ρ is the *density* and c_{p} is the specific heat of the material the body is constituted from. As greater or smaller is the thermal diffusivity of the substance, the greater or smaller is the speed in which the heat can spread through the body. Table 2.2 shows the values of thermal conductivity for some materials.

The ability of some substances to permit the passing of the energy flux is greater than that of other substances, as it occurs with respect to the electrical energy. Such an analogy is not incidental and is enough to justify the use of the electrical model to explain thermal conduction.

Let us have a board made of any material, with a surface area A and a thickness Δx. One of the faces of the board is at a temperature T_1, while the other face is at temperature T_2. The heat flux through the board will be proportional to its area and to the absolute value of the temperature differential, $|T_1 - T_2|$, but inversely

Table 2.2 Coefficients of thermal conductivity (k) for some materials

Material	Conductivity, $W \cdot m^{-1} \cdot {}^{\circ}C^{-1}$
Diverses	
Water (0°C)	0.562
Cotton	0.06
Aluminium	220
Sand (20 cm layer)	0.027
Asbestos cement board	0.63
Copper	386
Concrete	1.4
Cork	0.039
Stucco	0.72
Glass fibre	0.037
Polystyrene	0.025
Common brick	0.72
Glass	0.76
Wood	
Agglomerate	0.087
Plywood	0.12
Softwood	0.12
Hardwood	0.16
Animal tissues	
Subcutaneous fat (blubber)	0.18–0.25
Muscle	0.41
Skin (human)	0.037

From Rohsenow et al. (1998), Monteith and Unsworth (2008) and others

proportional to the thickness Δx. The thermal flux by conduction between both faces of the board will be given by

$$\mathbf{K} = KA(T_1 - T_2) \text{ W} \tag{2.26}$$

where K is the *thermal conductance* (W m^{-2} °C^{-1}) of the board. By taking into account the electric model, we must have the *thermal resistance*, r, of the board material; it is the inverse of the conductance, or $r = 1/K$. Then, after eliminating the term A from the formula 2.26 we have the thermal flux by conductance

$$\mathbf{K} = \frac{T_1 - T_2}{r} \text{ W m}^{-2} \tag{2.27}$$

2.2.2 Conduction in Multiple Layers

Suppose a board constituted into many layers of different materials and different thickness. Figure 2.6 shows the three-layer case.

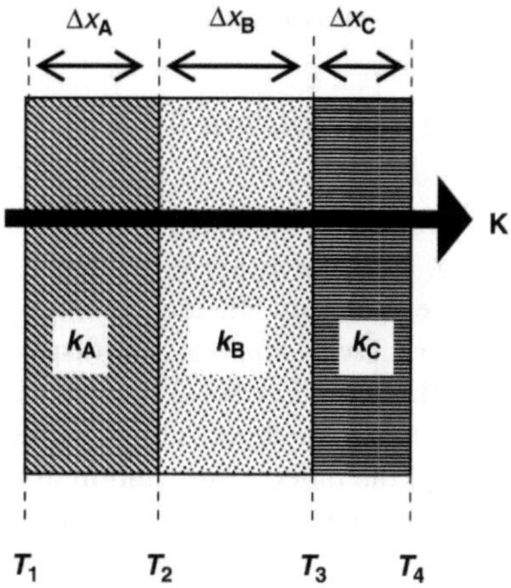

Fig. 2.6 Heat flux by conduction, **K**, through three layers of different thickness made from different material. Values T_1 to T_4 are temperatures of the surfaces limiting the layers, Δx_A to Δx_C are the respective layer thicknesses and k_A and k_C are the conductivity coefficients of the layers

By considering Eq. 2.27 we have:

$$\mathbf{K} = \frac{T_1 - T_2}{r_A} + \frac{T_2 - T_3}{r_B} + \frac{T_3 - T_4}{r_C}$$

where r_i (i = A, B, C) is the thermal resistance of the ith layer

$$r_i = \frac{\Delta xi}{k_i}$$

and Δx_i and k_i are the thickness and the thermal conductivity of the constituting material, respectively.

Considering the electrical model, the total resistance to heat flux through the three layers can be estimated by adding the three series partial resistances, $r_T = r_A + r_B + r_C$:

$$\mathbf{K} = \frac{(T_1 - T_2) + (T_2 - T_3) + (T_3 - T_4)}{r_A + r_B + r_C} = \frac{T_1 - T_4}{r_T}$$

Generalising for n layers, we can have a more practical formula:

$$\mathbf{K} = \frac{\rho c_p (T_1 - T_{n+1})}{\sum_{i=1}^{n} r_i} \text{ W m}^{-2} \tag{2.28}$$

where ρ is the density (g m^{-3}) and c_p is the specific heat (J g^{-1} °C^{-1}) of air at ambient temperature. Then the thermal resistance can be defined as

$$r_i = \frac{\rho c_p \Delta xi}{k_i} \text{ s m}^{-1} \tag{2.29}$$

2.3 Convection

2.3.1 Definition

Convection is the transfer of thermal energy by displacement of the air or any other fluid. If such a displacement is caused by a density differential – a consequence of the temperature difference – then the process is named as *free convection* or *passive convection*. If the displacement is caused by active forces, as pumps, ventilators or any wind source, we have a *forced* or *active* convection.

Let us have an object immersed into a fluid at a temperature that is lower than that of the object. The fluid in direct contact with the surface of the object will be heated to a temperature close to that of the surface. Even if the fluid is moving, there is always a layer of some thickness attached to the object's surface – the *boundary layer* – which is at a temperature higher than that of the fluid around it. As more turbulent is the fluid displacement, the less tick is this boundary layer.

Thermal energy is transferred by conduction from the object's surface into the boundary layer, which becomes less dense than the rest of the fluid involving it; there has a tendency for it to ascend against the gravity force and push away the surface. As it happens, the mass of heated fluid is displaced carrying itself the absorbed heat and is substituted by an equivalent volume of colder, denser fluid; this results into an increased temperature differential at the object's surface again, and the process is repeated.

Heat flux by convection depends on the following: (a) the temperature differential between the surface and the fluid in contact to it and (b) the *convection coefficient* of the fluid relative to the physical structure of the surface. Then

$$\mathbf{C} = h_c(T_s - T_a)\ \mathrm{W\,m^{-2}} \tag{2.30}$$

In the above equation, h_c is the *convection coefficient*, given by

$$h_c = \frac{k}{d}\frac{d}{z}\ \mathrm{W\,m^{-1}\,{}^{\circ}C^{-1}} \tag{2.30a}$$

where d is the *characteristic dimension* (m) of the surface, object or body; z is the thickness of the boundary layer (m). In practical terms, Eq. 2.30 can be re-written as:

$$\mathbf{C} = \frac{\rho c_p (T_s - T_a)}{r_H}\ \mathrm{W\,m^{-2}} \tag{2.31}$$

where ρ is the density (g·m^{-3}) and c_p the specific heat (J·g^{-1}·°C^{-1}) of the air at the given temperature; r_H is the resistance of the fluid to the heat transfer by convection,

$$r_H = \frac{\rho c_p d}{k N_u}\ \mathrm{s\,m^{-1}} \tag{2.32}$$

The value N_u in Eq. 2.32 is the *Nusselt number*, whose definition and calculation will be discussed as follows.

2.3.2 *Free Convection*

2.3.2.1 Nondimensional Numbers

The Nusselt number for free convection is defined in its general form as

$$N_u = a\,G_r^m P_r^n \tag{2.33}$$

where a, m and n are constants whose values vary according to the specific case; G_r is the *Grashof number*,

$$G_r = \frac{g\,d^3(T_s - T_a)}{v^2(T_a + 273.15)} \tag{2.34}$$

where g is the acceleration of gravity, d is the characteristic dimension of the surface, v is the kinematic viscosity of the air at temperature T_a (°C) and T_s is the surface temperature (°C).

Another nondimensional number to be considered is P_r or the *Prandtl number*, which is associated to the relation between actual thickness of the boundary layer and its thickness needed to perform the desired heat transfer. Then

$$P_r = \frac{\rho c_p v}{k} \tag{2.35}$$

where ρ, v, k and c_p are thermo-physical properties of the air at temperature T_a, as described in Chap. 1.

Some equations for the calculation of the Nusselt number will be given as follows, according to the geometrical conformation and the position of a specific surface or object.

Vertical Planes (Churchill and Chu 1975b)

$$N_u = \left\{0.825 + \frac{0.387(G_r P_r)^{1/6}}{\left[1 + (0.492/P_r)^{9/16}\right]^{8/27}}\right\}^2 \tag{2.36}$$

For the present case, the thermo-physical properties of the air must be calculates for the temperature $T_m = 0.5\,(T_s + T_a)$.

Vertical Cylinders (Monteith and Unsworth 2008)

$$N_u = 0.65\ G_r^{1/4}\, P_r^{1/3} \quad \text{for } 10^4 < G_r < 10^9 \tag{2.37a}$$

$$N_u = 0.123\ G_r^{1/3}\, P_r^{1/3} \quad \text{for } 10^9 < G_r < 10^{12} \tag{2.37b}$$

Equation 2.36 can be used also for this case if the cylinder has a relation diameter/length $\geq 35/G_r^{1/4}$.

Horizontal or Inclined Planes (Incropera et al. 2007)
When the plane has an inclination angle θ in relation to the vertical that is greater than or equal to 0, the g value in Eq. 2.34 must be substituted by $g.\cos\theta$; then use the N_u in Eq. 2.36. On the other hand, if the plane is almost horizontal, its characteristic dimension must be calculated as

$$d = (\text{Surface area})/\text{Perimeter m}$$

and N_u is calculated by using one of the equations:

$$N_u = 0.54\ G_r^{1/4}\, P_r^{1/4} \quad \text{for } 10^4 \leq G_r\, P_r \leq 10^7 \tag{2.38a}$$

$$N_u = 0.15\ G_r^{1/3}\, P_r^{1/3} \quad \text{for } G_r\, P_r > 10^7 \tag{2.38b}$$

Horizontal Cylinders (Churchill and Chu 1975a)

$$N_u = \left\{ 0.6 + \frac{0.387 (G_r P_r)^{1/6}}{\left[1 + (0.559/P_r)^{9/16}\right]^{8/27}} \right\}^2 \tag{2.39}$$

Spheres (Churchill 2002)

$$N_u = 2 + \frac{0.589 (G_r\, P_r)^{1/4}}{\left[1 + (0.469/P_r)^{9/16}\right]^{4/9}}. \tag{2.40}$$

2.3.3 *Forced Convection*

2.3.3.1 Nondimensional Numbers

Nusselt number for the forced convection is defined by the general equation

$$N_u = b\, R_e^p P_r^q \tag{2.41}$$

where R_e is the *Reynolds number*, given by

$$R_e = \frac{Ud}{v} \tag{2.42}$$

where U is the speed of the wind or the air displacement (m s^{-1}), d is the characteristic dimension of the surface or object (m) and v is the kinematic viscosity of the air at the given temperature (m^2 s^{-1}).

When a plane, smooth surface is exposed to a nonturbulent air stream, the boundary layer will become turbulent only in the case the Reynolds number exceeds the value $R_e = 10^5$; however, if the air stream is a turbulent one, this limit decreases to $R_e = 4{,}000$ (Monteith and Unsworth 2008). More details at this respect can be found in Gates (1980) and Rubesin et al. (1998).

Equations for the Nusselt number in the case of forced convection will be given according to the surface type.

Horizontal or Vertical Planes, Flux Parallel to the Surface (Incropera et al. 2007)

Assuming the direction of the flux as the same as that of the characteristic dimension d (m), we have

$$N_u = 0.332\, R_e^{1/2} P_r^{1/3} \tag{2.43}$$

Cylinders, Flux Perpendicular to the Axis (Churchill and Bernstein 1977)

$$N_u = 0.3 + \frac{0.62\, R_e^{1/2} P_r^{1/3}}{\left[1 + (0.4/P_r)^{2/3}\right]^{1/4}} \left[1 + \left(\frac{R_e}{282{,}000}\right)^{5/8}\right]^{4/5} \tag{2.44}$$

In this case, the characteristic dimension is $d =$ cylinder diameter (m), and the thermo-physical properties of the air must be calculated for the temperature $T_m = 0.5(T_s + T_a)$.

Cylinders, Flux Parallel to the Axis (Incropera et al. 2007)

$$N_u = 0.0296\, R_e^{4/5}\, P_r^{1/3} \tag{2.45}$$

where $d =$ cylinder length (m). According to Monteith and Unsworth (2008), for the case of body of a quadruped animal, it would be better to use the body volume V (m^3) to estimate the characteristic dimension as $d = V^{1/3}$, even for flux parallel to the axis or perpendicular to the axis.

Spheres (Incropera et al. 2007)

$$N_u = 2 + \left(0.4\, R_e^{1/2} + 0.06\, R_e^{2/3}\right) P_r^{0,4} \left(\frac{\mu}{\mu_s}\right)^{1/4} \tag{2.46}$$

where μ and μ_s are the coefficients of dynamic viscosity of the air at temperatures T_a and T_s, respectively. This equation can be eventually simplified by eliminating μ and μ_s.

2.3.4 Mixed Convection

In many circumstances, free convection occurs together with the forced one. A way that can be followed in order to solve this problem is to determine the Grashof and Reynolds numbers first and then calculate

$$\xi = \frac{G_r}{R_e^2} \tag{2.47a}$$

Lloyd and Sparrow (1970) studied the association of ξ value with that $N_u/R_e^{0.5}$ and observed that for P_r values about 0.71 (often determined for atmosphere under normal conditions) the critical value of ξ which eliminated the occurrence of forced convection was 0.08. On the other hand, the results of the cited paper showed also that convection was purely free when $\xi > 3$. According to Chapman (1987), those results involve an error less than 5 %. Then, we can apply the following criterion to decide what convection type is occurring:

If $\xi \leq 0.08$ there is forced convection;
If $\xi \geq 3$ there is free convection.

If $0.08 < \xi < 3$, the Nusselt number must be calculated for both convection types and then it is calculated the weighted value:

$$N_u = \left[\left(N_u^N\right)^p + \left(N_u^F\right)^p\right]^{1/p} \tag{2.47b}$$

according to Churchill (1977), and where:

N_u^N = Nusselt number for free convection.
N_u^F = Nusselt number for forced convection.
$p = 3$ for the general case, $p = 4$ for spheres, and $p = 3.5$ for horizontal cylinders.

2.3.5 Forced Convection in Tubes

In biological studies, it is eventually needed to determine heat transfer by convection within the respiratory ways. Thus, the knowledge of the principles involved in that mode of heat transfer can be of great interest.

Forced convection is generally predominant in such cases, in which there are considered tubes of small internal dimensions (diameter d and length z). The equation proposed by Sieder and Tate (1936) is considered yet as efficient enough to estimate the Nusselt number for those cases:

$$N_u = 1.86\left[\left(\frac{d}{z}\right)R_eP_r\right]^{1/3}\left(\frac{\mu}{\mu_s}\right)^{0.14} \tag{2.48}$$

where $\mu = \nu\rho$ is the dynamic viscosity of the air at the temperature of the flux within the tube and μ_s is the same property, for the temperature of the tube walls (in general, this temperature is the same as that of the body or rectal temperature).

Equation 2.48 refers to the case of a laminar flux. For a turbulent flux, which occurs in most of the cases, one can use the equations

$$N_u = 0.0395\ R_e^{0.75}\ P_r^{1/3} \quad \text{for } 10^4 < R_e < 5\times 10^4 \tag{2.48a}$$

$$N_u = 0.023\,R_e^{0.8}\,P_r^{1/3} \quad \text{for } 3\times 10^4 < R_e < 10^6 \tag{2.48b}$$

However, when the tube is very short with a relation of length to diameter as that $400 > z/d > 10$, it must be used the following equation proposed by Nusselt (1931):

$$N_u = 0.036\ R_e^{0.8}\ P_r^{1/3}\left(\frac{d}{z}\right)^{1/18} \tag{2.48c}$$

in which the atmospheric properties must be calculated for the temperature within the tube – which is that of the expired air, in animals. Practical application of those principles to animals will be discussed in details later in Chap. 4.

2.4 Mass Transfer

2.4.1 Definitions

Among organisms and the atmosphere there are constant transfers of mass, especially gases as O_2, CO_2 and water vapour. Those transfers are processed within the stomatal cavities of the plant leaves, while in animals, it occurs through the tissues of the respiratory system and the boundary layer at the body surface. Such transfers are generally turbulent ones; the interfaces between liquids and gases are zones in which the water molecules pass down from a physical state to another, and latent heat is exchanged along that process.

In order to understand the mass transfer process, let us suppose two points in the space, x_1 and x_2, each of them with a different concentration of water vapour, C_1 e C_2, respectively. In general terms, it is known that the concentration of a gas is given by

$$C = \frac{M\rho}{M_a\rho_a}$$

where M and ρ are the molecular mass and the density of that gas, respectively, while M_a and ρ_a refer to the air. Those two points are separated one from another by a layer with thickness dz that is permeable to the gas molecules, whose displacement is assumed to be at random. In those circumstances, there has equal probability of any molecule in x_1 be transferred to x_2 and *vice versa*.

Now, by applying Fick's law the mass flux can be estimated as

$$\Phi_{\mathrm{m}} = -D\frac{\mathrm{d}C}{\mathrm{d}z} = -\frac{D}{\mathrm{d}z}(C_1 - C_2) \tag{2.49}$$

where D is the gas diffusivity in the air and dC/dz is the concentration gradient.

2.4.1.1 Nondimensional Numbers

For the present purposes, the term mass transfer refers to the water vapour transfer, that is, to the evaporation process. The vapour transfer from surfaces exposed to air flux of a given intensity is analogous to the process of thermal convection, which was discussed before in this chapter (Sect. 2.3). Thus, the Nusselt number is used here also, but as a modified form – the *Sherwood number*, S_h – which is nondimensional also.

Now, by taking Eq. 2.49 again and eliminating its negative sign, we have

$$\mathbf{E} = \frac{D_{\mathrm{v}}}{d} S_h (C_1 - C_2) \tag{2.50}$$

where $\mathbf{E}$ is the flux density of thermal energy by evaporation, D_{v} is the diffusivity of water vapour in the air, d is the characteristic dimension of the surface, C_{s} is the concentration of water vapour at the surface and C_{a} is the concentration of water vapour in the atmosphere.

On the other hand, the Sherwood number can be defined as

$$S_h = \frac{\mathbf{E}}{(D_{\mathrm{v}}/d)(C_{\mathrm{s}} - C_{\mathrm{a}})} \tag{2.51}$$

and it represents the rate of vapour transfer that can occur if the same concentration differential would exist through an air layer of thickness d. In practical terms, it is possible to determine an equation for S_h in specific cases by merely calculating Grashof, Reynolds and Prandtl numbers for each case, in the same way as to determine the Nusselt number for convection. Then, it is found the N_u equation needed for the actual case and the Prandtl number, P_r, is substituted in that equation by the following nondimensional value:

$$S_c = \frac{\nu}{D_{\mathrm{v}}} \tag{2.52}$$

which is the *Schmidt number*, v is the kinematic viscosity of the air and D_v is the vapour diffusivity in the air at temperature T_a. Such a change of P_r by S_c is due to the differences in the effective thickness of the boundary layer, with respect to heat and mass transfer.

For example, if the equation chosen for the Nusselt number is

$$N_u = 0.332\, R_e^{1/2}\, P_r^{1/3}$$

then the corresponding equation for the Sherwood number will be

$$S_h = 0.332\, R_e^{1/2} S_c^{1/3}$$

2.4.2 Evaporation from Wet Surfaces

Suppose a wet surface, that is, one covered with a water film. The surface is at a temperature T_s (just that at which the water will evaporate) and the vapour pressure in the boundary layer over the surface is the saturation vapour pressure of the air at the same T_s temperature. On the other hand, the atmosphere is at a temperature T_a and partial vapour pressure P_v. If this system is adiabatic and there has no energy transfer by radiation or conduction, then at equilibrium, the sensible heat exchange by convection will be equal to the latent heat dissipation by evaporation, that is, $\mathbf{E} = \mathbf{C}$. The convective exchange is given by Eq. 2.31,

$$\mathbf{C} = \frac{\rho c_p (T_s - T_a)}{r_H}\ \mathrm{W\,m^{-2}}$$

where $r_H = \frac{\rho c_p d}{k N_u}\ \mathrm{s\,m^{-1}}$.

Similarly, the flux of heat by evaporation from a wet surface is given by

$$\mathbf{E} = \frac{\rho c_p [P_s(T_s) - P_v]}{P_a \gamma r_V}\ \mathrm{W\,m^{-2}} \qquad (2.53)$$

where ρ, c_p and γ are thermo-physical properties of the air at temperature T_a; P_a is the actual atmospheric pressure (kPa) and r_V is the resistance to mass transfer, given by

$$r_V = \frac{d}{D_V\, S_h}\ \mathrm{s\,m^{-1}} \qquad (2.53a)$$

It must be stressed that the saturation vapour pressure at the surface, $p_S(T_S)$, must be calculated for the surface temperature, T_S.

However, once evaporation and convection are functions of the wind velocity, Monteith and Unsworth (2008) suggested that Eq. 2.53 can be conveniently changed into

$$\mathbf{E} = \frac{\rho c_p [P_s(T_s) - P_v]}{\gamma^* r_H} \tag{2.54}$$

where

$$\gamma^* = P_a \gamma \left(\frac{r_V}{r_H} \right) \tag{2.54a}$$

Strictly speaking, Eqs. 2.53 and 2.54 are valid only when the surface is completely wetted and covered by a water film. As for the cutaneous surface of animals, such a condition would imply in sweating rates of 1,000 g·m^{-2}·h^{-1} or greater; this is true for humans and equines, but other animal species (as ruminants in general) lose heat by sweat evaporation without having a fully wet skin surface.

Then, if the surface is humid, but without a visible water film, it is needed to modify Eq. 2.54 as

$$\mathbf{E} = \frac{w \rho c_p [P_s(T_s) - P_v]}{\gamma^* r_H} \tag{2.55}$$

according to McArthur (1987) and in which the value w is the proportion of the actual evaporation rate to the rate that would be observed if the surface were wet (Gagge 1981).

The amount of thermal energy needed to convert 1 g of water at temperature T_s (surface) into 1 g of vapour at temperature T_a (atmosphere) is given by

$$\lambda - c_{pv}(T_s - T_a) \text{ Joules}$$

after Monteith (1972), in which λ is the latent heat of vaporisation of water at temperature T_s (Eq. 1.46) and c_{pv} is the specific heat of the water vapour at temperature T_a, estimated by Eq. 1.49. By relating the amount of effectively dissipated energy to that lost in the case that the surface was covered with a water film, it is possible to determine the correction factor shown in Eq. 2.53a and which is

$$w = \frac{\gamma^* r_H \mathrm{S} [\lambda - c_{pv}(T_s - T_a)]}{\rho c_p [P_s(T_s) - P_v]} \tag{2.56}$$

according to Silva (2000) and where S is the sweating rate (g·m^{-2}·s^{-1}).

2.5 Problems

Problem 2.1. The surface of a black body was at a temperature of 760 K. Calculate: (a) emissive power, (b) the wavelength at which the maximum emissive power occurs and (c) the power emitted at 1.1 μm and 6.5 μm wavelengths, respectively.

(a) *The Stefan-Boltzmann law is applied*:

$$\mathbf{R}_b = \sigma T^4 = 5.67 \times 10^{-8}(760)^4 = 18.9164\ \mathrm{W\,m^{-2}}$$

(b) *By the Wien's Law*:

$$\lambda_{max} = \frac{2,897.8}{T} = \frac{2,897.8}{760} = 3.81\ \mu\mathrm{m}$$

(c) *According to the Planck's law*:

$$\mathbf{R}_{\lambda,T} = \frac{a}{\lambda^5\left[e^{b/(\lambda T)} - 1\right]} = \frac{3.741775 \times 10^8}{\lambda^5\left[exp\left\{\frac{14,387.8}{\lambda T}\right\} - 1\right]}$$

$$\mathbf{R}_{1.1;760} = \frac{3.741775 \times 10^8}{(1.1)^5\left[exp\left\{\frac{14,387.8}{(1.1)(760)}\right\} - 1\right]} = 7.79\ \mathrm{W\,m^{-2}}$$

$$\mathbf{R}_{6.5;760} = \frac{3.741775 \times 10^8}{(6.5)^5\left[exp\left\{\frac{14,387.8}{(6.5)(760)}\right\} - 1\right]} = 1,853\ \mathrm{W\,m^{-2}}$$

Problem 2.2. A hog's body can be represented as a horizontal cylinder with hemispherical ends, 0.90 m length and 0.35 m average diameter. This animal is standing in a pen exposed to sun at an azimuth angle of 75°. The zenith angle of the sun is 7°, and the solar radiation was measured as S = 614 W m^{-2}. Calculate the direct solar radiation effectively received by the animal.

Data:

r = cylinder radius = 0.35/2 = 0.175 m
z = cylinder length = 0.9 m
β = sun elevation angle = 90 − θ = 90 − 7 = 83°
ω = azimuth angle of cylinder axis to sun = 75°
S = direct solar radiation = 614 W m^{-2}

By applying Eq. 2.20

$$F_c = \frac{(\sin\beta)^{-1}\left\{2rz\left[1-(\mathbf{cos}\beta)^2(\mathbf{cos}\omega)^2\right]^{1/2}+\pi r^2\right\}}{\mathbf{2\pi r(2r+z)}}$$

$$= \frac{1.00751\left\{2(0.175)(0.9)[1-(0.014852)(0.066987)]^{1/2}+\pi(0.175)^2\right\}}{2\pi(0.175)(0.35+0.9)} = 0.297$$

Animal's irradiance at the given conditions will be

$$\mathbf{S}_{\text{animal}} = F_c\,\mathbf{S}_{\text{dir}} = (0.297)(614) = 182.4\,\text{W m}^{-2}$$

Problem 2.3. There has a concrete wall 15 cm thick with external revetment of a 2.5-cm stucco layer and an internal one of 2-cm stucco plus 1-cm wood agglomerate. The external surface is exposed to 35°C temperature, while the internal one at 20°C. The atmospheric pressure is 98.5 kPa. Calculate the rate of thermal transfer by conduction through the wall.

Data:

n = *number of layers* = 4
Δx_1 = *layer 1 thickness* = 0.025 m
Δx_2 = *layer 2 thickness* = 0.15 m
Δx_3 = *layer 3 thickness* = 0.02 m
Δx_4 = *layer 4 thickness* = 0.01 m
Concrete conductivity = 1.4 $\text{W·m}^{-1}\,°\text{C}^{-1}$
Stucco conductivity = 0.72 $\text{W·m}^{-1}\,°\text{C}^{-1}$
Wood agglomerate conductivity = 0.087 $\text{W·m}^{-1}\,°\text{C}^{-1}$
T_1 = *external temperature* = 35°C
T_5 = *internal temperature* = 20°C

Atmospheric properties at T_a = 35°C = 308.15 K:

$$P_a = \text{atmospheric pressure} = 98.5\ \text{kPa}$$

$$\rho = \text{density} = \frac{3{,}484.358P_a}{T_a} = \frac{3{,}484.358(98.5)}{308.15} = 1{,}113.77\ \text{g m}^{-3}$$

$$c_p = \text{specific heat}$$
$$= 1.00522 + 0.0004577\ \exp\left\{\frac{T_a}{32.07733}\right\}\ \text{J g}^{-1}\,°\text{C}^{-1}$$
$$= 1.00522 + 0.0004577\ \exp\left\{\frac{35}{32.07733}\right\} = 1.006583\ \text{J g}^{-1}\,\text{C}^{-1}$$

Thermal resistances:

$$r_i = \frac{\rho c_p \Delta xi}{k_i}\ \mathrm{s\,m^{-1}}$$

$$\text{Concrete } (0.15\ \mathrm{m}) : r_1 = \frac{1{,}113.77(1.006583)(0.15)}{1.4} = 120.1181\ \mathrm{s\,m^{-1}}$$

$$\text{Stucco } (0.025\ \mathrm{m}) : r_2 = \frac{1{,}113.77(1.006583)(0.025)}{0.72} = 38.9272\ \mathrm{s\,m^{-1}}$$

$$\text{Stucco } (0.02\ \mathrm{m}) : r_3 = \frac{1{,}113.77(1.006583)(0.02)}{0.72} = 31.1417\ \mathrm{s\,m^{-1}}$$

$$\text{Agglomerate } (0.01\ \mathrm{m}) : r_4 = \frac{1{,}113.77(1.006583)(0.01)}{0.087} = 128.8623\ \mathrm{s\,m^{-1}}$$

$$\mathbf{K} = \frac{\rho c_p (T_1 - T_{n+1})}{\sum_{i=1}^{n} r_i}\ \mathrm{W\,m^{-2}}$$

$$\sum_{i=1}^{4} r_i = 120.1181 + 38.9272 + 31.1417 + 128.8623 = 319.0493$$

$$\mathbf{K} = \frac{\rho c_p (T_1 - T_{n+1})}{\sum_{i=1}^{n} r_i} = \frac{1{,}113.77(1.006583)(35-20)}{319.0493} = 52.71\ \mathrm{W\,m^{-2}}$$

Problem 2.4. In a location at 23°45′ south latitude and 189 m altitude there has a chamber which must be maintained at the temperature of 15°C, while in the outer, the temperature is as high as 38°C. The walls of the chamber were made of wood compensate 2.5 cm thick, and its inner surface is covered by a layer of polystyrene 5 cm thick. Calculate the thermal flux through that wall.

Data:

T_a = ambient temperature = 38°C = 311.15 K
T_{int} = chamber temperature = 15°C
L_t = Latitude = 23°45′ = 23.75°
z = Altitude = 189 m
Δx_1 = wood layer thickness = 0.025 m
Δx_2 = polystyrene layer thickness = 0.05 m
k_1 = thermal conductivity of the wood compensate = 0.12 s m^{-1}
k_2 = thermal conductivity of polystyrene = 0.025 s m^{-1}

Atmospheric properties at 38°C:

$$
\begin{aligned}
g &= 9.78013 + 8.18 \times 10^{-5} L_t + 1.168 \times 10^{-5} L_t^2 - 3.1 \times 10^{-6} z \\
&= 9.78013 + 8.18 \times 10^{-5}(23.75) + 1.168 \times 10^{-5}(23.75)^2 - 3.1 \times 10^{-6}(189) \\
&= 9.788075 \text{ m s}^{-2}
\end{aligned}
$$

$$
\begin{aligned}
P_a &= 101.325 \exp\left\{\frac{-zg}{287.04 T_a}\right\} \\
&= 101.325 \exp\left\{\frac{-189(9.788075)}{287.04(311.15)}\right\} = 96.65 \text{ kPa}
\end{aligned}
$$

$$
\begin{aligned}
c_p &= 1.00522 + 0.0004577 \exp\left\{\frac{T_a}{32.07733}\right\} \\
&= 1.00522 + 0.0004577 \exp\left\{\frac{38}{32.07733}\right\} = 1.006716 \text{ J g}^{-1}\,{}^\circ\text{C}^{-1}
\end{aligned}
$$

$$
\rho = \frac{3{,}484.358\, P_a}{T_a} = \frac{3{,}484.358\ (96.65)}{311.15} = 1{,}082.32 \text{ g m}^{-3}
$$

We have two layers, then
Compensate (0.025 m):

$$
r_1 = \frac{1{,}082.32(1.006716)(0.025)}{0.12} = 226.9972 \text{ s m}^{-1}
$$

Polystyrene (0.05 m):

$$
r_2 = \frac{1{,}082.32(1.006716)(0.05)}{0.025} = 2{,}179.1777 \text{ s m}^{-1}
$$

$$
\sum_{i=1}^{4} r_i = 226.9972 + 2{,}179.1777 = 2{,}406.1749 \text{ s m}^{-1}
$$

$$
\mathbf{K} = \frac{\rho\, c_p (T_a - T_{\text{int}})}{\sum_{i=1}^{n} r_i} = \frac{1{,}082.32(1.006716)(38 - 15)}{2{,}406.1749} = 10.42 \text{ W m}^{-2}
$$

If a more precise temperature control is desired within the chamber, the above calculations must be done using thicker polystyrene layers, until it is practical or economically available. For example, it is desired a thermal flux rate which is 10 % of that above calculated or 1.04 W m^{-2}:

$$
\mathbf{K} = \frac{\rho c_p (T_a - T_{\text{int}})}{\sum_{i=1}^{n} r_i} = \frac{1{,}082.32(1.006716)(38 - 15)}{226.9972 + r_2} = 1.04
$$

$$1.04(226.9972 + r_2) = 1{,}082.32\ (1.006716)(38 - 15)$$
$$\therefore r_2 = 10.6688698$$
$$r_2 = \frac{\rho c_p \Delta x}{k_2} = \frac{1{,}082.32(1.006716)\Delta x}{0.025} = 23{,}869.6795$$

$$43{,}583.468084\ \Delta x = 23{,}869.6795 \quad \therefore \Delta x = 0.548\ \text{m} = 54.8\ \text{cm}$$

In order to have a thermal flux as small as 10 W m^{-2} it would be needed a 55-cm polystyrene layer! Of course, it is not practical. Then a solution of compromise must be searched for, by establishing an acceptable thermal flux.

Problem 2.5. Suppose a horizontal plane with dimensions 0.4 × 0.95 m, whose surface is at 35°C. The ambient temperature is 30°C, and a 0.5 m s^{-1} wind blows perpendicularly to the shortest side of the plane. The respective location is at 21° latitude and 630 m altitude. Calculate the thermal exchange by convection.

Data:

T_a = air temperature = 30°C = 303.15 K
T_s = plane surface temperature = 35°C
U = wind velocity = 0.5 m s^{-1}
L_t = latitude = 21°
z = altitude = 630 m

$$\begin{aligned} g &= 9.78013 + 8.18 \times 10^{-5} L_t + 1.168 \times 10^{-5} L_t^2 - 3.1 \times 10^{-6} z \\ &= 9.78013 + 8.18 \times 10^{-5}(21) + 1.168 \times 10^{-5}(21)^2 - 3.1 \times 10^{-6}(630) \\ &= 9.785046\ \text{m s}^{-2} \end{aligned}$$

$$\begin{aligned} P_a &= 101.325 \exp\left\{\frac{-zg}{287.04\,T_a}\right\} = 101.325 \exp\left\{\frac{-630(9.785046)}{287.04(305.65)}\right\} \\ &= 94.449809\ \text{kPa} \end{aligned}$$

Thermal properties of atmosphere at temperature $T_m = 0.5(T_a + T_s) = 32.5°C$:

$$\begin{aligned} c_p &= 1.00522 + 0.0004577\ exp\left\{\frac{T_m}{32.07733}\right\} \\ &= 1.00522 + 0.0004577\ exp\left\{\frac{32.5}{32.07733}\right\} = 1.006481\ \text{J} \cdot \text{g}^{-1} \cdot{}^\circ \text{C}^{-1} \end{aligned}$$

$$\rho = \frac{3{,}484.358\,P_a}{T_a} = \frac{3{,}484.358(94.3951)}{305.65} = 1{,}076.0881\ \text{g m}^{-3}$$

$$k = \rho c_p(1.888 \times 10^{-5} + 1.324 \times 10^{-7} T_a)$$
$$= 1,076.0881(1.006481)[1.888 \times 10^{-5} + 1.324 \times 10^{-7}(32.5)]$$
$$= 0.025109$$

$$\nu = 1.32743 \times 10^{-5} + 9.22286 \times 10^{-8} T_a$$
$$= 1.32743 \times 10^{-5} + 9.22286 \times 10^{-8}(32.5) = 1.6272 \times 10^{-5}\ \mathrm{m^2 s^{-1}}$$

Non-dimensional numbers:
For forced convection in the present case $d = 0.95$; for free convection:

$$\mathrm{d} = \frac{0.4(0.95)}{2(0.4 + 0.95)} = 0.14074$$

By using Eqs. 2.35, 2.42 and 2.34, we calculate, respectively:

$$P_r = \frac{\rho\, c_p\, \nu}{k} = \frac{1076.0881(1.006481)(1.6272 \times 10^{-5})}{0.025109} = 0.701883$$

$$R_e = \frac{U\, d}{\nu} = \frac{0.5\ (0.95)\ V4}{1.6272 \times 10^{-5}} = 2,191.25$$

$$G_r = \frac{g\, d^3(T_s - T_a)}{\nu^2\, T_a} = \frac{9.785046(0.14074)^3(35 - 30)}{(1.6272 \times 10^{-5})^2(303.15)} = 1,699.206.2$$

But $G_r/R_e^2 < 0.08$, then it is a case of forced convection. By using Eq. 2.43, we have

$$N_u = 0.332\ R_e^{1/2} P_r^{1/3} = 0.332(29.191.25)^{1/2}(0.701883)^{1/3} = 50.4103$$

$$r_H = \frac{\rho\, c_p\, d}{k\, N_u} = \frac{1076.0881(1.006481)(0.95)}{0.025109(50.4103)} = 812.8834$$

$$\mathbf{C} = \frac{\rho\, c_p(T_s - T_a)}{r_H} = \frac{1076.0881(1.006481)(35 - 30)}{812.8834} = 6.7\ \mathrm{W\,m^{-2}}$$

Problem 2.6. A hog whose skin is at a 34°C temperature has the trunk as a horizontal cylinder with dimensions 0.90 × 0.35 m and is standing in an environment where temperature is 28.5°C, the atmospheric pressure is 94.5 kPa and there has a 1.1 m s^{-1} wind blowing perpendicularly at the body axis of the animal. Calculate the thermal exchange by convection.

Data:

T_a = ambient temperature = 28.5°C = 301.65 K
T_s = skin surface temperature = 34°C
U = wind velocity = 1.1 m·s^{-1}
P_a = atmospheric pressure = 94.5 kPa

As the wind blows perpendicularly at the animal's body axis, then the characteristic dimension of the body is $d = 0.35$. On the other hand, the wind velocity is $U > 1$ m·s^{-1}; thus we have a forced convection.

Properties of atmosphere at temperature $T_m = 0.5(T_a + T_s) = 31.25°C$:

$$c_p = 1.00522 + 0.0004577\ exp\left\{\frac{T_m}{32.07733}\right\}$$
$$= 1.00522 + 0.0004577\ exp\left\{\frac{31.25}{32.07733}\right\} = 1.006432\ \text{J} \cdot \text{g}^{-1} \cdot {}^{\circ}\text{C}^{-1}$$

$$\rho = \frac{3{,}484.358 P_a}{T_a} = \frac{3{,}484.358\ (94.5)}{301.65} = 1{,}091.57\ \text{g m}^{-3}$$

$$v = 1.32743 \times 10^{-5} + 9.22286 \times 10^{-8}\ \text{m}$$
$$= 1.32743 \times 10^{-5} + 9.22286 \times 10^{-8}(31.25) = 1.6156 \times 10^{-5}\ \text{m}^2\,\text{s}^{-1}$$

$$k = \rho c_p\left(1.888 \times 10^{-5} + 1.324 \times 10^{-7} T_m\right)$$
$$= \rho c_p\left[1.888 \times 10^{-5} + 1.324 \times 10^{-7}(31.25)\right] = 0.025287\ \text{W m}^{-2}\,{}^{\circ}\text{C}^{-1}$$

$$P_r = \frac{\rho\, c_p\, v}{k} = \frac{1091.57(1.006432)\left(1.6156 \times 10^{-5}\right)}{0.025287} = 0.701901$$

$$R_e = \frac{U d}{v} = \frac{1.1(0.35)}{1.6156 \times 10^{-5}} = 23{,}830.156$$

$$N_u = 0.3 + \frac{0.62\ R_e^{1/2} P_r^{1/3}}{\left[1 + (0.4/P_r)^{2/3}\right]^{1/4}}\left[1 + \left(\frac{R_e}{282.000}\right)^{5/8}\right]^{4/5}$$

$$= 0.3 + \frac{0.62(23.830.156)^{1/2}(0.701901)^{1/3}}{\left[1 + (0.4/0.701901)^{2/3}\right]^{1/4}}\left[1 + \left(\frac{23.830.156}{282.000}\right)^{5/8}\right]^{4/5}$$

$$= 87.422$$

$$r_H = \frac{\rho\, c_p\, d}{k N_u} = \frac{1{,}091.57(1.006432)\ (0.35)}{0.025287(87.422)} = 173.9341$$

$$\mathbf{C} = \frac{\rho\, c_p (T_s - T_a)}{r_H} = \frac{1091.57(1.006432)(34 - 28.5)}{173.9341} = 34.7\ \mathrm{W\,m^{-2}}$$

Problem 2.7. Let us have a horizontal cylinder with 0.72 m diameter and surface temperature of 36.5°C, which is suspended in a place where the acceleration of gravity is 9.7861 m s^{-2}, the atmospheric pressure is 99.641 kPa, dry bulb temperature is 32°C and a 0.9 m·s^{-1} wind blows perpendicularly at the cylinder axis. Determine the thermal exchange by convection.

Data:

g = acceleration of gravity = 9.7861 m·s^{-2}
P_a = atmospheric pressure = 99.641 kPa
T_a = dry bulb temperature = 32°C = 305.15 K
U = wind velocity = 0.9 m·s^{-1}
T_s = surface temperature of the cylinder = 36.5°C
d = diameter of the cylinder = 0.72 m

Properties of the air at temperature $T_m = 0.5\ (T_a + T_s) = 34.25$°C:

$$c_p = 1.00522 + 0.0004577\ exp\left\{\frac{T_m}{32.07733}\right\}$$
$$= 1.00522 + 0.0004577\ exp\left\{\frac{34.25}{32.07733}\right\} = 1.006553\ \mathrm{J\,g^{-1}\,°C^{-1}}$$

$$\rho = \frac{3{,}484.358\, P_a}{T_a} = \frac{3{,}484.358(98.7)}{307.4} = 1{,}129.5631\ \mathrm{g\,m^{-3}}$$

$$v = 1.32743 \times 10^{-5} + 9.22286 \times 10^{-8} T_m$$
$$= 1.32743 \times 10^{-5} + 9.22286 \times 10^{-8}(34.25) = 1.643 \times 10^{-5}\ \mathrm{m^2\,s^{-1}}$$

$$k = \rho c_p \left(1.888 \times 10^{-5} + 1.324 \times 10^{-7} T_m\right)$$
$$= \rho c_p \left[1.888 \times 10^{-5} + 1.324 \times 10^{-7}(34.25)\right] = 0.02667\ \mathrm{W\,m^{-2}\,°C^{-1}}$$

$$P_r = \frac{\rho\, c_p\, v}{k} = \frac{1{,}129.5631(1.006553)\left(1.643 \times 10^{-5}\right)}{0.02667} = 0.700425$$

$$R_e = \frac{Ud}{v} = \frac{0.9(0.72)}{1.643 \times 10^{-5}} = 39{,}440.0487$$

$$G_r = \frac{g\,d^3(T_s - T_a)}{\nu^2 T_a} = \frac{9.7861(0.72)^3(36.5 - 32)}{(1.643 \times 10^{-5})^2(305.15)} = 327{,}845{,}113.3$$

As $x = 0.08 < G_r/R_e{}^2 < 3$, then it is a mixed convection. By applying the principles previously studied, we calculate the Nusselt number for both convection types in horizontal cylinders as follows:

Free convection (Eq. 2.39):

$$N_u^N = \left\{ 0.6 + \frac{0.387(G_r P_r)^{1/6}}{\left[1 + (0.559/P_r)^{9/16}\right]^{8/27}} \right\}^2$$

$$= \left\{ 0.6 + \frac{0.387[(327{,}845{,}113.3)(0.701828)]^{1/6}}{\left[1 + (0.559/0.701828)^{9/16}\right]^{8/27}} \right\}^2 = 73.030933$$

Forced convection (Eq. 2.44):

$$N_u^F = 0.3 + \frac{0.62\,R_e^{1/2} P_r^{1/3}}{\left[1 + (0.4/P_r)^{2/3}\right]^{1/4}} \left[1 + \left(\frac{R_e}{282{,}000}\right)^{5/8}\right]^{4/5}$$

$$= 0.3 + \frac{0.62(39{,}440.0487)^{1/2}(0.701828)^{1/3}}{\left[1 + (0.4/0.701828)^{2/3}\right]^{1/4}} \left[1 + \left(\frac{39{,}440.048}{282{,}000}\right)^{5/8}\right]^{4/5}$$

$$= 72.187827$$

As the convection is a mixed one, we have

$$N_u = \left[\left(N_u^N\right)^{3.5} + \left(N_u^F\right)^{3.5}\right]^{\frac{1}{3.5}}$$

$$= \left[(73.030933)^{3.5} + (72.187827)^{3.5}\right]^{\frac{1}{3.5}} = 88.5156$$

$$r_H = \frac{\rho c_p d}{k N_u} = \frac{(1{,}129.5631)(1.006553)(0.72)}{(0.02667)(88.5156)} = 346.7664\ \mathrm{s\,m^{-1}}$$

$$\mathbf{C} = \frac{\rho\, c_p (T_s - T_a)}{r_H} = \frac{1{,}129.5631(1.006553)(36.5 - 32)}{346.7664} = 14.754\ \mathrm{W\,m^{-2}}$$

Problem 2.8. Consider the cylinder described in Problem 2.7 and the same environmental data. Suppose again that the surface of the cylinder is wetted and the atmospheric wet bulb temperature is 26°C. Determine the loss of latent heat from the cylinder surface.

Data:

g = *acceleration of gravity* = 9.78611 m·s^{-2}
P_a = *atmospheric pressure* = 99.641 kPa
T_a = *dry bulb temperature* = 32°C = 305.15 K
T_u = *wet bulb temperature* = 26°C
U = wind velocity = 0.9 m·s^{-1}
T_s = surface temperature of the cylinder = 36.5°C
d = diameter of the cylinder = 0.72 m
r_H = 346.7664 s m^{-1}

Properties of the air at T_a = 32°C:

$$c_p = 1.00522 + 0.0004577\, exp\left\{\frac{T_a}{32.07733}\right\}$$
$$= 1.00522 + 0.0004577\, exp\left\{\frac{32}{32.07733}\right\} = 1.006463 \text{ J g}^{-1}\,°\text{C}^{-1}$$

$$\rho = \frac{3484.358 P_a}{T_a} = \frac{3484.358(98.7)}{305.15} = 1,137.7523 \text{ g m}^{-3}$$

$$v = 1.32743 \times 10^{-5} + 9.22286 \times 10^{-8} T_m$$
$$= 1.32743 \times 10^{-5} + 9.22286 \times 10^{-8}(32) = 1.623 \times 10^{-5} \text{ m}^2\text{ s}^{-1}$$

$$k = \rho\, c_p\left(1.888 \times 10^{-5} + 1.324 \times 10^{-7} T_m\right)$$
$$= \rho\, c_p\left[1.888 \times 10^{-5} + 1.324 \times 10^{-7}(32)\right] = 0.0265 \text{ W m}^{-2}°\text{C}^{-1}$$

$$\lambda = \textit{latent heat of vaporization} \text{ (Eq. 1.46)}$$
$$= 2{,}500.788 - 2.37374(32) = 2,424.8284 \text{ J g}^{-1}$$

$$\gamma = \textit{psychrometric constant} \text{ (Eq. 1.47)}$$
$$= \frac{c_p}{0.6223\,\lambda} = \frac{1.006463}{0.6223(2424.8284)} = 6.6852 \times 10^{-4}\,°\text{C}^{-1}$$

$$D_v = \textit{water vapour diffusivity} \text{ (Eq. 1.48)}$$
$$= 2.12138 \times 10^{-5} + 1.4955 \times 10^{-7}(32) = 2.6 \times 10^{-5} \text{ m}^2\text{ s}^{-1}$$

$$P_S(T_u) = \textit{air saturation pressure at temperature } T_u \text{ (Eq. 1.50)}$$
$$= 0.61078 \times 10^{7.5T_u/(T_u+237.5)}$$
$$= 0.61078 \times 10^{7.5(26)/(26+237.5)} = 3.356779 \text{ kPa}$$

$$P_S(T_s) = \textit{air saturation pressure at temperature } T_s \text{ (Eq. 1.50)}$$
$$= 0.61078 \times 10^{7.5T_s/(T_s+237.5)}$$
$$= 0.61078 \times 10^{7.5(36.5)/(36.5+237.5)} = 6.094987 \text{ kPa}$$

$$P_v = \textit{partial vapour pressure} \text{ (Eq. 1.51)}:$$
$$= P_s(T_u) - P_a\gamma(T_a - T_u)$$
$$= 3.356779 - (99.641)(6.6852 \times 10^{-4})(32 - 26) = 2.957107 \text{ kPa}$$

Non-dimensional numbers:

$$R_e = \textit{Reynolds number} = \frac{U\,d}{v} = \frac{0.9(0.72)}{1.623 \times 10^{-5}} = 39{,}926.0629$$

$$S_c = \text{Schmidt number} = \frac{v}{D_v} = \frac{1.623 \times 10^{-5}}{2.6 \times 10^{-5}} = 0.624231$$

Sherwood number (the same Eq. 2.44 is used as for N_u, but substituting in it the value P_r by S_c):

$$S_h = 0.3 + \frac{0.62\,R_e^{1/2}S_c^{1/3}}{\left[1 + (0.4/S_c)^{2/3}\right]^{1/4}}\left[1 + \left(\frac{R_e}{282{,}000}\right)^{5/8}\right]^{4/5}$$

$$= 0.3 + \frac{0.62(39{,}926.0629)^{1/2}(0.624231)^{1/3}}{\left[1 + (0.4/0.624231)^{2/3}\right]^{1/4}}\left[1 + \left(\frac{39{,}926.0629}{282{,}000}\right)^{5/8}\right]^{4/5}$$

$$= 113.82$$

$$r_V = \frac{d}{D_v S_h} = \frac{0.72}{(2.6 \times 10^{-5})(113.82)} = 243.3 \text{ s m}^{-1}$$

$$\gamma^* = P_a\gamma\left(\frac{r_V}{r_H}\right) = 99.641\ (6.6674 \times 10^{-4})\left(\frac{243.3}{346.7664}\right) = 0.046612$$

$$\mathbf{E} = \frac{\rho\, c_p[P_s(T_s) - P_v]}{\gamma^*\, r_H}$$
$$= \frac{1137.7523(1.006463)(6.094987 - 2.957107)}{0.046612(346.7664)} = 222.3 \text{ W m}^{-2}$$

References

Campbell, G.S., Norman, J.M.: An Introduction to Environmental Biophysics, 2nd edn. Springer, New York (1998)

Chapman, A.J.: Fundamentals of Heat Transfer. Macmillan, New York (1987)

Churchill, S.W.: A comprehensive correlating equation for laminar assisting forced and free convection. AIChE J. **23**, 10–16 (1977) [*Apud* Rohsenow et al. 1998]

Churchill, S.W.: Free convection around immersed bodies. In: Heat Exchange Design Handbook. Begell House, New York (2002)

Churchill, S.W., Bernstein, M.: Correlating equation for forced convection from gases and liquids to a circular cylinder in crossflow. J. Heat Transf. **94**, 300–305 (1977)

Churchill, S.W., Chu, H.H.S.: Correlating equations for laminar and turbulent free convection from a horizontal cylinder. Int. J. Heat Mass Transf. **18**, 1049–1055 (1975a)

Churchill, S.W., Chu, H.H.S.: Correlating equations for laminar and turbulent free convection from a vertical plate. Int. J. Heat Mass Transf. **18**, 1323–1328 (1975b)

Clapperton, J.L., Joyce, J.P., Blaxter, K.L.: Estimates of the contribution of solar radiation to thermal exchanges of sheep. J. Agric. Sci. **64**, 37–49 (1965)

Gagge, A.P.: Rational temperature indices of thermal comfort. In: Cena, K., Clark, J.A. (eds.) Bioengineering, Thermal Physiology and Comfort. Elsevier, Amsterdam (1981)

Gates, D.M.: Biophysical Ecology. Springer, New York (1980)

Incropera, S., DeWitt, D.P., Bergman, T.L., Lavine, A.S.: Fundamentals of Heat and Mass Transfer, 6th edn. Wiley, New York (2007)

Lloyd, J.R., Sparrow, E.M.: Combined forced and free convection flow on vertical surfaces. Int. J. Heat Mass Transf. **13**, 434–440 (1970)

McArthur, A.J.: Thermal interaction between animal and microclimate: a comprehensive model. J. Theor. Biol. **126**, 203–238 (1987)

Monteith, J.L.: Latent heat of vaporization in thermal physiology. Nat. News Biol. **236**, 96 (1972)

Monteith, J.L., Unsworth, M.H.: Principles of Environmental Physics, 3rd edn. Academic, New York (2008)

Nusselt, W.D.: Wärmeraustausch zwischen Wand and Wasser in Roher. Forsch. Geb. Ingenieurwes. **2**, 309–314 (1931)

Rohsenow, W.M., Hartnett, J.P., Cho, Y.I.: Handbook of Heat Transfer, 3rd edn. McGraw-Hill, New York (1998)

Rubesin, M.W., Inouye, M., Parikh, P.G.: Forced convection, external flows. In: Rohsenow, W.R., Harnett, J.P., Cho, Y.I. (eds.) Handbook of Heat Transfer, 3rd edn. McGraw-Hill, New York (1998)

Sieder, E.N., Tate, E.G.: Heat transfer and pressure drop of liquids in tubes. Ind. Eng. Chem. **28**, 1429–1435 (1936)

Silva, R.G.: Um modelo para a determinação do equilíbrio térmico de bovinos em ambientes tropicais. Braz. J. Anim. Sci. **29**, 1244–1252 (2000)

Underwood, C.R., Ward, E.J.: The solar radiation area of man. Ergonomics **9**, 381–385 (1966)

Chapter 3
Thermal Balance and Thermoregulation

Abstract Aspects involved in thermal exchange between animals and environments are discussed, together with the use of concepts of stress and strain to explain those relationships. There are discussed the concepts of stress, strain, acclimatisation, acclimation and habituation; the classification of animals with respect to their adaptation to environment is discussed: homeothermy and heterothermy, endothermy and ectothermy, eurithermy and stenothermy and hibernation and estivation are explained. The concept of thermoregulation is discussed, and thermal balance is defined in its component parts. Mechanisms involved in metabolic thermogenesis are explained together with the endocrine regulation of metabolism. The methods of metabolism determination by indirect calorimetry are described. The relationship of metabolism with body mass and surface area is discussed for several animal species. There is explained the concept of thermal energy storage and its methods of estimation.

Keywords Adaptation • Heat storage • Heterothermy • Hibernation • Homeothermy • Metabolism • Strain • Stress • Thermal balance

3.1 Animals vs. Environment

3.1.1 Definitions and Terminology

Organisms in general are never independent of the environment in which they live. As a matter of fact, animals and their environments form complex systems in which each animal is affected by environmental factors and other animals, then reacting and affecting them. Because the extreme complexity and variability of the environment, animals are continuously in a process of reactions against it; such reactions often leave to genetic changes in the populations which are the main causes of the evolutionary process. Thus, evolution is a consequence of continuous adaptation of animals to variations in their environment.

R. Gomes da Silva and A.S. Campos Maia, *Principles of Animal Biometeorology*, Biometeorology 2, DOI 10.1007/978-94-007-5733-2_3,

3.1.1.1 Stress and Strain

In the study of the relationships of animals with their environment, the first phenomenon to be considered is the impact of the environmental factors on the organism. This impact is defined by the term *stress*, after the pioneering work of Selye (1936, 1950). It denotes the magnitude of forces external to the organism which tend to displace it from its equilibrium state by causing a series of reactions (Hatch 1963; Kerslake 1972). The book by Moberg (1985) gives a good introductory view on that matter.

On the other hand, *strain* is the result of the stress action, or the displacement of organic functions from their normal conditions, as a reaction to the stress.

For example, thermal stress must be defined in terms of environment only, while the organic strain is the consequence of the stress action, depending on the physiological characteristics of each animal's organism. Thus, when an animal has some symptoms from the action of environmental factors upon its organism, it is incorrect to say that the animal is "stressed"; the correct term would be "strained" or "tense".

Some terms describing the responses of animals to their environments will be defined as follows according to the IUPS Thermal Commission (2003).

3.1.1.2 Adaptation

It is a change in the organism which reduces the physiological strain caused by stressful components of the total environment. Adaptation that is a result of genetic selection (either natural or artificial) is a *genotypic adaptation,* which involves a permanent change in the genetic structure of a population in order to favour the survival in a given environment.

If adaptation occurs in the lifetime of an animal, it is said as *phenotypic adaptation*, whose objective is to reduce the physiological strain caused by stressful components of a given environment. Three types of phenotypic adaptation can be defined:

Acclimation. A term that aims to describe the adaptive changes that occur in an organism in response to experimentally induced changes in a given climatic factor, such as temperature, humidity or other.

Acclimatisation. Refers to the adaptive changes occurring in an organism in response to changes in the natural climate (e.g. seasonal or geographical). It may be specified as a particular phenotypic adaptation.

Habituation. Reduction of responses of the organism to the perception of a given stressful component of the climate, caused by repeated stimulation.

It must be explained that any physiological change occurred under the situations referred as acclimation, acclimatisation or habituation is potentially detrimental to the organism or parts of it. Such a change aims at the survival of the individual, and in its process, some less vital functions are affected or even damaged. For example, the process of acclimatisation of dairy cows often results in a serious reduction of its milk yield, for the survival of the cow is more important than its production.

Adaptation and Animal Classification

Thermal energy exchange is one of the most important aspects of the relationship between animals and environment; it involves several processes that affect morphological, physiological and behavioural characteristics of the animals. Animals differ in a great extent – by considering species, breeds, populations and even individuals – over the range of temperatures they can tolerate. However, the limits of thermal tolerance are neither fixed nor immutable and can eventually change with time.

Homeothermy and Heterothermy

Because the body temperature of birds and mammals is generally higher than that of the atmosphere, they have been traditionally named as "warm blood" animals, while amphibians, fishes and reptiles would be "cold blood" animals. However, if a rabbit and a lizard are placed together in an environment at mild temperature, it will be observed that the rabbit has its body temperature always higher than that of the lizard. On the other hand, if the ambient temperature is increased to a high level, it can be observed that the rabbit's body temperature changes in a small extent, while the lizard's temperature increases together with that of the air.

The rabbit is a *homeothermic* (from the Greek *homoios* = equal and *thermos* = heat) animal, whose internal body temperature tends to remain independent from the environmental variation. In other words, its temperature is maintained within a narrow range despite much larger variations in ambient temperature. Birds and mammals are homeothermic animals.

A *heterothermic* animal is that animal in which the variation of its internal body temperature varies within a relatively wide range either seasonally or nycthemerally (within a 24-h period); such a variation is generally dependent on the ambient temperature. Amphibians, fishes, reptiles and insects are heterothermic animals.

There has some species among the homeotherms in which the core temperature must be maintained within a very close range, about one standard deviation from the mean temperature of the species as measured in a thermo neutral environment. Those animals are named *cenothermic* and include the man and some superior mammals.

Endothermy and Ectothermy

In mammals and birds, most of the body heat is generated by the metabolic processes that are associated to the several organic functions; such animals are named as *endothermic* or *endotherms* (from Greek *endos* = inside). Those species are also known as *tachymetabolic* due to their high and controlled rates of metabolism, which give them body temperatures above that of the environment.

In many other species, as reptiles, amphibians and most fishes, the metabolic heat is produced at comparatively low rates; thus, in those animals, body temperature depends mainly on the behavioural control of the heat exchange between body and environment. They are *ectothermic* or *ectotherms* (from Greek *ectos* = outside). Several species of insects and some fishes can raise their body temperatures above that of environment through intense muscular activity. As a matter of fact, the heterotherms occupy an intermediary place between endotherms and pure ectotherms (Eckert et al. 1988).

Both endothermy and homeothermy present obvious advantages, for animals with those characteristics are able to maintain a balanced temperature in the nervous system, thus favouring stability of its functions. On the other hand, because their independence with respect to variations of the environmental temperature, such animals can adapt by themselves to a wider range of climatic environments than ectotherms can do. As a result, intertropical regions present equivalent occurrences of endothermic, heterothermic and ectothermic species; however, as the latitude increases, the homeotherms become increasingly prevalent, and there have neither reptiles nor amphibians native in regions near the polar circle (Schmidt-Nielsen 1990).

The high metabolic rates of the endotherms are due mainly to their greater mitochondrial content of the cells and also to the occurrence of two mitochondrial types in those animals. Their high rates of aerobic metabolism are favoured by high cell concentrations of mitochondria, leaving to a more stable body temperature and to a better performance of the locomotive performance, which is superior to that of ectotherms (Bennett and Ruben 1979; Bennett 1987).

Ectothermy was considered until recent times as a characteristic of less evolved organisms, with a slow life rhythm and low energy fluxes. However, as their body temperature tends to be close to that of the environment, the ectotherms have lower energy expenditure for the thermal exchange; with their low metabolic rates, such animals can apply most of their energy gain on growth and reproduction. Other advantages are:

(a) Due to their low metabolic rates, ectotherms need less food and can expend much time hidden inside shelters or burrows; thus, they are less exposed to predators.
(b) The high metabolic rates of the endotherms result into high energetic costs, involving metabolic rates per unit weight of body tissue 4–8 times higher than those of most ectotherms. Such high energy requirements imply greater food needs.
(c) Because their water needs are much below those of the endotherms, the ectotherms are able to live for a long time in dry environments, as deserts.

Eurythermy and Stenothermy

The tolerance by heterotherms of a wide range of changes in the thermal environment is named *eurythermy* (from the Greek *euros* = wide); such animals are able to adapt to several thermal environments. On the other hand, *stenothermy* (from the Greek *stenos* = narrow) refers to the organisms which must live in a narrow range of thermal environments and/or are intolerant to wide changes of their thermal environment. Many insect species are stenothermic and can survive only in stable thermal environments.

Hibernation and Estivation

Under certain conditions, some homeothermic animals can behave as heterotherms, and this is the case of many mammals of temperate regions, which enter into lethargy during winter; their metabolism is set to very low levels, just enough to sustain their lives without need of feeding. Such animals are named as *hibernators* (from the Latin *hibernare* = to pass winter). The condition of hibernation is needed because of the scarcity of food during the winter period, for example, for the bears. However, polar bears do not hibernate, because they remain on search for seals over the ice fields on high seas during the winter; in the summer months, they come back to the land.

On the other hand, some animals enter a state of lethargy during the summer, with reduction in their body temperature and metabolic levels; it is the *estivation* (from the Latin *aestivus* = in the summer). Such behaviour is presented by many desert-living species; in order to survive to the scarcity of water and the very high temperatures, the animals remain inactive inside deep burrows during the day time and search for food during night hours.

3.2 Thermoregulation

3.2.1 Definitions and Concepts

Thermoregulation is the process of thermal energy control in any physical system. Living organisms are physical systems in which thermal energy is produced along the metabolic processes associated to the life-supporting functions. At the same time, there are thermal energy exchanges – gains and losses – between the organisms and the external environment. As the body temperature is a measure of the thermal energy content of a given organism at a given moment, its variation is under the control of the existing thermoregulatory mechanisms. See Fig. 3.1.

The optimal thermal environment for a given animal is represented by the range of temperatures from the lower critical temperature (LCT) up to the upper critical

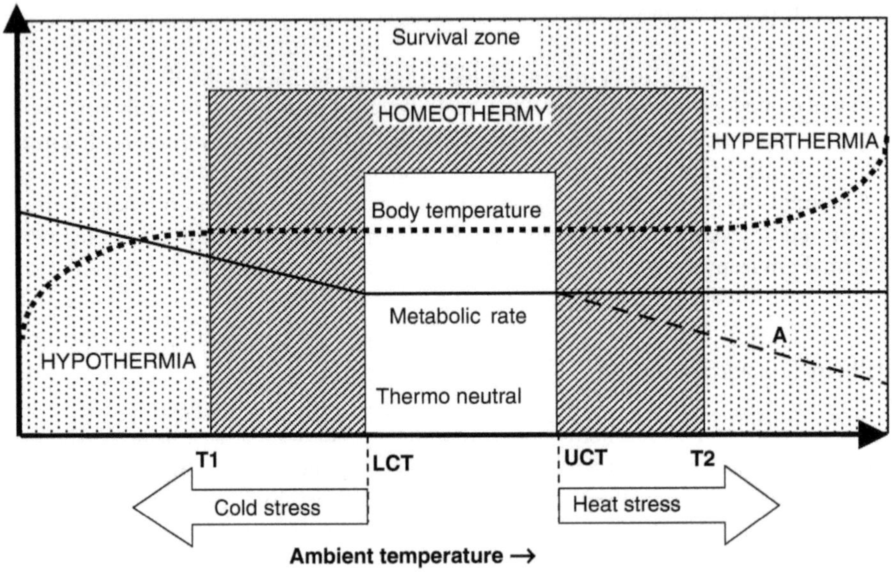

Fig. 3.1 Ambient temperature zones related to variations in metabolic rate and body temperature. *LCT* lower critical temperature, *UCT* upper critical temperature. See text for more explanation

temperature (UCT); such a range defines the *thermo neutral zone* (IUPS Thermal Commission 2003), at which temperature regulation is achieved without changes in the metabolic heat production or activation of the evaporative heat loss. In other words, when an animal is in the thermo neutral zone, its organism does not make any effort to gain or lose thermal energy. This concept does not apply to ectotherms.

As it is shown in Fig. 3.1, temperatures T_1 and T_2 mark the limits of the *zone of homeothermy*, within which it is possible for the animal to sustain its internal temperature in a relatively steady state, that is, within specified, narrow limits.

When the ambient temperature is lower than LCT, the organism is under *cold stress*; then it needs to replace the heat that was lost to the environment, by means of an increased metabolism rate together with behavioural changes. At an ambient temperature T_1, it attained a limit beyond which the organism is unable to sustain homeothermy even with an increased metabolism, and then the body temperature decreases (*hypothermia*) together with the ambient one until death. Inversely, when the ambient temperature is above T_2, the organism is under *heat stress*, a condition in which the heat loss physiological mechanisms are not enough to maintain thermal balance and then the body temperature progressively increases (*hyperthermia*[1]) until the survival limit is attained. As it is illustrated by the descending dashed

[1] *Note*: *Hyperthermia and fever cannot be confounded one to another. Hyperthermia is an increase of the body temperature that is related to thermoregulatory processes. Fever, on the other hand, is a hyperthermia caused by the reaction of the organism to a pathogen, aiming to establish a hostile environment against it.*

line A in Fig. 3.1, there has a decreasing need of metabolic heat; however, metabolic rate cannot be reduced below some limit, by which there has an increased action of the available heat loss mechanisms, together with thermal energy storage.

Man and even animals under adequate life conditions try to remain within their thermo neutral zone most of the time (Ivanov 1999). Clothing, housing, feeding and behaviour can provide adequate conditions for man, while the external coat (hairs, wool or feathers), subcutaneous fat, shelters and behaviour help animals to be in thermal comfort.

By definition, thermoregulation aims to sustain thermal balance in animals, and this balance can be described by the formula

$$\mathbf{S} + \mathbf{M} + \boldsymbol{\Gamma} \pm \mathbf{L} \pm \mathbf{C}_S - \mathbf{C}_R \pm \mathbf{K} - \mathbf{E}_\mathrm{S} - \mathbf{E}_\mathrm{R} = 0 \tag{3.1}$$

where **S** is the effective short-wave irradiance, **M** is the metabolic rate, $\boldsymbol{\Gamma}$ is the rate of thermal energy storage, **L** is the rate of heat exchange by long-wave radiation, $\mathbf{C}_\mathrm{S}$ is the rate of heat exchange by cutaneous convection, **K** is the rate of heat exchange by conduction; and $\mathbf{C}_\mathrm{R}$, $\mathbf{E}_\mathrm{S}$ and $\mathbf{E}_\mathrm{R}$ are the rates of heat loss by respiratory convection, cutaneous evaporation and respiratory evaporation, respectively.

The sensible heat flux by conduction (**K**) can be assumed as negligible for most terrestrial animals, since the physical contact between body and ambient surfaces is very limited in most cases, with exception for reptiles, amphibians and animals that lay down on the ground. In tropical environments, animals avoid contact with surfaces whose temperature is higher than that of the body; this would prevent any extra heat gain by conduction. However, it has been observed that many cows under the shade of trees remain lying down on the ground, while others remain upright; this may be evidence that the ground temperature is comfortable. On the other hand, conduction can be an important way of heat exchange in aquatic animals; marine mammals on land, for example, tend to have great contact surfaces with the ground, relative to their body surface area.

3.2.2 Object of Thermal Control

Despite the many studies on thermoregulation in animals carried out since more than 100 years ago, with several revisions published on that matter (e.g. Bligh 1973, 1985, 1998), many unsolved problems remain yet. In the old times, there was a belief that thermoregulatory mechanisms were integrated together in functional terms based on the hypothalamic temperature; in other words, the thermoregulatory centre would be located in the hypothalamus. However, changes observed in the thermal sensitivity of hypothalamus as responses to signals from skin thermo-sensors showed that the hypothalamic temperature could not be the controller element. Then, other thermo-physical parameters were considered, as the average body temperature, the body heat content, the thermal balance and the intensity and direction of the direction of heat fluxes through the skin (Burton and Edholm 1955; Gonzalez et al. 1978; Houdas and Guiev 1978; Bligh 1990; Webb 1995; Ivanov 1999).

The last of the above cited papers (Ivanov 1999) refers to a study with rabbits in a controlled thermal environment, measuring their temperatures in 18 different places of the body, the hypothalamus inclusive. By increasing the ambient temperature from 20°C onwards until the first thermoregulatory reaction was observed (vasodilatation in the ears) between 28 and 31°C, Ivanov recorded 23 thermal reactions in the animals, *without any change of the hypothalamic temperature* – at the same time the temperature of other body regions varied from 38 to 39°C; during this process, skin temperature increased 3–4°C. It was concluded that the hypothalamic temperature was not the only object of thermoregulation; in addition, the efficiency of the homeothermic process would depend on the ambient temperature, the skin temperature and the temperature of other body parts.

Sweating, for example, is an important thermoregulatory response in hot environments for man, cattle and other animals, and it is a function of the skin surface temperature (Berman 1971; Isabirye 1972; Finch 1973; Gatenby 1986; Silva and Maia 2011).

There exist thermo-sensors in every tissue of the body (Hensel 1981; Bligh 1990), and the temperature of the blood stream varies continuously, even under conditions of thermal comfort (Ivanov 1999); therefore, even if an individual is under thermal comfort, it is not sure that there has thermal homeostasis. In other words, thermoregulatory mechanisms work continuously under each and every condition.

3.2.3 Lower Critical Temperature

Blaxter (1977) proposed a method to estimate the lower critical temperature for several animal species, by means of the formula

$$LCT = T_r + I_c(\mathbf{E}_S + \mathbf{E}_R) - \mathbf{M}(I_t - I_c) \ \ {}^{\circ}\mathrm{C} \tag{3.2}$$

where T_r is the deep body temperature (°C), I_c is the thermal insulation of the hair coat (W^{-1} m^2 °C), I_t is the thermal insulation of the tissues (W^{-1} m^2 °C) and the other elements were previously defined. Values of LCT and UCT for some animals are given in Table 3.1.

It must be remembered, however, that most of the published data on this subject were obtained in temperate regions under the artificial conditions of climatic chambers; therefore, they are of little value for tropical climates, in which more studies are needed.

Table 3.1 Values of the lower (LCT) and upper (UCT) critical temperatures for some animals (°C)

Animal	Rectal temperature (°C)	LCT	UCT
Cattle			
European, adult	37.5–39.3 (38.3)	−6	27
European, calf		13	25
Zebu, adult	38.5–40.0 (39.1)	7	35
Sheep			
Adult	38.3–40.0 (39.1)	−2	20
Lamb		29	30
Wool-less breeds	38.9–40.5 (39.8)		
Swine			
Suckling		32	33
Adult	38.7–39.8 (39.2)	0	15
Goat			
Adult, Egyptian	38.7–40.7 (39.9)	10–20	25–30
Poultry	40.6–43.0 (41.7)	7	21
Chicken		13	27

Data from Bianca (1970), Hahn (1976), Salem et al. (1982) and non-published material from the present authors

3.3 Metabolism and Heat Storage

3.3.1 Thermogenesis

3.3.1.1 Definition

Thermogenesis is the transformation of chemical energy into heat and mechanical work by metabolic activities within the organism; it is usually expressed in terms of unit area of body surface (W m^{-2}), especially if we want to express metabolism in terms of body heat balance. Sometimes it is expressed in terms of body mass, W kg^{-1}.

3.3.1.2 Metabolic Rate

Published papers on animal metabolism often refer to a minimal metabolic level, the *basal metabolism.* However, according to Schmidt-Nielsen (1989), such a term is an incorrect one, because it assumes an ultimate lower level for the metabolic rate. However, when sleeping, a given individual has a metabolic rate below that considered as basal; it would be more correct to say it as *lowest observed metabolic rate* (LOMR). On the other hand, the metabolic rate of an animal that is resting in a thermo neutral environment is the *resting metabolism.*

3.3.1.3 Determination of the Metabolism

As the several physiological processes are carried out, different metabolic pathways are activated, thus producing thermal energy in direct proportion to the level of physical and chemical activity. Such an activity level is evaluated by the *metabolic rate*, which refers to the energy metabolism per unit time. It can be determined in three different ways:

1. *By calculating the difference between the energetic value of the ingested food and the energetic value of the excreta (urine and faeces).* As this method involves the assumption of no changes in the structural composition of the organism, then it cannot be used for growing animals or animals subjected to changes in the storage of fat or any other material.
2. *Direct calorimetry.* It is the determination of the total amount of heat produced and can be assumed, in principle, as the most accurate method to determine metabolism. The animal to be evaluated is placed into a chamber with thermally insulated walls; then, the amount of heat it eliminates by means of radiation, conduction, convection and evaporation is measured.
3. *Indirect calorimetry.* This is the most widely used method, but it cannot determine thermal energy produced along anaerobic processes, which occur also into the organism. The animal to be evaluated is placed into a closed chamber, in which there has a controlled air flux; this air is evaluated as for its gas content (O_2, CO_2, CH_4) before and after it enters the chamber. Urine is also collected and analysed for its nitrogen content.

Method 3 is based on the assumption that the heat produced by 1 L of O_2 during the oxidation processes of carbohydrates, fats or proteins is almost the same; as it is shown in Table 3.2, the highest heat production is that of carbohydrates (20.93 kJ L^{-1} of O_2), and the lowest one is that of proteins (18.34 kJ L^{-1} of O_2), a difference of 12.4% only. Therefore, it is usually considered an average value of 20.1 kJ L^{-1} for the thermal energy produced in relation to the O_2 consumed. The error resulting from the use of this average value is assumed as negligible because animal foods are mixtures of the three above-mentioned components.

However, it would be better to use the general formula

$$\mathbf{M} = a(\mathbf{O_2}) + b(\mathbf{CO_2}) + c(\mathbf{N}) + d(\mathbf{CH_4}) \text{ kJ} \tag{3.3}$$

where $\mathbf{O_2}$ is the volume of oxygen consumed (L), $\mathbf{CO_2}$ is the volume of carbon dioxide produced (L), **N** is the amount of nitrogen in the urine (g) and $\mathbf{CH_4}$ the volume of methane produced (L); *a*, *b*, *c* and *d* are coefficients used for each animal species and which are found in Table 3.3.

For ruminants, when there has no information about excreted N and produced CH_4, the following equation can be used using the coefficients given in Table 3.3:

Table 3.2 Thermal energy produced and O_2 consumed during metabolic processes

Component	O_2 consumed ($L\ g^{-1}$)	Heat produced kJ L^{-1} of O_2	Respiratory quotient
Carbohydrates	0.84	20.93	1.00
Lipids	2.00	19.68	0.71
Proteins (urea)	0.96	18.75	0.81
Proteins (uric acid)	0.97	18.34	0.74

Adapted from Schmidt-Nielsen (1990)
The values for proteins depend on the final product (urea, uric acid). The respiratory quotient is estimated as Q_R = (CO_2 produced)/(O_2 consumed)

Table 3.3 Coefficients used to estimate metabolism from O_2 consumption, CO_2 production, methane production and nitrogen excretion in the urine

Animal type	Consumed O_2 (kJ L^{-1})	Produced CO_2 (kJ L^{-1})	Excreted N in urine (kJ g^{-1})	Produced CH_4 (kJ L^{-1})
Ruminants[a]	16.18	5.16	−5.93	−2.42
Nonruminants[b]	16.57	4.50	−5.90	...
Birds[c]	16.20	5.00	−1.20	...

References: [a]Brouwer (1965); [b]Brockway (1987); [c]Farrell (1974)

$$\mathbf{M} = 1{,}000\,\Phi_{cap}\left(16.18\,O_{con} + 5.16\,CO_{exp}\right)\ W \tag{3.3a}$$

where Φ_{cap} is the air flux through the mask (L s^{-1}), O_{con} is the consumed oxygen (proportion) and CO_{exp} is the exhaled carbon dioxide (proportion).

Maia et al. (2011) determined the metabolic rate of goats exposed to sun and under shade in an equatorial semiarid region, by using the formula

$$\mathbf{M} = F_R V_T A^{-1}\left(0.75\ Q_O O_{con} + 0.25\ Q_{CO}\ CO_{exp}\right)\ W\ m^{-2} \tag{3.3b}$$

where F_R is the respiratory rate (breaths/s), V_T is the tidal volume (L/breath), Q_O and Q_{CO} are heat coefficients of O_2 and CO_2, respectively. The body surface area was estimated from the body weight (kg) as $A = 0.13\ W^{0.556}$ (m^2).

In many cases, we are able to measure oxygen consumption or CO_2 production only, and then we can use the following coefficients:

$$O_2 \text{ consumed} = 20.1\ kJ\ L^{-1}$$
$$CO_2 \text{ produced} = 24.1\ kJ\ L^{-1}$$

The values of the two above-mentioned coefficients are approximations and must be used only in the case of impossibility to apply formulas 3.3, 3.3a, and 3.3b. Determination of metabolism from CO_2 elimination only is a low precision approach, because the levels of normal activity of several physiological functions are subjected to wide variations (Schmidt-Nielsen 1990).

There are two reasons for that limitation: (a) The organism contains a considerable CO_2 storage and (b) the heat equivalent of a given volume of CO_2. The stored gas is eliminated in great amounts during respiratory hyperventilation, but this amount of CO_2 is not a measure of the metabolic rate. On the other hand, during an intense physical exercise, the lactic acid formed in the muscular tissue is transferred to the blood, and the CO_2 is eliminated in the lungs; however, this CO_2 is not also a measure of the metabolism. According to Schmidt-Nielsen (1990), CO_2 is less suitable than O_2 for estimating metabolism because different foodstuffs give rather different amounts of energy per litre of produced CO_2: fats, for example, give one-third more energy than do carbohydrates.

Another problem is the exchange of O_2 and CO_2 through the skin, despite the low rates in relation to those observed in the lung. In frogs, for example, the O_2 uptake in the lungs changes during the year, while the skin uptake is nearly constant throughout the year because it is related to the constant O_2 concentration in the atmosphere (Schmidt-Nielsen 1990). According to the same author, reptiles have a nearly impermeable skin, while in mammals, the oxygen uptake through the skin is barely measurable.

As for the skin CO_2 elimination, in humans it is less than 1% of that from the lungs (Alkalay et al. 1971). Results of a study with dairy cows (Loureiro 2004) showed that the CO_2 flux from the skin is directly related to the rate of cutaneous evaporation by the function

$$CO_2 = 0.065 + 0.0012\ \mathbf{E_S}\ \mathrm{L\ h^{-1}\ m^{-2}} \tag{3.4}$$

with a coefficient of determination $R^2 = 0.71$ and where $\mathbf{E_S}$ is the rate of cutaneous evaporation (W m^{-2}). The observed average rate of CO_2 uptake from the skin was 0.168 L h^{-1} m^{-2}, against that 128.9 L h^{-1} from the lungs.

3.3.1.4 Metabolism and Ambient Temperature

If the oxygen uptake is studied over a wide temperature range in an animal, it can be observed a negative relationship of O_2 consumption with the increasing temperature. For example, Finch (1985) evaluated Brahman, Shorthorn and crossbred cattle and observed that their O_2 consumption decreased significantly as the ambient temperature increased from 25 to 45°C. See Fig. 3.2. In the Shorthorns, this reduction began at 25°C, in crosses at 28°C and in Brahmans at 41°C.

3.3.1.5 Metabolism, Surface Area and Body Mass

Despite differences among body tissues as for their metabolic rates, it is possible to conceive an "average" metabolic rate for the whole body. Assuming that the thermal energy is exchanged between an animal's body and its environment through the skin surface, Rubner (1983) proposed that the metabolism of dogs

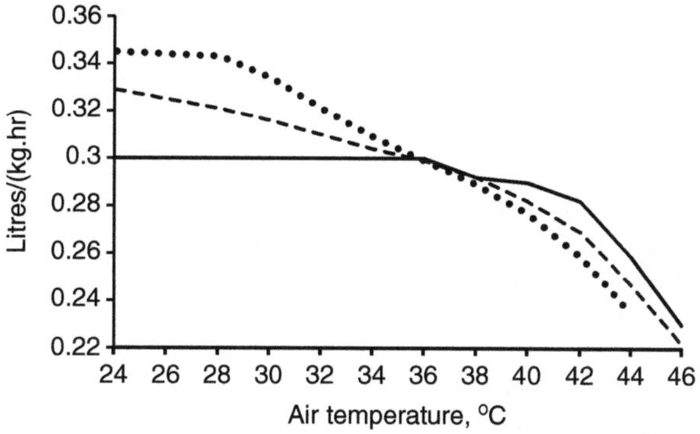

Fig. 3.2 Oxygen consumption by cattle of three groups: Brahman (*B*), Brahman x Hereford-Shorthorn crosses (*BHS*) and Shorthorn (*S*), according to the air temperature (Adapted from Finch 1985)

(3.2- to 31.2-kg body weight) was approximately constant when expressed in terms of unit area of body surface.

However, there are great difficulties of measuring surface area of animals, and the several formulas for predicting surface area from body weight give results differing by 70% for pigs and 30% for horses (Poczopko 1971). Some of those formulas will be discussed some paragraphs ahead.

According to Kleiber (1932), metabolism can be related directly to body mass (*W*, kg) as

$$\mathbf{M} = 3.552\ W^{0.739}\ \mathrm{W} \tag{3.5}$$

Brody et al. (1934) evaluated several species from mice to elephants, by using a slope of 0.734 in the above equation. In his book *The Fire of Life* (1961), Kleiber rounded off this slope to 0.75 or 3/4 and the coefficient 3.552–3.3922, in order to make calculations easier. However, Heusner (1982) suggested that Kleiber's 3/4 slope value could be a statistical artefact and that the correct value for any animal species would be 2/3, while the proportionality constant (the value 3.552 in the Eq. 3.5) would be different for each species. On the other hand, the exponent 2/3 would provide a better description of intraspecific variations (Feldman and McMahon 1983; Prothero 1984; Heusner 1985). Discussions about the matter are open yet, and the interested reader is addressed to the reviews by Hemmingsen (1960), McMahon (1973) and especially to that of Schmidt-Nielsen (1989). Table 3.4 shows some of the equations proposed for the estimation of resting metabolism in some animals.

The equations shown in Table 3.4 estimate the metabolic rate in terms of watts (W). However, in order to use the respective results to determine thermal balance by Eq. 3.1, we must change the units into W m^{-2}. Therefore, any equation from Table 3.4 must be divided by the body surface area of the respective animal. As an example, for a steer, we would have

$$\mathbf{M} = \frac{3.552\,P^{0.75}}{A}\ \mathrm{W\,m^{-2}} \tag{3.6}$$

Table 3.4 Equations for the estimation of resting metabolism as a function of body weight W (kg)

Animals	M (W)	References
Mammals		
Mouse (*Peromyscus*)	$8.0\ W^{2/3}$	Heusner (1982)
Cattle	$25.37\ W^{2/3}$	Heusner (1982)
Diverse species (2.4–100 g)	$W^{0.23}$	Bartels (1982)
Diverse species (110–260 g)	$W^{0.42}$	Bartels (1982)
Diverse species (>260 g)	$W^{0.76}$	Bartels (1982)
Rats, steers	$3.552\ W^{3/4}$	Kleiber (1961)
Birds		
Passerine (normal activity)	$589.9\ W^{0.704}$	Aschoff and Pohl (1970)
Passerine (resting)	$480.7\ W^{0.726}$	Aschoff and Pohl (1970)
Others (normal activity)	$381.0\ W^{0.729}$	Aschoff and Pohl (1970)
Others (resting)	$307.7\ W^{0.734}$	Aschoff and Pohl (1970)
Birds in general	$361.8\ W^{0.668}$	Schmidt-Nielsen (1989)
Reptiles		
Lizards (0.002–4.4 kg)	$28.64\ W^{0.62}$	Bartholomew and Tucker (1964)
Lizards (0.001–7 kg)	$32.7\ W^{0.83}$	Bennett and Dawson (1976)
Snakes (*Colubridae*)	$18.4\ W^{0.98}$	Galvão et al. (1965)
Snakes (*Boidae*)	$7.49\ W^{1.09}$	Galvão et al. (1965)
Serpents in general	$12.99\ W^{0.86}$	Galvão et al. (1965)
Amphibians		
Salamander (25°C)	$0.79\ W^{0.66}$	Ultsch (1973)
Marsupials		
Tasmanian devil	$194.8\ W^{0.739}$	MacMillen and Nelson (1969)
Kangaroos (9 a 54 kg)	$203.5\ W^{0.737}$	Dawson and Hulbert (1970)

In order to estimate body area, we must remember that as for geometrical bodies in general the surface area A is a function of the volume V, in the form $A = x{\cdot}V^{2/3}$, in which x is a constant depending on the geometric form: $x = 4.836$ for a sphere and $x = 6$ for a cube. A sphere is the form that has the least area in relation to its volume.

Animal bodies have complex forms, and the evaluation of their surface area can be very difficult, but by considering the above assumptions, Meeh (1879) proposed the formula

$$A = x W^{2/3}\ \mathrm{m}^2 \tag{3.6a}$$

for the estimation of the body surface area of mammals and in which W is the body weight (kg).

Values for the constant x in formula 3.6a were determined for several species, varying from 7.5 for porcupine to 12.5 for snakes (Benedict 1934). However, after a comprehensive review on this matter, Schmidt-Nielsen (1989) concluded that the search for more precise x values is a futility; he considered the average value $x = 10$ as adequate for every animal species.

Table 3.5 Body surface area as function of the body weight *W* (kg)

Animal	Surface area (m^2)	References
Sheep		
Merino (23–63 kg)	$0.171\ W^{0.5025}$	Bennett (1973)
Cattle		
Zebu	$0.15\ W^{5/9}$	Finch (1985)
European breeds	$0.13\ W^{5/9}$	Finch (1985)
Crossbreds	$0.14\ W^{5/9}$	Finch (1985)
Pigs	$0.09\ W^{2/3}$	Benedict (1934)
Birds	$0.1\ W^{2/3}$	Benedict (1934)

Bennett (1973) measured directly the total area of sheep hides, after correction for post-skinning shrinkage. The respective equations are given in Table 3.5. Finch (1985) proposed three equations for cattle, which are given also in Table 3.5.

Turning back to Eq. 3.6, let us estimate the metabolic rate of a European cattle breed. By taking the respective equation from Table 3.4 and substituting it for *A* value, we have

$$\mathbf{M} = \frac{3.552\ W^{0.75}}{A} = \frac{3.552\ W^{0.75}}{0.13\ W^{5/9}} = 27.323\ W^{0.1944}\ \mathrm{W\ m^{-2}} \tag{3.6b}$$

For sheep, we can use the formula given by Bennett (1973), also presented in Table 3.5 and which has a high coefficient of determination, $R^2 = 0.89$. Then

$$\mathbf{M} = \frac{3.552\ W^{0.75}}{A} = \frac{3.552\ W^{0.75}}{0.171\ W^{0.5025}} = 20.772\ W^{0.2475}\ \mathrm{W\ m^{-2}} \tag{3.7}$$

3.3.1.6 Circadian Variation of Metabolism

Metabolic activity barely remains at uniform rates along the time, even in strictly homeothermic animals. An evidence of this fact is the regular variation of the body temperature of homeotherms along time; this variation is synchronised with daily and seasonal changes of the light intensity, air temperature and probably other factors (Kleitman 1949; Beakley and Findlay 1955).

Among nocturnal rodents, for example, the body temperature and the metabolic rate show a well-evident circadian rhythm with peaks at night, while the inverse is observed for diurnal birds (Aschoff and Pohl 1970). Thermal conductance of the external body covering (hair coat in mammals and plumage in birds) varies also in accordance to circadian cycles (Aschoff 1981); such a variation can help the heat balance of the animals either in activity or resting.

Hahn et al. (1986) observed a circadian rhythm of the tympanic temperature in sheep, with maximum peaks at midnight and minimum at noon. Mendel and Raghavan (1964) recorded maximum body temperatures for sheep at evening (around 16:00 h) and minimum ones before dawn (around 04:00 h). Brown

(1971), Mohr and Krzywanek (1960) and Johnson (1991) determined also circadian variations of the body temperature in Merino sheep.

Silva and Minomo (1995) observed Corriedale sheep in a tropical environment, determining well-defined circadian and seasonal sine-shaped cycles of the rectal temperature. Maximum values occurred around 17:00 h in the summer and at 18:30 h in winter, while the minimum values were observed at dawn (05:00 h) in the summer and at 06:00 h in the winter. The body temperature peaks were out of phase with those of the ambient temperature; the difference was greater in the winter (4 h) than in the summer (3 h). Seasonal variation of the rectal temperature followed also a sine-shaped curve with maximum peak in December (average 39.8°C) and minimum one in July (average 39.0°C).

As for the metabolic rate, Blaxter and Boyne (1982) observed in several sheep breeds that the resting metabolism varied as a sine-shaped curve with minimal values in the midwinter and maximal ones in the midsummer. According to these authors, the observed increased metabolic rate in the summer would be due to factors other than changes in food intake.

Both circadian and seasonal cycles are hormonally regulated. There are strong evidences that the effect of light is transmitted to the hypothalamus from the pineal gland, which is located in the central area of the brain. This gland produces the hormone *melatonin*, as discovered by Lerner et al. (1959), and its activity is regulated by light/dark cycle; melatonin production is confined to the dark period, as was first observed by Wurtman et al. (1964).

In mammals, light is absorbed at the retinal surface of the eyes, and a signal is sent to the pineal gland; then, the hypothalamus is activated by way of the retinohypothalamic projection (Pickard 1982). Light causes suppression of the neural activity in the hypothalamic suprachiasmatic nucleus and then a reduction in melatonin synthesis. On the other hand, in the dark period, the suprachiasmatic nucleus is activated, thus increasing melatonin synthesis (Reiter 1991a). Besides, there are evidences that melatonin production can also be affected by the exposure of skin to light (Stumpf 1988).

According to Reiter (1991b), melatonin is secreted in the pineal gland probably in its active form, by elements known as *pinealocytes* (Reiter 1981) and then passed directly to the bloodstream. It is derived from the amino acid tryptophan by means of several reactions. Under the action of the enzyme tryptophan hydroxylase, the L-tryptophan is first changed into 5-hydroxy-L-tryptophan, which is then transformed into serotonin by means of decarboxylase; after a series of reactions, the end product is 5-methoxy-*N*-acetyltryptamine or melatonin (Reiter 1996, personal information).

Several tissues (e.g. the gastrointestinal tract) are able to convert tryptophan into serotonin, but the pineal gland is the only that has all the enzymes and substrates needed to melatonin synthesis. Melatonin secretion was initially thought to function exclusively as an intermediary between seasonal photoperiod changes and animal reproduction cycles (Reiter 1998), but other functions have been identified, as effects on the circadian rhythms (Reiter 1993a) and sleep mechanisms (Dawson and Encel 1993). Light intensity is an important factor to be considered in those cases, but the respective wavelength is also of importance; Brainard et al. (1985) observed that, at this respect, the blue light (0.475–0.525) μm is the most active band.

It is well accepted that melatonin has a major role in the neuroendocrine aspects of the reproductive physiology of animals (Reiter 1998); photoperiod variation is translated by the pineal gland into variations of the melatonin secretion, with a subsequent action on the hypothalamic-pituitary system that controls reproduction (Robinson 1987; Reiter 1993b). Other functional aspects of melatonin have been identified, as those related to the immune function (Maestroni et al. 1989; Maestroni 1993) and antioxidative processes.

According to Reiter (1998), there is also evidence supporting a physiological role for melatonin in thermoregulation. It has been observed an increase in core temperature after nocturnal exposure to light (Badia et al. 1990), resulting also in increased alertness and psychomotor performance. In humans, melatonin secretion causes a decrease of the body temperature across the sleep period. It has been suggested that at least half the diurnal variation in the core temperature is the result of melatonin secretion.

3.3.2 Categories of Thermogenesis

According to Himms-Hagen (1983), it would be two categories of thermogenesis. *Obligatory thermogenesis* is that associated with those metabolic reactions essential to the cells, those related with endothermic reactions and those associated with the processing of food. It occurs in all the body tissues and is controlled basically by thyroid hormones.

On the other hand, *facultative thermogenesis* may or may not be occurring at any given time; it may be brought on by cold, by exercise or by diet. It involves two mechanisms: (a) muscle activity, which is controlled by the nervous system through mediation by acetylcholine and (b) direct thermogenesis in brown adipose tissue, measured by noradrenaline.

Diet-Induced Thermogenesis. It occurs mainly in the brown adipose tissue (BAT) and is directly mediated by noradrenaline and also eventually by glucocorticoids, insulin or thyroid hormones. It depends on the amount of food ingested, being associated primarily with overeating.

Cold-Induced Thermogenesis. Occurs in two sites: (a) in the BAT under the control by glucocorticoids, by insulin and by thyroid hormones and (b) in the skeletal muscle, through shivering mediated by acetylcholine or, eventually, by glucocorticoids and catecholamines (adrenaline and noradrenaline). It occurs only at ambient temperatures below thermo-neutrality.

Exercise-Induced Thermogenesis. The site is the skeletal muscle (mechanical work), and it is directly mediated by acetylcholine or, eventually, by glucocorticoids and catecholamines.

Metabolic reactions associated to the obligatory thermogenesis are endocrine mediated, and then the changes in endothermic thermogenesis are usually slow and require hours or days to be completed. On the other hand, the reactions involved in the facultative thermogenesis occur principally in two organs (skeletal muscle and BAT) and can be switched on and off in minutes rather than hours. Its objective is to supply any heat need during cold stress, when the organism is under temperatures below its LCT.

The role of BAT in facultative thermogenesis has been long recognised (Smith and Horwitz 1969); it was later shown by Foster and Frydman (1978) that the BAT is the major site of heat production in cold-acclimated rats, which presented three- to fourfold increases in thermogenesis after exposure to cold. It is interesting to note that animals under such conditions of cold stress tend to a twofold increase in food intake, as compared to those maintained under 28°C (Leung and Horwitz 1976; Van Hardeveldt et al. 1979); that extra food is just the energy source that allows the BAT to produce more heat.

The BAT is a very important source of heat for newborn calves (Alexander et al. 1975), but it decreases rapidly after the first weeks from birth (Monteith and Mount 1974). This tissue is also of importance for thermoregulation in newborn of other animal species as man, sheep, dogs, cats and rodents in general (Alexander 1962; Alexander and Williams 1968; Thomson and Jenkinson 1969; Alexander et al. 1970; Slee 1985).

The BAT is located especially between the shoulders just below the neck and is strongly vascularised. Its cells have a great content of mitochondria (in which cell respiration takes place) whose considerable amount of cytochrome oxidase gives the characteristic brown colour to BAT. The thermogenesis is mediated by the binding of noradrenaline to the BAT receptors, and at this point, there has two options:

(a) There has in the cells an increased ATP uptake, which is hydrolysed to give work and heat.
(b) The ATP is resynthesised from ADP, thus resulting into the releasing of phosphorus (in general linked to H^+) which passes through the mitochondrial membrane. Once within the mitochondria, H^+ is linked to O_2 in order to give H_2O and heat.

The BAT is heated rapidly along those processes, and the resulting heat is carried out to other tissues by the blood. Such a heating process is also of importance for hibernating animals, especially when they are awakening at the end of the winter.

3.3.3 Endocrine Regulation of Metabolism

Thyroid hormones have a very important role in the control of metabolism. It has been long known that metabolic rate increases after administration of thyroid hormones, while a decrease is observed after thyroidectomy. However, it is not easy to interpret the results of experiments as for the role of thyroid hormones, because they have great influence on the action of other hormones involved in thermogenesis (as the growth hormone and the catecholamines) and have an attenuating action on the sympathetic nervous system (Axelrod 1975; Gibson 1981). According to Fregly et al. (1979), the thermogenic effect of catecholamines is increased by hyperthyroidism and decreased by hypothyroidism.

The main active thyroid hormone, 3,5,3′-tri-iodothyronine or T_3, is only in part secreted by the thyroid gland. Nearly 96% of it is formed in several tissues through deiodination of thyroxine or T_4 (3,5,3′,5′-tetra-iodothyronine), which is the principal product of the gland, through the action of the enzyme 5′-deiodinase (Eisenstein et al. 1978; Norman and Litwak 1987; Kahl et al. 1995; Randall et al. 1997).

Thyroid hormones present thermogenic action by three ways: (a) by changing the properties of the mitochondria, leaving to an increase of its respiration rate; (b) by increasing tissue content in mitochondria; and (c) by increasing action of the enzyme $N_a^+K^+$-ATP-ase in the tissues, thus affecting sodium pumping (Smith and Edelman 1979). Although those mechanisms are not mutually exclusive, there is no agreement as to their relative importance.

It has been observed that animals under thermal stress present changes in both T_3 and T_4 levels. Thomson (1973) reviewed data on the variation of thyroid function in cattle and concluded that acclimation to hot environments was associated with an increased body temperature and decreased thyroid activity. Similar results were found in other species, as sheep, camel, horses, goats and buffaloes (McFarlane et al. 1974; Valtorta et al. 1982; El-Nouty et al. 1978).

It is a matter of fact that in animals under heat stress, there has quite a decrease in the food intake which would cause a reduced thyroid activity, as it has been observed (Valtorta et al. 1982; Christopherson et al. 1979; Ingram and Kasciuba-Uscilko 1977; Abdullah and Falconer 1977); however, the food intake was not a primary effect on the thyroid function as was observed in the referred papers. Magdub et al. (1982) carried out a study with dairy cows by evaluating the effects of environmental temperature variation on T_3 and T_4 content of blood plasma, urine, faeces and milk. All those contents (except urine T_3) were reduced after 5 days under exposure to 31.2°C. Some of the observed results are shown in Table 3.6. These results show a considerable decrease of the thyroid hormone levels after exposure of the animals to high ambient temperatures, especially in the milk. According to Johnson (1985), the decreased thyroid hormone levels of heat stressed animals would be the result of a decreased TSH release by the pituitary gland. In fact, the study by Magdub et al. (1982) showed that TSH release was reduced from 2.6 to 2 ng mL^{-1} during exposure of the animals to 31.2°C. However, TSH variation only cannot explain the considerable decrease of thyroid hormone levels.

Nascimento et al. (1996) observed Corriedale sheep in a tropical environment, correlating their T_3 and T_4 blood serum contents with air temperature and air humidity levels measured daily until 15 days before blood sampling. There were determined positive correlation coefficients (0.200–0.586) of T_4 levels with the environmental temperature until 6–7 days before blood sampling; as for 8–15 days before, the correlations were negative with a peak $r = -0.471$ in the 12th day

Table 3.6 Effect of environmental temperature on the thyroid hormone (T_3 and T_4) content of milk and blood plasma

Hormone	Ambient 17.6°C	Ambient 31.2°C
T_4		
Plasma (ng mL^{-1})	79.1 ± 2.1	66.1 ± 3.9
Milk (μg day^{-1})	30.2 ± 1.4	16.8 ± 1.4
T_3		
Plasma (ng mL^{-1})	1.46 ± 0.20	0.62 ± 0.01
Milk (μg day^{-1})	21.10 ± 0.82	9.24 ± 0.79

Source: Magdub et al. (1982)

before sampling. Correlations between environmental temperature and T_3 showed the same trends as those for T_4, but the coefficients were lower. Correlations of the air humidity with T_4 until the 7th day before blood sampling were positive (0.168–0.738); with respect to other days, the coefficients were negative (−0.173 to −0.509, with a peak at the 12th day); the values for T_3 were also lower than those for T_4. The results are illustrated by Fig. 3.3. It could be deduced from those data that when an animal is exposed to heat stress, the thyroid hormone levels increase in the subsequent 5 days, then decreasing. It is evident that, if the heat stress is constant, the T_4 level will be maintained below the normal level. Some authors (Gregerman and Crowder 1963; Collins and Weiner 1968; Yousef and Johnson 1968; Sanchez and Evans 1972; Guerrine and Bertchinger 1983) observed also increased circulatory levels during the initial phase of the heat stress, in proportion to the magnitude of exposure. Table 3.7 shows some on data that were published on thyroid hormones of sheep.

3.3.4 Thermal Energy Storage

3.3.4.1 Concepts

In general, homeotherms increase metabolism rate under ambient temperatures that were below the respective LCT. However, there has no reverse action when temperature is above HCT, because at this point the metabolism is at its lowest level. In other words, the rate of resting metabolism cannot be reduced to a level below a minimum limit, even if the metabolic heat is excessive as for the prevailing circumstances. In such a case, the excess heat must be get rid of by means of the several mechanisms available. It was shown previously (Part 2: Basic Physical Mechanisms) that under high ambient temperatures, the heat exchange by conduction and convection may become inoperative, due to a very narrow temperature differential between organism and environment. On the other hand, radiation heat exchange depends on the nature of the surfaces involved and the ambient sources of thermal radiation; this can be a serious problem in tropical regions. Therefore, evaporation becomes the main way by which the excess heat gets rid of animal's body, and, as a consequence, animal survival and well-being in hot environments require adequate water supply.

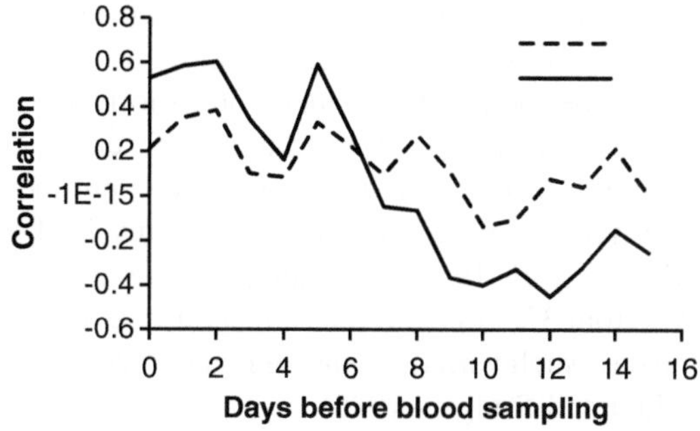

Fig. 3.3 Variation of correlation coefficients of thyroid hormones (T3 and T4) blood serum levels with ambient temperature measured several days before blood sampling. Data recorded along 12 months in Corriedale sheep under temperatures from 19 to 45°C (Adapted from Nascimento et al. 1996)

Table 3.7 Thyroid hormone levels in sheep according to several authors (values in ng mL^{-1})

Animals	Conditions	T3	T4	Author(s)
Romney Marsh				[1]
Ewes		1.69	40.5	
Lambs		4.41	51.0	
Blackbelly	19–34.5°C		67.6	[2]
Dorset			90.7	
Crossbred			113.7	
Corriedale				[3]
Females	2.6–6.2°C	1.23	39.5	
Males		1.38	67.5	
Corriedale				[4]
Females	Tropical	0.83	77.4	
Males		0.40	56.6	
Rambouillet	15°C	0.98	78.7	[5]
Patanwadi	Hot dry	1.65	163.7	[6]
Rambouillet cross		1.48	128.0	
Merino cross		1.60	130.8	
Corriedale	19–45°C	0.60–1.19	70.3–90.0	[7]

Authors: [1] Millar and Albyt (1985), [2] Ross et al. (1985), [3] Canola (1982), [4] Canola (1987), [5] Anderson et al. (1988), [6] Patel et al. (1992), [7] Nascimento et al. (1996)
T3 = tri-iodothyronine, T4 = thyroxine

In some *habitats*, as the deserts and tropical semiarid regions atmospheric temperature may largely exceed that of an animal body, such a condition is sometimes made worse by the intense solar radiation and water shortage. Animal species living in those regions need physiological mechanisms adequate to protect organism against excess water losses; therefore, such animals are likely to present some of the following characteristics:

1. Body surface covered by a thick, keratinized epithelium.
2. Sweating glands low both in number and productivity; they are concentrated in the less exposed regions of the body.
3. Skin covered by wool or a thick coat of long, thin hairs.
4. Ability to recover water that was evaporated from the respiratory tract.
5. Ability to recover water from the excreta (urine and faeces).

As for example, the Bedouin goats from Africa and the Middle East, when they are well watered, are likely to show an average water loss of 1.08 g kg^{-1} body weight per hour, while under water shortage conditions, the loss is 0.55 g kg^{-1} h^{-1} only (Louw 1993). Dromedaries with easy access to water lose 9.1 kg of water during the ten hottest hours of the day, but the same animals under water deprivation have average losses of 2.8 kg only in the same period of the day (Schmidt-Nielsen et al. 1957). African elephants increase its body temperatures during the day from 32.5 to 37.5°C as a response to air temperature differences of 10°C along the day; such variation was related to mechanisms of water and energy conservation (Elder and Rodgers 1975).

Those mechanisms of water loss control may be eventually insufficient or inadequate, and then some species – which can be named as *xeromorphic* ones – developed an additional mechanism that is the ability to store thermal energy during the day hours, in order to minimise the need of water evaporation for cooling. Thus, their body temperature can be increased along the day until maximum heat storage is reached; this thermal energy is eliminated later during the night hours, when the ambient temperature is much below that of the body. This mechanism was well studied by Schmidt-Nielsen et al. (1957) in dromedaries, for which they observed body temperature increases up to 6°C during the day; the stored heat was eliminated by convection and radiation during the cold night.

Ground squirrels are little rodents found in the deserts of North America, which live under extreme levels of air temperature and shortage of water, which is not normally found by them in the liquid state. Nevertheless, such animals have intense diurnal activity at the ground surface even during the hottest hours of the day. Chapell and Bartholomew (1981) refer that those rodents are able to tolerate body temperatures up to 42°C, when they enter their burrows and remain there until the body temperature decreases to normal levels and then return back to the ground surface. With such behaviour, they need quite a little water supply.

Laboratory studies carried out by McLean et al. (1982, 1983a, b) suggest that cattle maintained under wide variations of the ambient temperature (about 10°C) tends also to store heat in the body. However, in great mammals, the changes of the deep body temperature are very little, even if they were exposed to wide variations in the ambient temperature. On the other hand, superficial tissues (those more close to the body surface) tend to change their temperature in response to variations of the ambient one. Therefore, if the body mass is sufficiently big, there has a great potential of heat storage, even if there has little change in the body core temperature. A possible consequence of those facts would be that the cooling of surface tissues during the night and its heating during the day may leave to a lower production of metabolic heat (McLean et al. 1983b).

Thermal storage is symbolised here as $\mathbf{\Gamma}$ and is positively related to the body mass in such an order that the heaviest individuals tend to store greater amounts of thermal energy and need less production of metabolic heat. The following analogy will help to understand that point:

If we put a 50-g stone under the strong solar radiation together with another stone 1 kg weight, after some time, we will observe that the big stone is colder than the little one. The explanation is that a little object has surface area wider than that of a biggest one, in relation to the respective volumes. As for example, consider two cubes, one sided 5 cm and the other 50 cm. We have the following measurements:

5 cm *cube*: *6 faces with* 5 × 5 cm
Surface area = 6 × (5 × 5) = 150 cm^2
Volume = 5 × 5 × 5 = 125 cm^3
Area/volume relationship = 1.2

50 cm *cube*: *6 faces with* 50 × 50 cm
Surface area = 6 × (50 × 50) = 15,000 cm^2
Volume = 50 × 50 × 50 = 125,000 cm^3
Area/volume relationship = 0.12

Therefore, it is easy to conclude that as great the body is as smaller will be its surface in relation to the volume. This relation of area to volume is extremely important as for animal adaptation to its environment, especially when we consider water economy and elimination of heat by evaporation. According to Schmidt-Nielsen (1991), if a man exposed to the intense heat is able to lose 1 L h^{-1} of sweat in order to maintain his thermal balance (about 0.6 L m^{-2} h^{-1}), it is possible to use the same relationship to estimate the amount of sweat for other animals under the same conditions, in terms of percentage of body weight per hour:

Man 70 kg	1.4
Dromedary 500 kg	0.8
Donkey 95 kg	1.3
Rabbit 2.5 kg	4.7
Rat 0.1 kg	12.8

Small rodents weighing 10–100 g would evaporate water at a rate of 15–30% of their body weight per hour. But it is impossible, because a water loss amounting 10–20% of the body weight is deadly for all mammals. Consequently, due to their little body size, most of the desert rodents lodge in deep burrows during the day hours (estivation behaviour) in order to avoid body water loss.

On the other hand, the dromedaries are advantageous to be used in the desert just because the combination of a great body with a great ability to economise water. According to Schmidt-Nielsen et al. (1957), a dromedary that drinks water every day in order to become fully hydrated has body temperature changes about 2% in the daytime, but if the animal is not watered, the circadian variation of its temperature will be greater: 34°C in the morning and 41°C in the later evening; such a difference points to the heat body storage. For a 500-kg animal, the above temperature difference corresponds to 12,142 kJ of thermal energy storage or an economy of 5-L water. During the night, the animal easily eliminates the excess heat by radiation and convection. High body temperatures during the hottest day hours are advantageous because the fact that when the core temperature increases the temperature difference between the body and the environment is reduced, and then it is lower the heat gain.

3.3.4.2 Determination of Heat Storage

According to McLean et al. (1982, 1983a), changes in the mean body temperature (ΔT_b) of cattle can be estimated from the difference between the amount of heat produced (*M*) and the amount of heat lost (*H*) within a given time interval *t* (s):

$$\Delta T_b = (M - H)t/(W\,c_{pt}) \tag{3.8}$$

where *W* is the body weight (kg) and $c_{pt} = 3.47$ J g^{-1} °C^{-1} is the mean specific heat of body tissues.

For the case of animals in a fluctuating thermal environment, McLean et al. (1983b) proposed the formula

$$\Delta T_b = \alpha\,\Delta T_c + (1 - \alpha))\Delta T_s \tag{3.9}$$

where T_c is the body core temperature (which is estimated by the rectal temperature, T_r), T_s is the body surface temperature and $\alpha = 0.85$ is a weighting factor. The value of α can be calculated as

$$\alpha = (\Delta T_b - \Delta T_s)(\Delta T_c - \Delta T_s)^{-1} \tag{3.10}$$

However, if the temperature differences are small, the resulting value of α can be inaccurate. In addition, it seems that the value of α increases with increasing environmental temperature.

Heat storage can be evaluated as the difference between the amount of heat produced and the amount of heat lost within a given time interval:

$$\Gamma = c_{pt}(T_1 - T_2)\ \mathrm{W} \tag{3.11}$$

where c_{pt} is the average specific heat of the body tissues, while T_1 and T_2 are body temperatures taken at different times. Some authors used the value $c_{pt} = 3.47$ J g^{-1} °C^{-1} for the great mammals, but more recent studies have got the value $c_{pt} = 3.34$ J g^{-1} °C^{-1}; however, fat individuals with about 50% body fat content present values as low as $c_{pt} = 2.73$ J g^{-1} °C^{-1}.

Blaxter (1989) suggested the following formula for the estimation of heat storage in animals:

$$\Gamma = \frac{c_{pt}\Delta T_r\,W}{t}\ \mathrm{W} \tag{3.12}$$

where ΔT_r is the rectal temperature differential for the time period t (s) and W is the body weight (g). In order to have coherence with other elements of the heat balance equation, Eq. 3.12 must be divided by the body surface area (m^2), thus obtaining

$$\mathbf{\Gamma} = \frac{c_{pt}\Delta T_r W}{A\,t} = \frac{3.34\,\Delta T_r\,W}{A\,t}\ \mathrm{W\,m^{-2}} \tag{3.13}$$

By considering the case of cattle, for example, it is possible to use some of the respective equations given in Table 3.5 in order to estimate the surface area. After some adjustments, we have

$$\mathbf{\Gamma} = \frac{3.34\Delta T_r\,W}{\phi W^{5/9} t} = \frac{3.34\,W^{0.444}\Delta T_r}{\phi t}\ \mathrm{W\,m^{-2}} \tag{3.14}$$

where ϕ is a coefficient specific for the animal type: $\phi = 0.15$ for Zebu, $\phi = 0.13$ for European breeds and $\phi = 0.14$ for crossbreds.

For sheep

$$\mathbf{\Gamma} = \frac{3.34\Delta T_r\,W}{0.171\,W^{0.5025} t} = \frac{19.532\,W^{0.4975}\Delta T_r}{t}\ \mathrm{W\,m^{-2}} \tag{3.15}$$

Finally, an alternative way to estimate heat storage is to apply the heat balance Eq. 3.1 by making

$$\mathbf{\Gamma} = \mathbf{S} + \mathbf{M} \pm \mathbf{L} \pm \mathbf{C_S} \pm \mathbf{K} - \mathbf{C_R} - \mathbf{E_S} - \mathbf{E_R}\ \mathrm{W\,m^{-2}} \tag{3.16}$$

3.4 Problems

Problem 3.1. Body weight was recorded for each of seven Holstein cows in a subtropical environment. Estimate their metabolic rates.

Data:

Cow #	Body weight, kg
1	550
2	520
3	590
4	430
5	500
6	570
7	540

Solution: By using Eq. 3.6 we obtain for cow #1:

$$\mathbf{M} = 27.323\ W^{0.1944} = 27.323\ (550)^{0.1944} = 93.17\ \mathrm{W\ m^{-2}}$$

The results for the seven animals are:

Cow #	Body weight, kg	**M** (W m^{-2})
1	550	93.17
2	520	92.16
3	590	94.45
4	430	88.81
5	500	91.46
6	570	93.81
7	540	92.83

Problem 3.2. Five ewes of the Morada Nova breed were evaluated for the rectal temperature twice a day, at 06:00 and 14:00 h, respectively. Calculate the heat storage.

Data:

		Rectal temperature (°C)	
Animal	Weight (kg)	06:00 h	14:00 h
1	18.0	36.6	41.7
2	18.2	36.8	41.2
3	19.7	37.6	40.7
4	24.0	37.7	41.3
5	20.5	37.4	41.9

By using Eq. 3.12, we have for animal n° 1

$$\boldsymbol{\Gamma} = \frac{19.532\ W^{0.4975}\Delta T_{\mathrm{r}}}{t} = \frac{19.532(18)^{0.4975}(41.7-36.6)}{28{,}800} = 0.014569\ \mathrm{W\ m^{-2}}$$

where t = 14:00–06:00 h = 8 h = 28,800 s

The results for all the five animals will be

Animal	**Γ** (W m^{-2})
1	0.014569
2	0.012638
3	0.009262
4	0.011866
5	0.013714
Mean	0.012410

Problem 3.3. Considering the sheep data of Problem 3.2, determine the metabolic heat production of the animals.

By using Eq. 3.7,

$$\mathbf{M} = \frac{3.552\,W^{0.75}}{0.171\,W^{0.5025}} = 20.772\,W^{0.2475}\ \mathrm{W\ m^{-2}}$$

we have for animal #1

$$\mathbf{M} = 20.772\ W^{0.2475} = 20.772(18)^{0.2475} = 42.48\ \mathrm{W\ m^{-2}}$$

and so on for all the animals.

In some instances, it would be interesting to consider the metabolism in terms of body weight, so we can use Eq. 3.5 divided by the body weight

$$\mathbf{M} = \frac{3.552\ W^{0.75}}{W} = 3.552\ W^{-0.25}\ \ \mathrm{W\ kg^{-1}}$$

Animal	Weight (kg)	**M** (W m^{-2})	**M** (W kg^{-1})
1	18.0	42.48	1.72
2	18.2	42.59	1.72
3	19.7	43.44	1.69
4	24.0	45.61	1.60
5	20.5	43.87	1.67

It is clear that the metabolic rate increases directly with the surface area, but decreases with the body weight. Thus, the greater the body mass, the lower the metabolic rate.

References

Abdullah, R., Falconer, I.R.: Responses of thyroid activity to feed restriction in the goat. Aust. J. Biol. Sci. **30**, 207–210 (1977)

Alexander, G.: Temperature regulation in the newborn lamb. V. Summit metabolism. Aust. J. Agric. Res. **13**, 100–105 (1962)

Alexander, G., Williams, D.: Shivering and non-shivering thermogenesis during summit metabolism in young lambs. J. Physiol. (Lond.) **198**, 251–254 (1968)

Alexander, G., Bell, A.W., Williams, D.: Metabolic response of lambs to cold. Effect of prolonged treatment with thyroxine and of acclimation to low temperatures. Biol. Neonat. **15**, 198–200 (1970)

Alexander, G., Bennett, J.W., Gemonel, R.T.: Brown adipose tissue in the newborn calf. J. Physiol. (London) **244**, 227–230 (1975)

Alkalay, I., Suegutsu, S., Constantine, H., Stein, M.: Carbon dioxide elimination across human skin. Am. J. Physiol. **220**, 1434–1436 (1971)

Anderson, R.R., Nixon, D.A., Akasha, M.A.: Total and free thyroxine and triiodothyronine in blood serum of mammals. Comp. Biochem. Physiol. A **89**, 401–404 (1988)

Aschoff, J.: Thermal conductance in mammals and birds: its dependence on body size and circadian phase. Comp. Biochem. Physiol. A **60**, 611–619 (1981)

Aschoff, J., Pohl, H.: Rhythmic variations in energy metabolism. Fed. Proc. **29**, 1541–1552 (1970)

Axelrod, J.: Relationship between catecholamines and other hormones. Recent Prog. Horm. Res. **31**, 1–27 (1975)
Badia, P., Culpepper, J., Myers, B., Boecker, M., Harsh, J.: Psychophysiological and behavioral effects of bright and dim light. Sleep Res. **19**, 387 (1990)
Bartels, H.: Metabolic rate of mammals equals the 0.75 power of their body weight. Exper. Biol. Med. **7**, 1–11 (1982)
Bartholomew, G.A., Tucker, V.A.: Size, body temperature, thermal conductance, oxygen consumption, and heart rate in Australian varanid lizards. Physiol. Zool. **37**, 341–354 (1964)
Beakley, W.R., Findlay, J.D.: The effect of environmental temperature and humidity on the rectal temperature of calves. J. Agric. Sci. **45**, 339–345 (1955)
Benedict, F.G.: Die Oberflächenbestimmung verschiedener Tiergattungen. Ergeb. Physiol. **36**, 300–346 (1934)
Bennett, J.W.: Regional body surface area of sheep. J. Agric. Sci. **81**, 429–432 (1973)
Bennett, A.F.: Evolution of the control of body temperature: is warmer better? In: Dejours, P., Bolis, L., Taylor, C.R., Weibel, E.R. (eds.) Comparative Physiology: Life in Water and on in Land. Liviana Press, Padova (1987)
Bennett, A.F., Dawson, W.R.: Metabolism. In: Gans, C., Dawson, W.R. (eds.) Biology of the Reptilia, Physiology A, vol. 5. Academic, New York (1976)
Bennett, A.F., Ruben, J.A.: Endothermy and activity in vertebrates. Science **206**, 649–654 (1979)
Berman, A.: Thermoregulation in intensively lactating cows in near-natural conditions. J. Physiol. **215**, 477–489 (1971)
Bianca, W.: Animal response to meteorological stress as a function of age. Biometeorology **4**, 119–122 (1970)
Blaxter, K.L.: Environmental factors and their influence on the nutrition of farm livestock. In: Haresign, W., Swan, H., Lewis, D. (eds.) Nutrition and the Climatic Environment. Butterworts, London (1977)
Blaxter, K.L.: Energy Metabolism in Animals and Man. Cambridge University Press, Cambridge (1989)
Blaxter, K.L., Boyne, A.W.: Fasting and maintenance metabolism in sheep. J. Agric. Sci. **99**, 611–620 (1982)
Bligh, J.: Temperature Regulation in Mammals and Other Vertebrates. North Holland, Amsterdam (1973)
Bligh, J.: Temperature regulation. In: Yousef, M.K. (ed.) Stress Physiology in Livestock, vol. 1. CRC Press, Boca Raton (1985)
Bligh, J.: Cell, cell-talk and mammalian homeothermy. In: Bligh, J., Voigt, K. (eds.) Thermoreception and Temperature Regulation. Springer, Berlin (1990)
Bligh, J.: Mammalian homeothermy: an integrative thesis. J. Thermal Physiol. **35**, 941–972 (1998)
Brainard, G.C., Lewy, A.J., Menaker, M., Frederickson, R.H., Miller, L.S., Weleber, R.G., Cassone, V., Hudson, D.: Effect of light wavelength on the suppression of nocturnal plasma melatonin in human volunteers. Ann. N. Y. Acad. Sci. **453**, 376–378 (1985)
Brockway, J.M.: Derivation of formulae used to calculate energy expenditure in man. Clin. Nutr. **41C**, 463–471 (1987)
Brody, S., Procter, R.C., Ashworth, U.S.: Basal metabolism, endogenous nitrogen, creatinine and neutral sulphur excretions as functions of body weight. Missouri Agric. Exp. Sta. Res. Bull. **220**, 1–40 (1934)
Brouwer, E.: Report of subcommittee on constants and factors. In: Blaxter, K.L. (ed.) Energy Metabolism. Academic, London (1965)
Brown, G.D.: Thermal study of sheep at pasture in Western New South Wales. Aust. J. Agric. Res. **22**, 797–808 (1971)
Burton, A.C., Edholm, O.G.: Man in a Cold Environment. E. Arnold, London (1955)
Canola, J.C.: Determinação por radioimunoensaio dos níveis séricos de triiodotironina e tiroxina em ovinos da raça Corriedale no Rio Grande do Sul [Determination by radio-immuno-assay of

the serum levels of tri-iodothyronine and thyroxine of Corriedale sheep in Rio Grande do Sul]. M.Sc. dissertation, Universidade Federal de Santa Maria, Santa Maria, RS, Brazil (1982)
Canola, J.C.: Determinação por radioimunoensaio da variação nos níveis séricos dos hormônios tireotrófico (TSH), triiodotironina (T3) e tiroxina (T4) em ovinos da raça Corriedale [Determination by radio-immuno-assay of the variation of the serum levels of thyroxine stimulating hormone (TSH), tri-iodothyronine (T3) and thyroxine (T4) of Corriedale sheep]. Doctoral thesis, São Paulo State University, Botucatu, SP, Brazil (1987)
Chapell, M.A., Bartholomew, G.A.: Activity and thermoregulation of the antelope ground squirrel, *Ammospermophilus leucurus*, in winter and summer. Physiol. Zool. **54**, 215–223 (1981)
Christopherson, R.J., Gonyou, H.W., Thomson, J.R.: Effects of temperature and feed intake on plasma concentration of thyroid hormones in beef cattle. Can. J. Anim. Sci. **59**, 655–658 (1979)
Collins, K.J., Weiner, J.S.: Endocrinological aspects of exposure to high environmental temperatures. Physiol. Rev. **48**, 785–839 (1968)
Dawson, D., Encel, N.: Melatonin and sleep in humans. J. Pineal Res. **15**, 1–12 (1993)
Dawson, T.J., Hulbert, A.J.: Standard metabolism, body temperature, and surface areas of Australian marsupials. Am. J. Physiol. **218**, 1233–1238 (1970)
Eckert, R., Randall, D., Augustine, G.: Animal Physiology: Mechanisms and Adaptations, 3rd edn. Freeman, New York (1988)
Eisenstein, Z., Hagg, S., Vagenakis, A.G., et al.: Effect of starvation on the production and peripheral metabolism of 3,3′,5′-triiodothyronine in eutheroid obese subjects. J. Endocrinol. Metab. **47**, 889–893 (1978)
Elder, W.H., Rodgers, D.H.: Body temperature in the African elephant as related to ambient temperature. Mammalia **39**, 395–399 (1975)
El-Nouty, F.D., Elbanna, I.M., Davis, T.P., Johnson, H.D.: Aldosterone and ADH response to heat and dehydration in cattle. J. Appl. Physiol. **48**, 249–253 (1978)
Farrell, D.J.: General principles and assumptions of calorimetry. In: Morris, T.R., Freeman, B.M. (eds.) Energy Requirements of Poultry. British Poultry Science Ltd, Edinburgh (1974)
Feldman, H.A., McMahon, T.A.: The ¾ mass exponent for energy metabolism is not a statistical artifact. Resp. Physiol. **52**, 149–163 (1983)
Finch, V.A.: Comparison of non-evaporative heat transfer in different cattle breeds. Aust. J. Agric. Res. **36**, 497–508 (1985)
Finch, V.A.: Thermoregulation and heat balance in some East African herbivores. Ph.D. thesis, University of Nairoby, Kenya (1973)
Foster, D.O., Frydman, M.L.: Non-shivering thermogenesis in the rat. II. Measurements of blood flow with microspheres point to brown adipose tissue as the dominant site of calorigenesis induced by noradrenaline. Can. J. Physiol. Pharmacol. **56**, 119–122 (1978)
Fregly, M.J., Field, E.P., Katovich, M.J., Barney, C.C.: Catecholamine-thyroid hormone interaction in cold-acclimated rats. Fed. Proc. **38**, 2162–2169 (1979)
Galvão, P.E., Tarasantchi, J., Guertzenstein, P.: Heat production of tropical snakes in relation to body weight and body surface. Am. J. Physiol. **209**, 501–506 (1965)
Gatenby, R.: Exponential relation between sweat rate and skin temperature in hot climates. J. Agric. Sci. (Cambridge) **106**, 175–183 (1986)
Gibson, A.: The influence of endocrine hormones on the autonomic nervous system. J. Auton. Pharmacol. **1**, 331–358 (1981)
Gonzalez, R.R., Nichi, Y., Gagge, A.P.: Mean body temperature and effective temperature as indices of human thermoregulatory response. In: Houdas, Y., Guieu, J.D. (eds.) Trends in Thermal Physiology. Masson, Paris (1978)
Gregerman, R.I., Crowder, S.E.: Estimation of thyroxine secretion rate in the rat by the radioactive thyroxine turnover technique: influence of age, sex, and exposure to cold. Endocrinology **72**, 383 (1963)
Guerrine, V.H., Bertchinger, H.: Effect of exposure to a hot-humid and a hot-dry environment on thyroid hormone values in sheep. Br. Vet. J. **139**, 119–128 (1983)

Hahn, G.L.: Shelter engineering for cattle and other domestic animals. In: Johnson, H.D. (ed.) Progress in Animal Biometeorology, vol. 1. Swets and Zeitlinger, Amsterdam (1976)
Hahn, G.L., Nienaber, J.A., Klemcke, H.G., Gose, G.L.: Body temperature fluctuation in meat animals. Presentation to the American Society of Agricultural Engineers, Saint Joseph, MI (Paper n° 86-4009) (1986)
Hatch, T.F.: Assessment of heat stress. In: Hardy, J.D. (ed.) Temperature, its Measurement and Control in Science and Industry, vol. 3, pp. 307–318. Reinhold, New York (1963)
Hemmingsen, A.M.: Energy metabolism as related to body size and respiratory surfaces, and its evolution. Rep. Steno Meml Hosp (Copenhagen) **9**, 1–110 (1960)
Hensel, H.: Thermoreception and Temperature Regulation. Academic, London (1981)
Heusner, A.A.: Energy metabolism and body size. I. Is the 0.75 mass exponent of Kleiber's equation a statistical artifact? Resp. Physiol. **48**, 1–12 (1982)
Heusner, A.A.: Body size and energy metabolism. Ann. Rev. Nutr. **5**, 267–293 (1985)
Himms-Hagen, J.: Thyroid hormones and thermogenesis. In: Girardier, L., Stock, M.J. (eds.) Mammalian Thermogenesis, pp. 141–177. Chapman & Hall, London (1983)
Houdas, Y., Guiev, J.D.: New Trends in Thermal Physiology. Masson, Paris (1978)
Ingram, D.L., Kasciuba-Uscilko, H.: The influence of food intake and ambient temperature on the rate of thyroxine utilization. J. Physiol. (Lond.) **270**, 431–435 (1977)
Isabirye, J.S.J.: The sweat gland response to solar radiation in three species of domestic animals. Nairobi, Kenya. M.Sc. dissertation, Nairoby University (1972)
IUPS, International Union of Physiological Sciences: IUPS Thermal Commission glossary of terms for thermal physiology, 3rd edn. J. Thermal Biol. **28**,75–106 (2003)
Ivanov, K.P.: Subject of temperature control and the main function of thermoregulation of an organism. J. Thermal Biol. **24**, 415–421 (1999)
Johnson, H.D.: Physiological responses and productivity of cattle. In: Yousef, M.K. (ed.) Stress Physiology in Livestock, vol. 1. CRC Press, Boca Raton (1985)
Johnson, K.G.: Body temperature and respiratory rates of free ranging Merino sheep in and out of shade during summer. Aust. J. Agric. Res. **42**, 1347–1357 (1991)
Kahl, S., Capuco, A.V., Bineli, M., et al.: Comparison of growth hormone releasing factor and somatotropin: thyroid status of lactating, primiparous cows. J. Dairy Sci. **78**, 2150–2158 (1995)
Kerslake, D.M.: The Stress of Hot Environments, 316 p. Cambridge University Press, Cambridge (1972)
Kleiber, M.: Body size and metabolism. Hilgardia **6**, 315–353 (1932)
Kleiber, M.: The Fire of Life: An Introduction to Animal Energetic. Wiley, New York (1961)
Kleitman, N.: Biological rhythms and cycles. Physiol. Rev. **29**, 1–30 (1949)
Lerner, A.B.., Case, J.D., Heinzlman, R.V.: Structure of melatonin. J. Am. Chem. Soc. **81**, 6084 (1959)
Leung, P.M.B., Horwitz, B.A.: Free-feeding patterns of rats in response to changes in environmental temperature. Am. J. Physiol. **231**, 1220–1224 (1976)
Loureiro, C.M.B.: Dioxido de carbon liberado pela respiração e pela epiderme de vacas Holandesas expostas ao ambiente tropical (Carbon dioxide released by respiration and across the skin of Holstein cows in a tropical environment). Monograph, São Paulo State University (2004)
Louw, G.: Physiological Animal Ecology, 288 p. Longman Scientific, London (1993)
MacMillen, R.E., Nelson, J.E.: Bioenergetics and body size in dasyurid marsupials. Am. J. Physiol. **217**, 1246–1251 (1969)
Maestroni, G.J.M.: The immune-neuroendocrine role of melatonin. J. Pineal Res. **14**, 1–10 (1993)
Maestroni, G.J.M., Conti, A., Pierpaoli, W.: Melatonin, stress, and the immune system. Pineal Res. Rev. **7**, 203–226 (1989)
Magdub, A., Johnson, H.D., Belyea, R.L.: Effect of environmental heat and dietary fiber on thyroid physiology of lactating cows. J. Dairy Sci. **65**, 2323–2326 (1982)

Maia, A.S.C., Domingos, H.G.T., Araujo, J.P.F.Q., Chiquitelli-Neto, M., Silva, R.G.: Thermoregulation in goats managed in semiarid region: a study of production, gain and loss of heat. In: Proceedings of the 19th International Congress of Biometeorology, Auckland, New Zealand. Paper nr.237, 4–8 December 2011

McFarlane, W.V., Howard, B., Good, B.F.: Tracers in field measurements of water, milk and thyroxine metabolism of tropical ruminants. In: Tracer Techniques in Tropical Animal Production. International Atomic Energy Agency STI/Publ. n° 360, Vienna (1974)

McLean, J.A., Downie, A.J., Watts, P.R., Glasbey, C.A.: Heat balance of ox steers (*Bos taurus*) in steady-temperature environments. J. Appl. Physiol. **52**, 324–332 (1982)

McLean, J.A., Downie, A.J., Jones, C.D.R., Stombaugh, D.P., Glasbey, C.A.: Thermal adjustments of steers (*Bos taurus*) to abrupt changes in environmental temperature. J. Agric. Sci. **100**, 305–314 (1983a)

McLean, J.A., Stombaugh, D.P., Downie, A.J.: Body heat storage in steers (*Bos taurus*) in fluctuating thermal environments. J. Agric. Sci. **100**, 315–322 (1983b)

McMahon, T.: Size and shape in Biology. Science **179**, 1201–1204 (1973)

Meeh, K.: Oberflächenmessungen des menschlichen Körpers. Zeitschr. Biol. **15**, 425–458 (1879)

Mendel, V.E., Raghavan, G.V.: A study of diurnal temperature patterns in sheep. J. Physiol. (Lond.) **174**, 206–216 (1964)

Millar, K.R., Albyt, A.T.: Evaluation of RIA kits for the measurement of thyroid hormones in sheep and cattle sera. N. Z. Vet. J. **33**, 116–117 (1985)

Moberg, G.P.: Animal Stress, 324 p. Waverly Press, Baltimore (1985)

Mohr, E., Krzywanek, H.: Variation of core temperature rhythms in unrestrained sheep. Physiol. Behav. **48**, 467–473 (1960)

Monteith, J.L., Mount, L.E.: Heat Loss from Animals and Man. Buttherworths, London (1974)

Nascimento, M.R.B.M., Silva, R.G., Arduino, A.R.: Variation of the serum levels of thyroid hormones in Corriedale sheep bred in a tropical environment. In: Proceedings of the 14th International Congress of Biometeorology, Ljubljana, Slovenia, 1–8 September, 2, pp. 333–340 (1996)

Norman, A.W., Litwack, G.: Hormones, 806 p. Academic, San Diego (1987)

Patel, J.S., Mehta, V.M., Tajane, K.R., Kulkarni, V.V., Radadia, N.S.: Effects of thermal exposure on some of the hormones in Patanwadi and its crosses. Indian J. Anim. Sci. **62**, 89–90 (1992)

Pickard, G.E.: The efferent connections of the suprachiasmatic nucleus of the golden hamster with emphasis on the retinohypothalamic projection. J. Comp. Neurol. **211**, 65–83 (1982)

Poczopko, P.: Metabolic levels in adult homeotherms. Acta Theriol. **16**, 1–21 (1971)

Prothero, J.: Scaling of standard energy metabolism in mammals. I. Neglect of circadian rhythms. J. Theoret. Biol. **106**, 1–8 (1984)

Randall, D., Burgreen, W., French, K.: Animal Physiology: Mechanisms and Adaptations, 4th edn., 727 p. Freeman, New York (1997)

Reiter, R.J.: The mammalian pineal gland: structure and function. Am. J. Anat. **162**, 287–313 (1981)

Reiter, R.J.: Pineal gland: interface between the photoperiodic environment and the endocrine system. Trends Endocrinol. Metab. **2**, 13–19 (1991a)

Reiter, R.J.: Dynamic organization of endocrine networks: the neuroendocrine-reproductive axis and the pineal gland as examples. Ann. N. Y. Acad. Sci. **618**, 11–27 (1991b)

Reiter, R.J.: The melatonin rhythm: both a clock and a calendar. Experientia **49**, 654–664 (1993a)

Reiter, R.J.: The pineal gland: from last to first. Endocrinologist **3**, 425–431 (1993b)

Reiter, R.J.: Melatonin, active oxygen species and neurological damage. Drug News Perspect. **11**, 291–296 (1998)

Robinson, J.E.: Photoperiodic and steroidal regulation of the luteinizing hormone pulse generator in ewes. In: Crowley, W.F., Hofler, J.G. (eds.) The Episodic Secretion of Hormones, pp. 159–166. Wiley, New York (1987)

Ross, T.T., Goode, L., Linnerud, A.C.: Effects of high ambient temperature on respiration rate, rectal temperature, fetal development and thyroid gland activity in tropical and temperate breeds of sheep. Theriogenology **24**, 259–269 (1985)

Rubner, M.: Über den Einfluss der Körper grösse auf Stoffund Kraftwechsel. Zeitschr. Biol. **19**, 535–562 (1983)

Salem, M.K., Yousef, M.K., El-Sherbiny, A.A., Khalil, M.H.: Physiology of the sheep and goats in the tropics. In: Yousef, M.K. (ed.) Animal Production in the Tropics. Praeger, New York (1982)

Sanchez, O., Evans, J.W.: Ovine thyroid adaptation to heat exposure. Environ. Physiol. Biochem. **2**, 201–208 (1972)

Schmidt-Nielsen, K.: Animal Physiology: Adaptation and Environment, 4th edn. Cambridge University Press, New York (1990)

Schmidt-Nielsen, K.: Animal Physiology: Adaptation and Environment, 4th edn. Cambridge University Press, New York (1991)

Schmidt-Nielsen, K.: Scaling: Why is Animal Size so Important? 241 p. Cambridge University Press, Cambridge (1989)

Schmidt-Nielsen, K., Schmidt-Nielsen, B., Jarnum, S.A., Houpt, T.R.: Body temperature of the camel and its relation to water economy. Am. J. Physiol. **188**, 103–112 (1957)

Selye, H.: A syndrome produced by diverse nocuous agents. Nature (London) **138**, 32 (1936)

Selye, H.: Stress. Acta, Montreal (1950)

Silva, R.G., Maia, A.S.C.: Evaporative cooling and cutaneous surface temperature of Holstein cows in tropical conditions. Braz. J. Anim. Sci. **40**, 1143–1147 (2011)

Silva, R.G., Minomo, F.R.: Circadian and seasonal variation of the body temperature of sheep in a tropical environment. Int. J. Biometeorol. **39**, 69–73 (1995)

Slee, J.: Physiological responses and adaptations of sheep. In: Yousef, M.K. (ed.) Stress Physiology in Livestock, vol. 2. CRC Press, Boca Raton (1985)

Smith, T.J., Edelman, I.S.: The role of sodium transport in thyroid thermogenesis. Fed. Proc. **38**, 2150–2153 (1979)

Smith, R.E., Horwitz, B.A.: Brown fat and thermogenesis. Physiol. Rev. **49**, 330–425 (1969)

Stumpf, W.E.: The endocrinology of sunlight and darkness. Naturwissenschaften **75**, 247–251 (1988)

Thomson, G.E.: Review of the progress of dairy science. Climate physiology of cattle. J. Dairy Sci. **40**, 441–450 (1973)

Thomson, G.E., Jenkinson, D.M.: Non-shivering thermogenesis in the newborn lamb. Can. J. Physiol. Pharmacol. **47**, 249–252 (1969)

Ultsch, G.R.: A theoretical and experimental investigation of the relationships between metabolic rate, body size, and oxygen exchange capacity. Resp. Physiol. **18**, 143–160 (1973)

Valtorta, S., Hahn, G.L., Johnson, H.D.: Effect of high ambient temperature (35°C) and feed intake on plasma T_4 levels in sheep. Proc. Soc. Exp. Biol. Med. **169**, 260–265 (1982)

Van Hardeveldt, C., Zuidwijk, M.J., Kassenaar, A.A.H.: Studies in the origin of altered thyroid hormone levels in the blood of rats during cold exposure. I. Effect of iodine intake and food consumption. Acta Endocrinol. **91**, 473–483 (1979)

Webb, P.: The physiology of heat regulation. Am. J. Physiol. **268**, R838–R850 (1995)

Wurtman, R.J., Axelrod, J., Fisher, J.E.: Melatonin synthesis in the pineal gland: effect of light mediated the sympathetic nervous system. Science **143**, 1328–1329 (1964)

Yousef, M.K., Johnson, H.D.: Effects of heat and feed restriction during growth on thyroxine secretion rate of male rats. Endocrinology **82**, 477–484 (1968)

Chapter 4
Heat Exchange Between Animals and Environment: Mammals and Birds

Abstract Heat exchange between the body surface of animals and the environment is explained, together with the methods of determination. The influence of body surface structures is discussed: skin, hair coat, sweat glands and sweating. Importance of skin and hair coat pigmentation is discussed: melanin and tropical adaptation. Aspects of the thermal insulation at body surface and heat transfer through the hair coat are also discussed. Methods for the measurement of convection and evaporation at the body surface and hair coat radiative properties and the radiant heat exchange with environment are described. Calculation of the total heat flux through the hair coat is discussed. Heat transfer at the respiratory system, which includes respiratory convection in mammals and birds and respiratory evaporation, is also discussed.

Keywords Convection • Evaporation • Hair coat • Radiation • Respiratory surfaces • Skin surface

4.1 Body Surface

4.1.1 Body Surface and Heat Exchange

4.1.1.1 Introduction

As it was well described by Cena and Monteith (1975a), … *animals are irregular three dimensional objects whose orientation with respect to sun and wind can change from minute to minute according to their pattern of behaviour… whereas the metabolic heat production of animals and the mechanisms for maintaining a constant deep body temperature play a major role in determining heat loss to the environment.*

R. Gomes da Silva and A.S. Campos Maia, *Principles of Animal Biometeorology*, Biometeorology 2, DOI 10.1007/978-94-007-5733-2_4,

In fact, despite the great importance of the coat with respect to the regulation of thermal exchanges between animals and environment, the surface irregularities and its relative mobility sometimes make difficult the evaluation of thermal exchange.

The outer surface of the body is the main frontier between organism and the environment, the other border being represented by the lung and respiratory tissues of terrestrial animals and aquatic mammals and the branchial tissues of aquatic organisms in general. Such a borderline condition determines the characteristics of the surface as functions of the environment and the nature of the organism. Thus, animals living in deserts and extremely dry environments must be protected against water losses and the intense solar radiation; those living in cold regions must have an adequate thermal insulation in order to prevent loss of heat; in very hot regions, the animals must be able to transfer the body excess heat to the atmosphere and, at the same time, to reduce the input of heat from environment. Other animals live in regions characterised by wide climatic variations and in which extreme cold occurs alternately with extreme heat; the organisms of such animals must be able to get along with those conflicting conditions.

The thermal energy going out the inner body is transferred through the body surface (a) by conduction through the superficial, dermal and epidermal tissues; (b) by conduction along the fibres of the coat (hairs, wool or feathers); (c) by conduction through the air and the water vapour remaining between coat fibres; (d) by free or forced convection; (e) by evaporation over the skin; and (f) by radiation. Those mechanisms will be discussed separately.

4.1.1.2 Body Surface Structures

Considering its whole, the environment around an animal is rather complex and likely to change continuously in its different aspects; therefore, the outer surface of the body must have characteristics mutable as well and able to be adjusted to environmental changes. In other words, animal's body surface cannot be merely an assemblage of different, fixed, stable structures but a dynamic, adjustable whole. Its nature is directly related to the animal's standard of living and to its energy exchanges with environment. The main types of body surface cover are the following:

Mammals:	Hairs
	Wool
	Bare skin
Birds:	Feathers
Fishes:	Scales
	Hide
Reptiles:	Scales
Amphibians:	Bare skin

Hair coat is the most common skin cover in mammals and constitutes a thermal insulation by means of its physical structure and hair type; besides, the air trapped among the fibres is of great importance for its thermal insulation properties. In a review carried out by Marston (1955), there were considered two main functional groups of skin covers: *mechanical protection* (scales, spines, bristles and mane) and *thermal protection* (wool, hairs and plumage). This last group only will be discussed here.

Several wild mammals and even those furry ones – as the *nutria* or *ratão-do-banhado*,[1] only to mention a South American species – have an outer cover of long, thick hairs and an inner one of very thin, silky hairs. Some Asian and European sheep breeds have thick, long hairs covering the wool, as those of the Shetland Islands and of the Great Britain highlands (Swaledale, Welsh Mountain, Herdwick). Its hair-covered fleece seems to be related to a protection against the cold rain that is frequent in those regions (Yeates 1965); the water runs down easily over the hairs, thus avoiding wool to get soaking wet; in such a case, the thermal insulation would be destroyed. The fleece is a type of cover that may have a high density as that of sheep or less dense as that of other animals – for example, the camels. Its nature is very different of that of hairs and is a protective cover used by animal species both of cold and of hot, dry climates.

The bird plumage presents several feather types that vary as for the size, form and structure. The so-called contour feathers are the predominant type and give the body shape and form; its upper end is rigid, while at the lower end there has a plumed structure. The *fluffs have* long, loose filaments that are often longer than its axis; they are located at the basis of the contour feathers, and its function seems to be thermal insulation. Finally, there are the *semiplumes*, which constitute an intermediary type between feathers and fluffs.

The scales that occur in reptiles and fishes were of different nature, but they were not yet studied as for their heat transfer properties. On the other hand, animals covered by a bare skin only had been extensively studied at this respect; notwithstanding, the major part of those studies refers to human beings.

4.1.1.3 Hair Coat Structures

Number of Hair Follicles

Hair follicles are developments from the epidermis itself. In cattle, they are formed by around the 78th day of gestation and are well defined at the 166th day (Lyne and Heideman 1959). Therefore, a calf is born with a definitive number of hair follicles which is genetically established.

However, Turner et al. (1962) observed in cattle that the number of follicles per unit area decreases with age, as it is shown in Table 4.1. By their results, the relative

[1] *Myocastor coypus bonariensis*. A rodent like to the beaver which lives in the marshes of southern Brazil, Uruguay and Argentina.

Table 4.1 Average number of hair follicles per unit surface area in cattle at different ages, adjusted for body weight

Breed group	Age	Body weight kg	Number of follicles cm^{-2}
European	Foetus	3.7	23,300
	Newborn	31.3	5,186
	2–10 months	133.4	2,061
	12–22 months	218.6	1,413
	6 years	406.0	887
Zebu	Newborn	35.8	5,652
	2–10 months	134.7	2,446
	12–22 months	271.7	1,530

Adapted from Turner et al. (1962)

number of follicles tends to stabilise in animals at about 24 months of age or at the end of its body growth. Such a relationship with age was explained by the expansion of the skin body growth, without any change in the actual number of follicles.

In cattle, there has one hair follicle type only, although its structure varies from thick hairs with medulla to thin, medulla-less ones. Sheep have two types: *primary follicles*, those producing hairs and develop first, and *secondary follicles*, which produce wool fibres developing around the primary ones. Those follicles are irrigated at their bases by many blood vessels; Ryder (1958) established the great importance of such irrigation for the active growth of both hair and wool.

Growth Cycles of the Hair Follicles

Once a follicle is completely developed, its growth occurs in cycles or alternate periods of quiescence. The hair growth is a result of the intense mitotic activity of the basal cells in the bulb that receive direct blood irrigation through the papilla. As new cells are formed, they migrate upwards until keratinisation begins; the cells reach its maximum size at this point (Matolstsy 1958). There are growth cycles of the hair follicles, each cycle having three successive phases: the *anagenic* or active, in which the bulb cells are intensively divided; the *catagenic* or intermediary phase, when the cell division ceases; and the telogenic of quiescent phase, in which the follicle becomes inactive.

The follicle growth cycles may be seasonal, with one or two hair changes during the year, as it is observed in horses, cattle, wild sheep and other species. There are also waving changes, in which the follicle growth occurs in activity waves with well-defined patterns, as in fur animals. The mechanism of those changes is not clearly understand yet, but the hair growth begins just after an increase of the blood irrigation at the bulb (Johnson 1965), probably induced by the hormonal changes associated to variations in the photoperiod. It has been observed in European cattle breeds that animals present in general long, thin hairs during the winter, while in the summer the hairs are short, thick and with medulla. According to Dowling and Nay (1960) and Hayman and Nay (1961), European cattle has two hair changes during

Table 4.2 Seasonal variation of the hair coat characteristics of 381 Holstein and 345 Jersey cows in the state of São Paulo, Brazil

Animals and season or month	Hair length (mm)	Coat thickness (mm)	Number of hairs (hairs cm^{-2})	References
Holstein				Pinheiro and Silva (1998)
Spring	12.93	4.36	557	
Fall	14.67	3.06	776	
Jersey				Silva et al. (1988)
Jan–Mar	14.75	3.17		
Apr–Jun	16.58	4.55		
Jul–Sep	17.49	5.48		
Oct–Dec	12.80	3.23		

the year, one in the spring (when the summer hair coat is formed) and the other in the fall (when it is formed the winter coat).

In tropical regions, there were observed also two hair coat changes per year. Silva et al. (1988) studied two Jersey herds in São Paulo, Brazil, concluding that both the hair coat thickness and the hair length increased in April and reduced almost suddenly in October, as it is shown in Table 4.2. Nicolau (1993) obtained similar results for Caracu cattle in the mountains of Minas Gerais. As for Holstein cattle, Pinheiro and Silva (1998) reported significant differences ($P < 0.01$) between spring and fall measurements for the hair coat thickness (4.36 and 3.06 mm, respectively), hair length (12.93 and 14.67 mm) and number of hairs per unit area (557 and 776 hairs cm^{-2}); this study involved 381 cows from two different regions in the state of São Paulo (see Table 4.2). Ryder (1966, 1971) studied the seasonal variation of the fleece in sheep and the hair coat in goats, determining also changes in the spring and in the fall which were attributed to variations in the photoperiod.

4.1.1.4 Sweat Glands

Ruminants in general have *apocrine* sweat glands (Findlay and Yang 1950; Dowling 1955; Hafez et al. 1955; Nay and Hayman 1956) that are directly associated with hair follicles. The cells discharge their secretion into the gland lumen, which unloads later at the hair just over the skin surface. It is interesting to remember that until 1955, it was largely believed that animals in general did not sweat, except the horses. Notwithstanding, the occurrence of active sweating glands was observed by several authors long time ago in cattle (Gurlt 1835; Ellenburger 1906; Yamane and Ono 1936; Findlay and Yang 1950; Villares and Berthet 1951) and in sheep (Lee and Robinson 1941; Rieck et al. 1950; Knapp and Robinson 1954). A detailed study on the morphology of sweat glands in many cattle breeds can be found in the paper of Jenkinson and Nay (1973).

As for the physiological importance of sweating for cattle, it was first shown by Ferguson and Dowling (1955), Dowling (1955) and Taneja (1959). It is actually believed that the importance of sweating for livestock in general is in decreasing order: horses, donkeys, cattle, goats, sheep and swine (McDowell 1972). However, there has a great variation among breeds.

The sweat gland of European cattle breeds present coil-shaped structures lower in diameter (up to 100 μm) in comparison to that of Zebu cattle, which is a sack-like structure 180–200 μm in diameter (Nay 1959). However, Jersey cattle – which is a European breed easily acclimatised in hot environments – has sweat glands much like those of the Zebu, though smaller in size. It has been also referred (Amakiri 1974; Turner et al. 1962) that cattle of tropical origin has a number of sweat glands per unit area (1,400–2,200 glands cm^{-2}) higher than that of European cattle (<1,000 glands cm^{-2}) on the average.

In sheep, the sweat glands are also of the apocrine type, associated with the primary follicles (not those producing wool), as it was observed by Bligh (1961) and Robertshaw (1966). These glands seem to be not high-producing ones; Silva et al. (1990) observed Polwarth ewes under temperatures from 25 to 46°C and recorded an average sweating rate of 90.5 ± 5.5 g m^{-2} h^{-1}. Other authors as Brook and Short (1960a, b) and Alexander and Williams (1962) recorded high sweating rates for sheared animals in hot environments (40°C) and suggested the importance of sweating for thermoregulation in sheep. On the other hand, it has been observed that for sheep with thick fleece exposed to high temperatures, the heat loss by respiratory evaporation is more important than the cutaneous evaporation, which in such a case amounts up to 35% only of the total evaporative losses (Hofmeyr et al. 1969).

Sweating is not continuous but intermittent both in cattle and in sheep; it occurs when the gland lumen is full of secretion and there has a contraction of the muscular fibres in the gland walls; the secretion is then driven out to the skin surface. Johnson (1973) observed that in sheep, the sweat discharges are more frequent under high temperatures, especially for sheared animals. During intervals between the discharges, there has a water loss from the skin by means of insensible perspiration.

Insensible perspiration is not easily separated from sweating because both of them occur simultaneously. Experimental results by Silva and Maia (2011) suggest that for Holstein cows in a tropical environment, the heat loss through insensible perspiration amounts 48 W m^{-2} on the average. However, insensible perspiration is often confounded with sweating; for example, rabbits lose about 40% of their latent heat over the skin (Hales 1974), though this species has no sweating glands (Jenkinson 1970).

With respect to their distribution on the body surface, the number of active sweat glands varies according to the body region; those with more glands are trunk, neck, dewlap, belly and extremities in the decreasing order. According to Silva and Maia (2011), the flank and the neck are the most representative body regions as for the sweating rate in cattle.

Latent heat transfer by evaporation is one of the most important thermoregulatory mechanisms for animals in general and especially the homeotherms, although there has a great variation among species as for the relative importance of respiratory evaporation and cutaneous evaporation. The amount of latent heat loss by cutaneous evaporation depends on the following factors:

(a) Nature and activity of the sweat glands
(b) Skin permeability to fluids coming from the inner body by insensible perspiration
(c) Nature and structure of the skin coat and its permeability to water vapour

In some animals – as the water buffalo and the pachyderms in general – there has a little number of sweat glands, and their epidermis is bare or sparsely covered by coarse hairs or bristles. Such animals find in the environment the water they need to moisten the skin and thus to lose heat by cutaneous evaporation. It is the *wallowing behaviour,* sometimes observed even in cattle.

4.1.1.5 Sweating Rate

Despite considerable individual variation of the sweating rate, there have significant differences between livestock breeds at this respect, according to the environment in which animals live. Besides, sweating is not homogeneous in the whole body surface and varies according to the body region; in cattle, it is greater in the flank and the neck; it seems to be more intense also in those skin areas covered with thinner hair coat (Schleger and Turner 1965; Pan et al. 1969).

The study by Salimos (1980) was probably the first to be carried out in Brazil on the sweating rate of cattle. This author recorded the sweating rate of 32 Jersey and Holstein cows along 1 year and observed that Jerseys had average sweating rate greater (156 g m^2 h^{-1}) than that of black-spotted (124 g m^2 h^{-1}) or red-spotted Holsteins (125 g m^{-2} h^{-1}). Cows at the lactation peak, 100–200 days, presented figures higher than those cows in other lactation periods; this was attributed to their higher metabolic rate and higher heat production.

Silva et al. (1988) evaluated the sweating rate of 523 Jersey cows in São Paulo, Brazil, obtaining an average of 123.3 ± 2.7 g m^{-2} h^{-1} during the summer under temperatures from 24.5 to 33°C. In the USA, Allen et al. (1970) recorded averages between 28 and 438 g m^{-2} h^{-1} for Jerseys also. Bertipaglia (2007) investigated the sweating rate of 1,607 Braford cows in Mato Grosso do Sul, Brazil; the animals were under field conditions, with air temperatures 21–33°C, and their average sweating rate was 320.0 ± 0.2 (89.2–591.5) g m^{-2} h^{-1}. Other results for beef cattle are those for Hereford × Shorthorn and Brahman × Shorthorn crosses in Australia, with 488 and 500 g m^{-2} h^{-1}, respectively (Schleger and Turner 1965), and the study by McMannus et al. (2005) with several local breeds in Brazil, which averaged 269.1 g m^{-2} h^{-1}.

Other animal species have less information about sweating function, and there have often problems as for the comparison of results because different measurement criteria are used. In fact, some authors use determinations in g m^{-2} h^{-1}, while others use g h^{-1} kg^{-1}; in the last case, it is alleged the difficulty to evaluate body surface area. However, this last variable can be estimated from the body mass, as it was explained in Chap. 3. Table 4.3 shows the data recorded in some studies.

Table 4.3 Average sweating rates recorded for four animal species

Animals	Conditions	$g\ m^{-2}\ h^{-1}$	References
Sheep			
Wool-less	25–35°C	217.2–240.3	Silva et al. (1990)
Wool-less	39.6°C	235.4	Brook and Short (1960a, b)
Merino	Capsule	32	Ames et al. (1971)
Suffolk	Colorimetric	35–265	Nogueira et al. (1987)
Chokla	41–43°C	79.5–112	Rai et al. (1979)
Polwarth	25–46.5°C	90.5	Silva et al. (1990)
Corriedale	21–41°C	1.303*	Silva and Starling (2003)
Cheviot	30°C	0.345*	Brockway et al. (1965)
Blackface	38°C	0.77–0.90*	Hofmeyr et al. (1969)
Donkeys			
	Resting	240–270	Bullard et al. (1970)
	Resting	300	Soholt et al. (1977)
	Walking at sun	460–510	Yousef et al. (1970)
	Walking at sun	330–594	Dill et al. (1979)
	40°C, resting	166	Allen and Bligh (1969)
	45°C, resting	195	Yousef (1985)
Horses			
Breton	Colorimetric	486.9	Titto et al. (1994)
Anglo-Arab	Colorimetric	369.8	Titto et al. (1994)
Mangalarga	Colorimetric	271.2	Titto et al. (1994)
Goats			
Saanen	27–35.6°C	104–115	Ligeiro et al. (2006)
Alpine	27–35.6°C	99	Ligeiro et al. (2006)
Crossbreds	27–35.6°C	145	Ligeiro et al. (2006)
Saanen	22–33°C	179	Oliveira et al. (2006)
Bedouin	Sun	185.1	Finch et al. (1980)
Bedouin	Shade	133.2	Finch et al. (1980)
Crossbreds	Semiarid	123	Costa (2011)

Values with an asterisk (*) refer to measurements done in $g\ h^{-1}\ kg^{-1}$

As for the evaluation of sweating rate, the colorimetric method of Schleger and Turner (1965) was used for a long time, but now it is preferred the direct determination of the amount of water evaporated from the skin surface by means of a ventilated capsule such as that developed by Gatenby (1980). This and other special methods will be described in Chap. 7.

4.1.2 Skin and Hair Coat Pigmentation

4.1.2.1 Evolutionary Aspects

Colour is one of the most remarkable characteristics of the body surface. At least in the case of vertebrates, the colour has been object of much speculation among zoologists, geneticists and animal scientists. It is a rather complex matter especially as for the livestock, which are subjected to both natural selection and the human selection pressure. However, the objectives of the man are often contrary to those of the nature, and many studies of animal colour are plagued by misleadingly anthropocentric bases.

In general terms, the colour of the outer body surface has the following known natural functions:

(a) Relationship to solar radiation
(b) Camouflage
(c) Signalising and sexual attraction

These three functions are not mutually exclusives as for a given species in a given environment, but they can combine one to another in order to prevail the function which gives greater advantage for the survival of the species or of the individual itself. An interesting example is that of the polar bear, which has a thick, dense hair coat over a highly pigmented skin. It is very clear that the white colour contributes a lot for the success of the bear as a predator, in an environment almost fully white during most part of the year. On the other hand, such a body covering has a high reflectance for the solar radiation, resulting in a poor absorption of thermal energy – a disadvantage in an environment with very low temperatures. Grojean et al. (1980) proposed the hypothesis that the hairs of the polar bear, which have a large medulla and are almost transparent, act rather as optical fibres that drive solar radiation into the skin (which has a high absorptance coefficient). Therefore, the skin pigmentation has the function of absorbing thermal energy from the solar radiation and thus heating the body.

Another illustrative example is a full black seabird like to the cormorant and known in Brazil as *biguá*. It builds the nest in the coast rocks and remains exposed to direct sun along the day, absorbing great amounts of thermal energy; its response to the heat stress is an increased respiratory rate. When seeking for food, it dives deep into the cold water, by which its plumage is permeable to water; on its return back to the nest, the bird is under cold stress because its feathers are fully soaked wet, then it remains exposed to sun and the black plumage allows fast body heating and drying. On the other hand, the gulls get nesting and feeding together with the biguás, but they are white and their plumage are hydrophobic as they do not dive and get resting by floating on the water surface; their white colour avoids excess heating of the body by the solar radiation.

The two examples above mentioned show how great is the importance to the zoologist or animal scientist of the degree to which thermal radiation will penetrate through the coats of different colours and thereby affect the thermal balance of animals.

Complex colour standards of many species, as the several ungulates that live in the African savannas, were developed under natural selection. Those animals have a light yellowish-brown hair coat that has a reasonable reflectance coefficient for the solar radiation but which gives also a camouflage within the savannas' brushes. An example is the South African springbok (*Antidorcas marsupialis*), whose rump and ventral surface are white; there has also a dark-brown stripe along the flank. According to Hofmeyr and Louw (1987), on cold mornings the springbok orientates its body axis to the sun in order to allow the dark stripe to absorb solar radiation and

Table 4.4 Heat exchange by black and white Bedouin goats in the conditions of the Negev desert

Exposure (at noon)	Sun		Shade	
Animal colour	Black	White	Black	White
Absorption of short waves (0.3–2.5 μm)	288	153	53	38
Absorption of long waves (>2.5 μm)	463	457	463	464
Body heat gain	104	49	24	15
Heat loss by evaporation	177	123	92	89

Values in W m^{-2} according to data from Finch et al. (1980)

give some body heating; as air temperature rises, the animal orientates its body parallel to the sunbeam, thereby reducing the area exposed to solar radiation; at the same time, the white ventral region reflects the short-wave solar radiation coming in from the ground surface by reflection.

As for the black and white stripes of zebras, there has much discussion yet, and convincing explanations of it are still awaited. There has no thermoregulatory advantages, and it is more difficult yet to explain them as a camouflage, since they make animals remarkably conspicuous in the savannas' environment. On the other hand, the stripes of the tiger are important for its camouflage within the high grass fields and bamboo groves of Southern Asia.

In the Middle East and particularly in Palestine region, the Bedouins have favoured black-coloured goats for centuries; black goats are just the predominant livestock in the Sinai and Negev (Shkolnik et al. 1972). These authors referred that in the winter (when food availability is minimal in the Negev), the metabolic rate of Bedouin goats exposed to sun is 25% lower than that of the white goats; such a difference would be attributed to the fact that black goats absorb more the solar radiation and spend less energy to produce heat under low air temperatures. On the other hand, the black colour would be a disadvantage during the summer, but the black goats are able to lose more heat through cutaneous and respiratory evaporation than the white goats; such ability was experimentally confirmed by Finch et al. (1980) whose results are summarised in Table 4.4.

The results presented in Table 4.4 show that the net gain of heat by short-wave radiation by black animals was about twofold that gained by the white ones. The excess heat gained by the blacks was eliminated through evaporative cooling (cutaneous + respiratory); however, there had almost no difference between the two animal types under shade. There had also very low differences as for long-wave absorption.

4.1.2.2 Melanin

The term "melanin" refers to black and brown pigments of epidermis, hairs, feathers, eyes, nervous system and other animal tissues. According to Riley (1997), its main – or even the only – known function in vertebrates is to protect body surface against the ultraviolet radiation; however, some of its intermediary metabolites present antibiotic

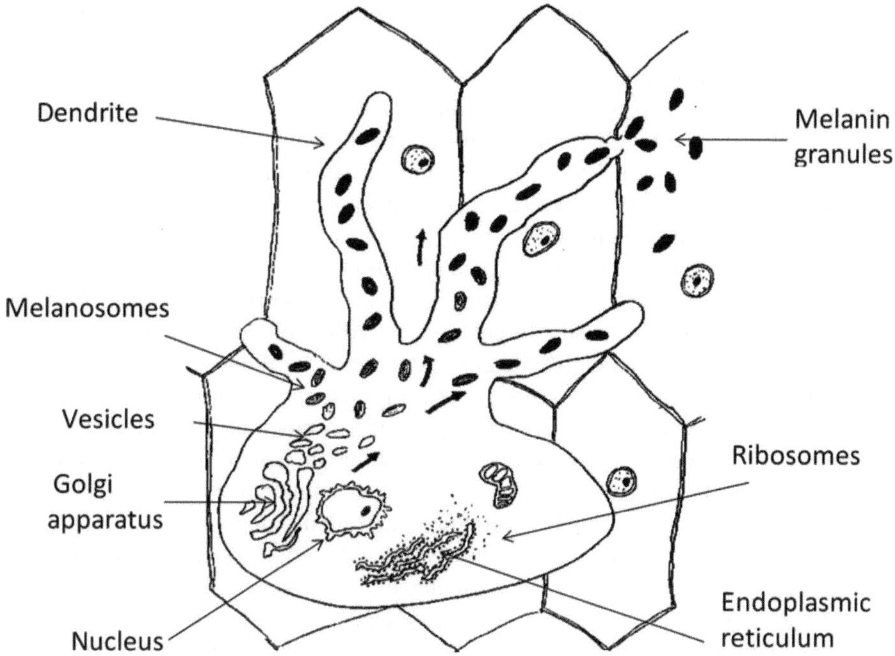

Fig. 4.1 Schematic representation of a melanocyte

properties, and this would be an explanation for its presence in nervous tissues. Melanin contributes also to strengthen the cell walls of plants and of insect cuticles.

Melanin pigmentation of the epidermis and of the outer body coat had been long studied as for its paper in the exchange of radiant energy by animals (Schleger 1962; Cowles 1967; Hamilton and Heppner 1967; Hutchinson and Brown 1969; Hutchinson et al. 1975; Dawson and Brown 1970; Kovarik 1973; Hamilton 1973; Cena and Monteith 1975a; Walsberg 1978; Finch et al. 1984; Hansen 1990; Gebremedhin et al. 1997; Hillman et al. 2001; Silva et al. 1988, 2003, 2010). It is generally known that melanin pigmentation of skin is important to protect deep tissues from damage by ultraviolet radiation, which leaves eventually to neoplasm both in animals and in humans (Nishimura and Frisch 1977; Bhume et al. 1992; Gardiner et al. 1972; Urbach 1997). The comprehensive review by Urbach (1993) is a good source of references on the effects of ultraviolet radiation upon several animal species.

The composition of melanin was studied by Thompson (1962), and it results from the polymerisation of oxidation products of ortho-dihydroxy-phenol compounds originated from the amino acid tyrosine. In short, melanin can be defined as an insoluble brown or black pigment formed by an enzymatic oxidation of tyrosine by tyrosinase in the cytoplasm of specialised cells, the *melanocytes*. In Fig. 4.1, there is a schematic illustration of a melanocyte.

The melanocyte is a distinctive cell with dendrites which are present at the epidermal-dermal junction of skin (basal layer), in the hair bulbs, in the retinal epithelium of the eye and in other tissues. Its cytoplasm contains small granules in different stages of maturity that migrate into the dendrites as true melanin granules; these dendrites are later phagocytised by other cells in which the melanin granules are dispersed throughout. The intensity of pigmentation of the skin, for example, is not due to the number of melanocytes but to the amount of melanin granules into the cytoplasm of the epithelium cells.

The vesicles produced by the Golgi apparatus contain tyrosine of dietary origin; in the case of an eventual lack of tyrosine, it can be synthesised by the liver from the phenylalanine, which is easily found in most food sources. According to Seiji (1967), the vesicles contain also the enzyme tyrosinase produced by the ribosomes in the endoplasmic reticulum (see Fig. 4.1); this enzyme remains inactive until it is activated by the ultraviolet radiation (UV) coming throughout the skin layers. Once tyrosinase is activated, there begins an oxidation process of tyrosine, which is successively transformed into 3,4-dihydroxy-phenylalanine (DOPA), DOPA-quinone, DOPA-chrome, Indole-5,6-quinone and finally into melanin.

As it is formed, melanin is set in layers within the vesicles, which change successively into pre-melanosomes, melanosomes and melanin granules. In this last stage, all the tyrosine content was already polymerised, and the granules that are in the extreme end of the dendrites become full black or dark brown. The next step is the introduction of the granules within the cytoplasm of other cells in the proximity of the melanocyte.

The cells migrate to the skin surface as the outer (and older) cells are destroyed by the use and new ones are produced in the basal layer. If the individual were not exposed again to the solar radiation, melanogenesis is decreased and the new epithelial cells will contain lower cytoplasmic amounts of melanin granules. Therefore, the epidermis loses with time its pigmentation as its cells are substituted for others with lower melanin content.

In the individuals having a permanently dark skin, there has a genetically driven mechanism which maintains a high level of melanogenesis independently of any variation of the exposure to UV radiation. Such a mechanism is just developed by natural selection under environments with high levels of UV radiation; that is the explanation for the dark skin of most human and animal populations autochthonous of tropical regions.

The higher the melanocyte activity, the greater the melanin content of the basal cells in the epidermis. However, in the human black races, the number of melanocytes per unit skin area is nearly the same as that of the white races or about one melanocyte for each four to ten epidermal cells. According to some authors in the cattle breeds of tropical origin – which have skin always much darker than that of European breeds – there has also the same number of melanocytes per unit area (Yang 1952; Hafez et al. 1955; Amakiri 1979).

European spotted cattle breeds as Holstein-Friesian, Simmental, Hereford and others have the same number of melanocytes in either colour spot, but in the white spots, the melanocyte action is genetically inhibited (Schleger and Bean 1973).

Finally, the tyrosinase synthesis is controlled by a dominant gene; if it is substituted by its recessive allele, there has the condition known as *albinism*, in which the melanogenesis is blocked and there has a complete absence of pigmentation; the pink-like colour of the skin and the eyes is due to the effect of light reflection on the surface blood veins. Both human and animal albinos have shorter lifespan than the normal individual and are susceptible to blindness because of the absence of melanin in the iris that protects eyes against UV radiation.

4.1.2.3 Pigmentation and Tropical Adaptation

Animals with dark-coloured outer surface – either with or without hair coat – are generally considered as more affected by heating through the solar radiation than those light-coloured ones (Riemerschmidt 1943; Stewart 1953; Finch et al. 1984; Hansen and Lander 1988). Therefore, it has been suggested by some authors that white animals would be the most adequate ones for tropical environments (Goodwin et al. 1995, 1997).

However, inconsistent results had been presented by several of those studies due to the fact that both hair coat structure and skin pigmentation were not taken into account. In addition, many of those studies were carried out in temperate regions, where the animals are not exposed to the intense solar radiation that occurs in the tropical ones. In the above cited studies by Goodwin and co-workers, the criterion they considered to justify the supposed superiority of white-hair-coated cattle was based on the milk yield, not on thermal-exchange-based arguments. On the other hand, most Brazilian dairy cattle breeders have long observed that predominantly white Holstein cows are disadvantageous in the low-latitude regions, where solar radiation is very intense.

As a matter of fact, it was suggested by Kovarik (1964) that short-wave solar radiation can be transmitted throughout the hair coat and that such transmission would be more intense in white hair coats. Such a hypothesis was later experimentally confirmed by Hutchinson and Brown (1969), Kovarik (1973), Cena and Monteith (1975a), Walsberg (1978), Gebremedhin et al. (1997) and Silva et al. (2003). The UV radiation is transmitted through the white hair coat and strikes the non-pigmented skin surface that is always associated to that coat colour, then giving rise to damages in the deep tissues.

For example, Nikula et al. (1992) observed Beagle dogs as for dermatosis and skin cancer caused by exposure to solar radiation. They recorded cases of haemangioma (6.5%), haemangio-sarcoma (4.8%) and squamous carcinoma (6.4%) in a total of 991 exposed animals. There was found a significant positive correlation of the incidence of dermatosis and neoplasia with sun exposure of non-pigmented skin areas. The greatest incidence was observed in locations of greater altitude and higher UV radiation.

Kimura and Doi (1994) observed significant increases of the skin thickness in hairless dogs exposed during 6 days to UV radiation at 20 W m^{-2} for 2 h day^{-1}. The epidermis presented irregular groups of cells with degenerative alterations, many of

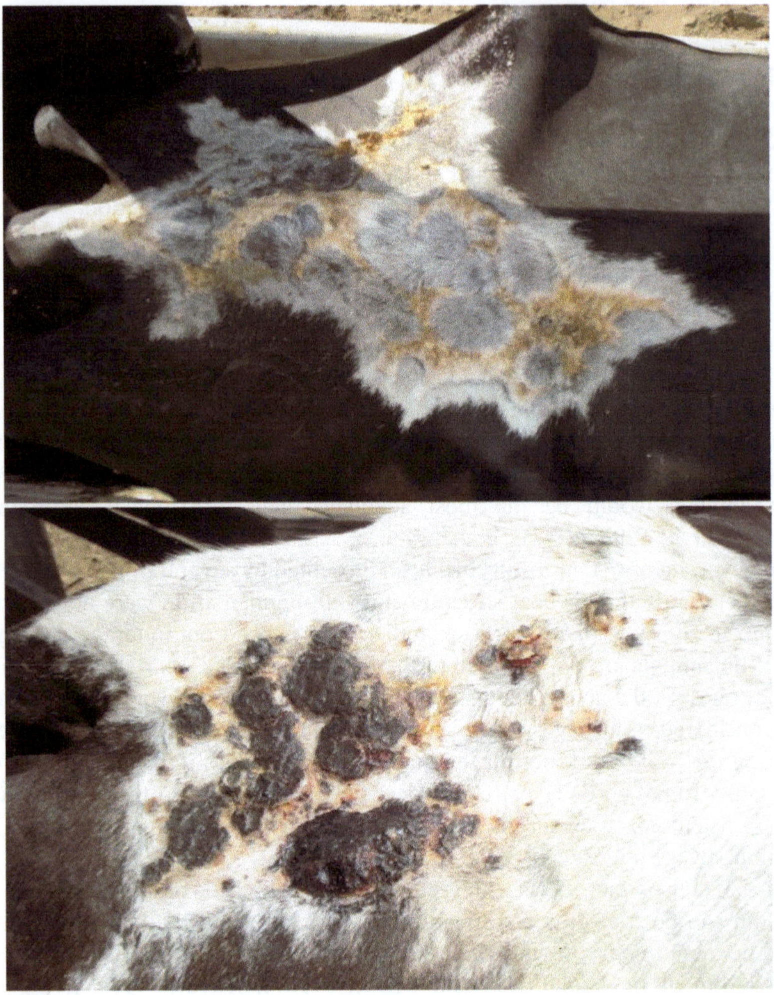

Fig. 4.2 Intense skin burning by solar radiation in the white spots of Holstein cows in an equatorial region (Ceará, Brazil) (Photos by R.G. Silva 2010)

them with picnotic nucleus. Such alterations are often associated to cancerous processes, and squamous carcinoma can be experimentally induced on animal skin by UV radiation of the same type that causes sunburn (Gruijl and Van der Leun 1995).

Intense burning of the skin is often observed in Holstein cows exposed to sun in tropical regions, with destroying of deep tissues (Fig. 4.2). In such cases, the problem occurs in the white spots only where the skin is non-pigmented.

4.1.3 Thermal Exchange at the Body Surface

4.1.3.1 Skin Surface

Thermal Insulation

Thermal insulation is defined as the inverse of conductivity, which can have values from 0.410 to 0.426 W m^{-2} °C^{-1} for cattle (Holmes 1998). It depends on the blood flux at the surface; for example, the temperature at the surface of a rabbit's ear increases rapidly from 15 to 35°C as the deep body temperature increases from 39.5 to 40°C; then it becomes stabilised, despite a subsequent increase of the body temperature to 42.5°C (Eckert et al. 1988). Such an increase of the cutaneous temperature is due to an increased blood flow after a dilatation of the capillary vases at the surface.

The increased blood flow from the inner body results into a greater thermal energy transfer to the body surface. In other words, tissue thermal insulation is reduced with vasodilatation and increased with vasoconstriction.

Sensible heat exchanges through the body surface tissues can be described by the formula

$$\mathbf{K} = \frac{\rho\, c_{\mathrm{p}}(T_{\mathrm{b}} - T_{\mathrm{s}})}{r_{\mathrm{t}}}\ \mathrm{W\ m^{-2}} \tag{4.1}$$

where r_t is the thermal resistance of cutaneous tissues (s m^{-1}), T_b is the deep body temperature (°C) and T_s the skin temperature. The value of r_t has been given by some authors (e.g. Turnpenny et al. 2000) as being proportional to $W^{1/3}$, where W is the body weight (kg). Table 4.5 gives the values suggested by several authors for the thermal resistance of the peripheral tissues in animals, while in Table 4.6, there are the respective thermal conductivity values (k, W m^{-1} °C^{-1}).

4.1.3.2 Conduction and Convection Through the Coat

It has been suggested by some authors as Blaxter (1989) that the thermal conduction along the hairs (or feathers, wool fibres) is negligible and that most of the energy is transferred by conduction through the air among the fibres. However, at least part of that energy is transported by the fibre mass itself. According to Hammel (1955), the free convection within the hair coat would be associated to the conduction, though the vapour diffuses among the fibres and the water absorbed by them would cause some increase in the thermal transfer, but such an effect of the moisture would be not great.

Data recorded by Cena and Monteith (1975a, b) showed that all the three mechanisms radiation, conduction and convection are involved together in the process of heat transfer throughout the hair coat of animals. On the other hand, animals chronically exposed to heat stress which need prompt heat elimination towards the

Table 4.5 Thermal resistance (s m^{-1}) of peripheral tissues in some animals, considering cutaneous vasoconstriction and vasodilatation

Animal	Vasoconstriction	Vasodilatation	References
Cattle			
European, adult	171	49	Webster (1974)
Adult, fat	342	–	Blaxter (1967a)
Steer	156	50	Blaxter (1967b)
Calf	111	50	Blaxter (1989)
¾ Hereford-Zebu	–	15	Gatenby (1986)
Hereford × Holstein	–	32	Gatenby (1986)
Zebu	–	18	Blaxter (1989)
Jersey	$225 - 5.44\ T_s$	29	McArthur (1987)
Swine			
Adult	178	–	Webster (1974)
Suckling	100	60	Blaxter (1967a, b)
Suckling	98	37	Blaxter (1989)
Newborn	25	30	Webster (1974)
Sheep			
Adult	90	–	Blaxter (1967a, b)
Adult	163	–	Webster (1974)
Birds			
Poultry	110	–	Blaxter (1989)

T_s is the skin temperature (°C)

Table 4.6 Thermal conductivity (k) of some animal tissues

Tissue	k (W m^{-1} °C^{-1})	References
Blood (human)	0.505	Dumas and Barozzi (1984)
Muscle		
Cattle	0.410–0.426	Chato (1968)
Swine	0.518–0.618	Chato (1968)
Fat		
Cattle	0.185–0.233	Chato (1968)
Swine	0.317–0.402	Chato (1968)
Minke whale	0.200–0.280	Kvadscheim et al. (1996)
Seal	0.190	Kvadscheim et al. (1994)
Porpoise	0.100	Worthy and Edwards (1990)
Dolphin	0.1025	Silva (2004)
Skin		
Crocodile (trunk)	0.432	Drane (1981)
Crocodile (tail and belly)	0.334	Drane (1981)
Swine	0.209	Cohen (1977)

atmosphere tend to have thick hairs, as compared to those living in cold environments. This is an evidence that there has thermal conduction within the hairs per se.

Though the heat transfer through the hair coat is processed by several physical mechanisms and not by molecular conduction only, the term *thermal insulation* is generally used for comparative purposes:

Table 4.7 Thermal insulation (I) and resistance (r_K) values for the body coat of some animals, per unit of thickness (cm)

Animal	Coat type	Condition	I (W^{-1} m^2 °C)	r_K (s m^{-1})
Cattle	Hair coat			
Diverse		$U = 0.2$ m s^{-1}	0.198	239.7
Diverse		$U = 0.7$ m s^{-1}	0.151	182.8
Ayrshire		Settled	0.099	120.0
Ayrshire		Erect	0.066	80.0
Galloway			0.074	90.0
Sheep	Fleece			
Diverse		$U = 0.3$ m s^{-1}	0.195	236.1
Diverse		$U = 1.1$ m s^{-1}	0.166	201.0
Diverse		$U = 1.9$ m s^{-1}	0.134	162.2
Diverse		$U = 3.4$ m s^{-1}	0.132	159.8
Diverse		$U = 4.3$ m s^{-1}	0.102	123.5
Merino			0.231	280.0
Down			0.157	190.0
Blackface			0.124	150.0
Cheviot			0.124	150.0
Others	Hair coat			
Husky dog			0.240	290.0
Polar bear			0.108–0.172	131.0–208.4
Rabbit			0.221–0.261	267.3–315.4
Air		Still	0.360	435.8

Data from Blaxter (1967b, 1977), Hammel (1955) and Scholander et al. (1950)
The values were estimated from the atmospheric air properties at 20°C. U = wind speed (m s^{-1})

$$I = \frac{\text{Temperature gradient}}{\text{Flux of sensible heat}} (\text{Surface area})\ W^{-1}\,m^2\,°C$$

However, it is most practical to use the thermal resistance instead of the insulation:

$$r_K = \rho\, c_p I \text{ s m}^{-1} \tag{4.2}$$

where ρ and c_p are, respectively, the density and the specific heat of the air at the given temperature. Table 4.7 shows the values for I and r_c that were estimated by several authors for the body coat of some animals.

As for the birds, there has not much information. However, it was referred by Wathes and Clark (1981) that the thermal resistance of the plumage of poultry increases directly with the plumage mass per unit area, according to the function

$$r_K = 150 + 650\ W_p \text{ s m}^{-1} \tag{4.3}$$

where W_p is the plumage mass per unit surface area of the body (kg m^{-2}); those authors estimated W_p values from 0.05 to 0.8 kg m^{-2}, and the average thermal

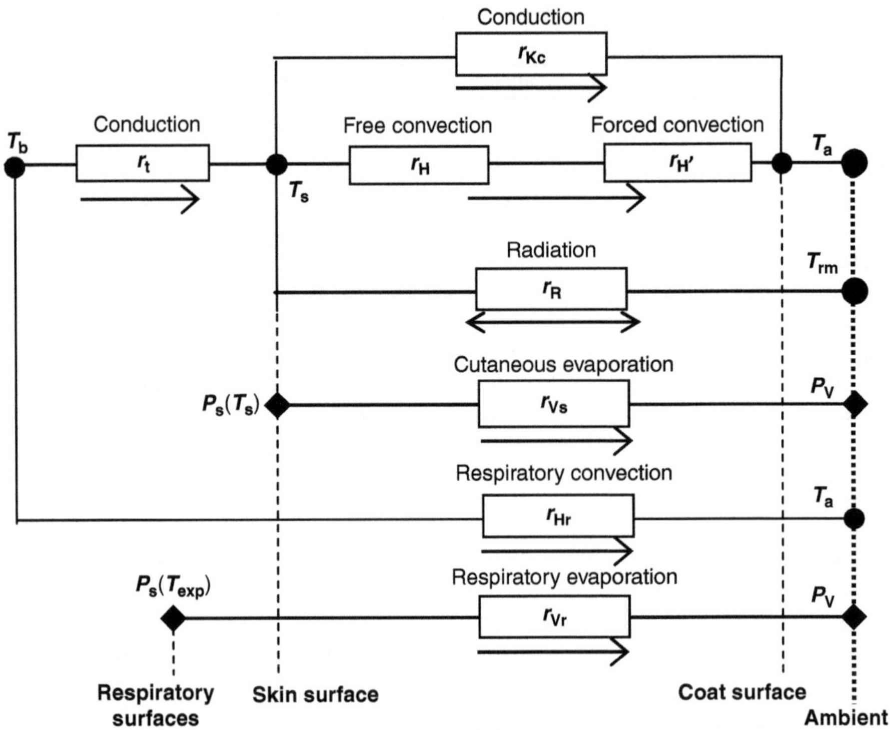

Fig. 4.3 Electrical analogue for sensible and latent heat transfer through the hair coat of animals. T_b = deep body temperature, T_s = skin temperature (Modified from Silva 2000)

resistance was 180 s m^{-1} per cm of coat thickness, close to that found for the sheep fleece (see Table 4.7).

McArthur and Monteith (1980) described the sensible heat transfer through the hair coat of animals in terms of analogy to an electric model, in which the total flux results from the partial fluxes by branches each one corresponding to a different physical mechanism. It can be illustrated as in Fig. 4.3.

The concept of thermal conductivity as described previously seems to be of less utility for the present case because the several physical mechanisms involved and considering also that the hair coat is not constituted by such a homogenous material a specific property as the k coefficient would be attributed thereto. However, there has sensible heat transfer among the air molecules that remain within the hair coat; then, the thermal resistance of the coat would be defined by the formula

$$r_{Kc} = \frac{\rho\, c_p z}{k} \text{ s m}^{-1} \quad (4.5)$$

where z is the coat thickness (m) and k the thermal conductivity of the air within the hair coat.

Table 4.8 Thermal insulation (W^{-1} m^2 °C) of the hair coat of some animals according to the wind speed

Wind (m s^{-1})	Rabbits	Horses	Swines
0	0.341	0.149	0.109
3.5	0.199	0.100	0.044
8.0	0.149	0.077	0.030

Adapted from Treggear (1965)

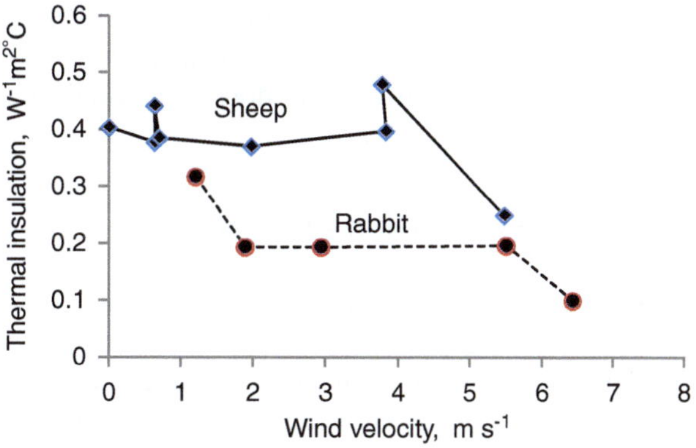

Fig. 4.4 Variation of thermal insulation of sheep fleece 4 cm thick and of the rabbit hair coat 1.8 cm thick according to the wind speed (Modified from Cena and Monteith 1975b)

Experimental observations by Cena and Monteith (1975b) revealed that for some mammals, the temperature gradient between skin and the outer surface of the hair coat was smaller than that expected one in the case the heat transfer was due either to molecular diffusion or to conduction and radiation combined together. As the heat transfer rate increased directly with the temperature gradient, it was concluded that free convection was the main thermal transfer mechanism working within the hair coat, so long as the air remained still within the coat.

On the other hand, the situation changes when the animal is exposed to the wind stream; several papers have been published on the effect of wind upon the thermal insulation of the hair coat in many species. For example, Treggear (1965) studied the conductance of the coat in rabbits, horses and pigs exposed to wind speeds from zero to 8 m s^{-1}; a summary of his results is given in Table 4.8.

Cena and Monteith (1975b) evaluated experimentally sheep fleece samples 4 cm thick, concluding that their thermal conductivity was not affected by wind blowing parallel to the skin surface with speed from zero to 6 m s^{-1}. However, when the conductivity data was converted into thermal insulation, these I values decreased from 0.402 to 0.256 W^{-1} m^2 °C as the wind speed increased up to 6 m s^{-1}. See Fig. 4.4.

The effect of wind was stronger on the hair coat of rabbits than that on the sheep fleece, an expected result because rabbit's hairs are more loose than the fleece and can be easily separated by the wind. When the wind blows perpendicularly to the skin surface, its effect becomes more accentuated, as was observed by Joyce and Blaxter (1964) for sheep; the thermal insulation of the fleece under those conditions decreased in proportion to the increase in wind speed. Similar results were obtained by Campbell et al. (1980), according to whom the thermal resistance of the fleece decrease linearly with the wind speed.

McArthur and Monteith (1980) proposed functions for both body sides relatively to the wind[2]:

$$r_{Hc} = 2961.7\,U^{-1.0548}\ \text{s m}^{-1},\quad \text{windward side}$$
$$r_{Hc} = 2843.7\,U^{-0.7658}\ \text{s m}^{-1},\quad \text{leeward side} \tag{4.6}$$

In order to apply the above formulas, the wind speed U must be always evaluated at the windward side. The thermal resistance of the hair coat of a given animal will be the average of the measurements done in the windward and in the leeward body sides.

4.1.3.3 Convection at the Body Surface

For the calculation of thermal exchange by convection between the body surface of an animal and the atmosphere, there were used the formulas presented in Chap. 2 for the Nusselt number of cylinders and spheres, according to the circumstances and the general body form. Many authors take into consideration the trunk only because it represents most part of body volume, then neglecting the head and the paws; however, it would be advisable to include these parts, and, in that case, the convective heat exchange must be calculated separately for each part and added later to that of the trunk. Figures 4.5 and 4.6 illustrate the procedure to calculate forced and free convection in those cases, respectively.

After considering the irregularities of an animals' body surface, Monteith and Unsworth (2008) suggested that its characteristic dimension would be estimated by the cubic root of the body volume. For example, considering the trunk as a cylinder, we have its length (z) and its radius (r); then $d = (\pi z r^2)^{1/3}$ metres. For a sphere with radius r,

$$d = \left(\frac{3}{4}\pi r^3\right)^{1/3}\ \text{m}$$

A cylindrical model was proposed by McArthur and Monteith (1980) for sheep using the following formulas to give the Nusselt number for the forced convection:

$$N_u = 0.106\,R_e^{0.68}\,P_r^{1/3}\quad \text{Sheared animals} \tag{4.6a}$$

$$N_u = 0.126\,R_e^{0.88}\,P_r^{1/3}\quad \text{Normal fleece} \tag{4.6b}$$

within the limits $2 \times 10^4 \le R_e \le 3 \times 10^5$.

---

[2] Windward: the direction from which the wind blows. Leeward: direction to which the winds goes, or the side that receives no wind upon.

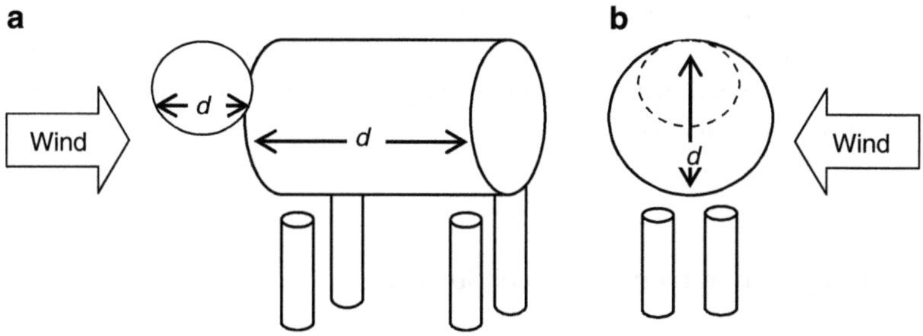

Fig. 4.5 Cylindrical model used to calculate the forced convection in four-legged mammals. As for the case of wind blowing parallel to the body axis (**a**), the characteristic dimension *d* will be the body length. In the case the wind blows perpendicularly to the body axis (**b**), the *d* value will be the average body diameter. In either case, the characteristic dimension of each paw and that of the head will be the respective diameter. Measurements in metres (m)

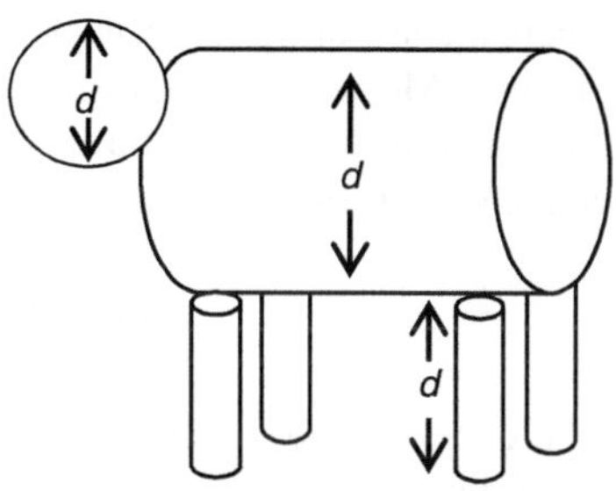

Fig. 4.6 Cylindrical model used to calculate free convection in four-legged mammals. The characteristic dimension (*d*) of the trunk and the head will be the respective average diameter, while that of each paw will be its length. Measurements in metres (m)

The exchange of heat by convection on the body surface of dairy cows was evaluated by Maia et al. (2005a, b) under the assumption of a horizontal cylindrical model whose characteristic dimension was the average trunk diameter.

The case of birds was studied by several authors; in particular, Wathes and Clark (1981) used as models combinations of a sphere (for the trunk) and cylinders (for neck and paws), considering the same principles described for mammals to determine the characteristic dimension. However, the body of most birds is like rather an elongated ellipsoid with diameter x and length z; by using the cubic root technique, we would calculate the characteristic dimension of a bird as

$$d = \left[\frac{3}{4}\pi\left(\frac{x+z}{4}\right)^3\right]^{1/3} \quad \text{m} \tag{4.7}$$

As for the Nusselt number for forced convection in birds, we can use the formula experimentally established by Wathes and Clark (1981):

$$\mathrm{N_u} = 2 + 0.886\,\mathrm{R_e^{0.48}}\,\mathrm{P_r^{1/3}} \tag{4.8}$$

In all of those cases, it is important that the body surface temperature be evaluated by means of a radiometer or infrared thermometer without any contact with the surface; if the animal is covered by plumage, hair coat or fleece, such a temperature will be that of the outer surface of the coat. Once obtained the Nusselt number, the thermal flux by convection can be calculated by using Eq. 2.39, considering definition of r_H given by formula 2.40.

4.1.3.4 Evaporation from the Body Surface

Estimation of the latent heat loss from the skin surface can be done for some animal species by first obtaining its sweating rate and then multiplying it by the latent heat of vaporisation of water at the skin temperature:

$$\mathbf{E_S} = \lambda S \ \mathrm{W\,m^{-2}} \tag{4.9}$$

where the sweating rate S is given in g m^{-2} s^{-1} and the latent heat of vaporisation λ in J g^{-1}. Monteith (1972) considered that in such cases, it is needed to adjust λ for the temperature differential between the skin and the atmosphere; then, Eq. 4.9 is changed as follows:

$$\mathbf{E_S} = S\left[\lambda - c_{\mathrm{pv}}(T_{\mathrm{s}} - T_{\mathrm{a}})\right] \ \mathrm{W\,m^{-2}} \tag{4.10}$$

where c_{pv} is the specific heat of the water vapour (J g^{-1} °C^{-1}), as given by Eq. 1.44a.

Although both expressions 4.9 and 4.10 were valid, they have no structural compatibility with equations for the other heat exchange mechanisms; such compatibility is necessary for the calculation of the whole thermal equilibrium of the animal. Then, the following equation can be considered instead of the two above-mentioned ones:

$$\mathbf{E_S} = \frac{\rho c_{\mathrm{p}}[P_{\mathrm{s}}(T_{\mathrm{s}}) - P_{\mathrm{v}}]}{P_{\mathrm{a}}\gamma r_{\mathrm{Vs}}} \ \mathrm{W\,m^{-2}} \tag{4.11}$$

which is similar to that 2.59 and where $P_{\mathrm{s}}(T_{\mathrm{s}})$ is the saturation vapour pressure at the skin surface temperature T_{s}; that r_{Vs} value is the resistance to the vapour exchange at the body surface, as calculated by formula 2.53; all the air properties must be calculated for the temperature T_{s}.

Fortunately, several precision devices are now available for the determination of evaporation at any surface, as, for example, the VapoMeter.[3] Ventilated capsules can also be easily used, as those described by Gatenby (1986) and Maia et al. (2005a); for their use (which will be described later in the chapter on Special Methods), it is considered the general formula

[3] *Delfin Technologies, Kuopio, Finland*, www.delfintech.com

$$E_s = \frac{\Phi_{cap}\lambda(\Psi_{cap} - \Psi_{atm})}{A_{cap}} \quad \mathrm{W\,m^{-2}} \tag{4.11a}$$

where Φ_{cap} is the flux of air throughout the capsule ($m^3\,s^{-1}$), λ is the latent heat of vaporisation of water ($J\,g^{-1}$) at the surface temperature and A_{cap} is the cutaneous surface area (m^2) encircled by the capsule; Ψ_{cap} and Ψ_{atm} are the levels of absolute humidity ($g\,m^{-3}$) of the air within the capsule and that of the atmosphere, respectively.

Maia et al. (2005a) determined the cutaneous evaporation of Holstein cows under temperatures from 10 to 36°C and observed that E_s became the main way of heat loss from animals at 36°C, reaching a rate up to 350 W m^{-2} at the same time the sensible heat exchange decreased with the increase of the environmental temperature. Similar results were obtained by Finch (1985), McLean (1963a) and Gebremedhin et al. (1981). Hillman et al. (2001) observed E_s values as high as 240 W m^{-2} in Holstein cows maintained under shade, while the rate at the sun reached 500 W m^{-2}.

It was recorded by Gatenby (1986) a positive association of the cutaneous evaporation with the skin surface temperature for Holstein cows in the UK and Kenya. This association was later confirmed by Silva and Maia (2011) for the same breed in Brazil; those last authors proposed a predictive equation for E_s from the surface temperature T_s, that is,

$$E_s = 31.5 + exp\{(T_s - 27.9)/2.19115\} \quad \mathrm{W.m^{-2}} \tag{4.12}$$

with determination coefficient as high as $r^2 = 0.68$. Such equation gave more precise results in Brazil than that of Gatenby (1986), which showed significant underestimates for such conditions.

4.1.4 Radiation and Hair Coat Pigmentation

4.1.4.1 Radiative Properties

In the past, it was supposed that solar radiation was absorbed into the hair coat of animals by its outer surface only, as it can be deduced from the studies by Riemerschmidt and Elder (1945), Stewart (1953), Priestley (1957) and Joyce et al. (1966). However, solar radiation can be transmitted throughout the hair coat, as it was suggested first by Kovarik (1964) and experimentally confirmed by Hutchinson and Brown (1969) and Cena and Monteith (1975a). The radiation coming in from the environment onto the outer surface of an animal (hair coat, fleece or plumage) may be reflected and absorbed, but part of it passes through the fibre layer and attains the skin surface; this depends on the physical properties of the coat, on its morphological structure and on the nature of the incident radiation.

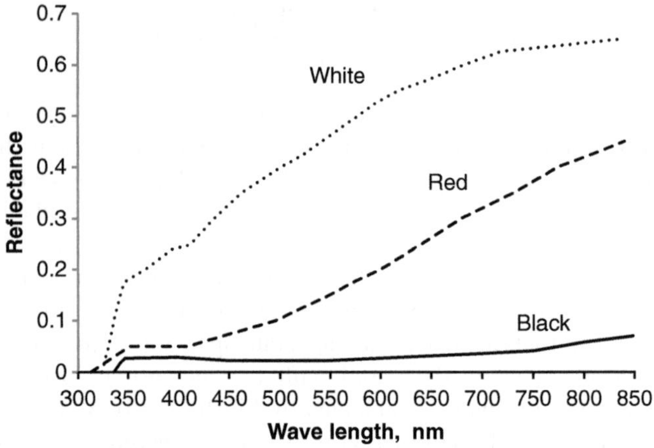

Fig. 4.7 Spectral reflectance of the hair coat of cattle for wavelengths from 300 to 850 nm (0.3–0.85 μm), according to the colour and a 90° incidence angle (Built with data from Silva et al. 2003)

Silva et al. (2003) evaluated cattle skin samples of several different colours together with their hair coat; they determined their respective radiative properties by spectrophotometry for wavelengths from 0.3 to 0.85 μm. The results showed in Fig. 4.7 for the reflectance of white, red and black hair coats are close to those given by Hutchinson et al. (1975). The radiative properties of the skin below the hair coat were also evaluated by Silva et al. (2003), and some of the results are shown in Table 4.9.

On the other hand, the amount of thermal radiation (long wave) emitted by the skin and transmitted through the hair coat to the ambient is greater for low-density hair coats, those with small number of fibres per unit area; it is the case of swine, the pachyderms in general and the buffalo. In tropical deer, the hair coat is not also much dense.

The physical characteristics of the hair coat are of great importance, as for the thermal radiation. According to Cena and Monteith (1975a), the *effective reflectance* of some hair coat or fleece is a function of (a) individual fibre reflectance, (b) physical structure of the coat and (c) reflectance of the skin surface below the coat. Those authors defined the following equations for the effective reflectance, effective transmittance and effective absorptance, respectively:

$$\rho* = \frac{[\rho_c - \rho_s(\rho_c + \alpha_c)]\sinh x + y\,\rho_s \cosh x}{[\alpha_c + \rho_c(1 - \rho_s)]\sinh x + y\cosh x} \tag{4.13}$$

$$\tau* = \frac{y}{[\alpha_c + \rho_c(1 - \rho_s)]\sinh x + y\cosh x} \tag{4.14}$$

$$\alpha* = 1 - \rho * - \tau * (1 - \rho_s) \tag{4.15}$$

Table 4.9 Radiative properties of the skin of some animals, for wavelengths from 0.3 to 0.85 μm (Silva et al. 2003)

Skin colour	Animal	ρ	α	τ
Black	Holstein cow	0.06	0.93	0.01
	Brangus cow	0.07	0.92	0.01
	Buffalo	0.23	0.77	0
Red	Holstein cow	0.44	0.37	0.19
	Simmental cow	0.29	0.54	0.17
	Pantanal deer	0.31	0.56	0.13
Grey	Nelore cow	0.04	0.91	0.05
Light brown	Canchim cow	0.66	0.27	0.07
Non-pigmented	Holstein cow	0.53	0.43	0.04

ρ reflectance, α absorptance, τ transmittance

where

$$x = yz\{dn\tan[\arccos(z/c)]\}$$

$$y = \left(\alpha_c^2 + 2\rho_c\,\alpha_c\right)^{1/2}$$

d = average diameter of the hairs (m)
n = number of hairs per unit area (m^{-2})
c = average length of the hairs (m)
ρ_s, α_s = reflectance and absorptance of the skin, respectively
ρ_c, α_c = reflectance and absorptance of the coat, respectively
sinh(x) and cosh(x) are the hyperbolic sine and cosine of x

The above defined effective radiative properties do not correspond to those measurements of reflectance, transmittance and absorptance values usually found in the literature, which were often evaluated on hair samples extracted from the skin; its measurement was done without taking into consideration the morphological structure of the coat. Exceptions at this respect are the data by Gebremedhin et al. (1983) and Silva et al. (2003).

Studies on the solar radiation effect on animals must take also into consideration the structural characteristics of their hair coat under the normal conditions at a given environment and not the formal radiation laws (Cena and Monteith 1975a; Walsberg 1978). For example, Walsberg (1978) observed in pigeons that the absorption of thermal radiation was higher for black than for the white plumage, when the birds were in a still atmosphere; however, the reverse happened when the wind speed exceeded 4 m s^{-1}. The explanation for such effect was that the wind permeates into the plumage deeply and exposes the skin to the incoming radiation. However, the heat gained in such a process can be compensated by the convective cooling, especially under a strong wind. Hutchinson and Brown (1969) and Cena and Monteith (1975a) observed also that radiation enters more easily through the white than the black coats under the same conditions.

The case of the polar bears referred by Grojean et al. (1980) is very illustrative also. According to these authors, the white, transparent hairs of the bears act as solar short-wave radiation conductor onto the skin surface, which is highly pigmented; therefore, great amounts of thermal energy arrive at the cutaneous surface at the same time the thermal insulation properties of the coat are maintained in order to reduce body heat loss.

The above facts show that it is possible for a researcher to reach erroneous conclusions about the thermal exchanges by radiation, unless he takes into consideration the circumstances and the physical and morphological characteristics of the hair coat of a given animal.

Finally, it must be stressed that the body surface colour does not affect significantly both absorption and emission of long-wave thermal radiation, as the emissivity coefficient of such a surface varies barely from 0.94 to 0.99 for the different animal species (human inclusive) and independently of the colour (Gates 1980). In the deserts and arid terrains in general, the soil absorbs great amounts of solar radiation, then heating a lot; the thermal energy so accumulated is later emitted upwards as long-wave radiation. The animals living in those environments are much affected by that upward radiation, and then their body colour is of less importance as for thermoregulation; it is important with respect to the solar short-wave radiation only.

4.1.4.2 Thermal Radiation Through the Hair Coat

A detailed study on the heat transfer by radiation through the hair coat of animals was carried out by Davis and Birkebak (1974), who proposed a model based on the coat structure and the physical properties of the respective component fibres. If it is assumed that the coat fibre layer is an isothermal one and that its emissivity is ε_c, the radiant energy flux within the coat would be described by the equation

$$\mathbf{R}_c = \varepsilon_c \sigma \left(T_m^4 - T_c^4\right) \text{ W m}^{-2} \tag{4.16}$$

where σ is the Stefan-Boltzmann constant and T_m and T_c are the average temperatures within the coat and on its surface, respectively (degrees K). A linear approximation of the above equation can be obtained by the following procedure (Monteith and Unsworth 2008):

$$\begin{aligned}\mathbf{R}_c &= \varepsilon_c \sigma \left[T_m^4 - (T_m + \Delta T)^4\right] \\ &= \varepsilon_c \sigma \left[4T_m^3 \Delta T + 6T_m^2 \Delta T^2 + 4T_m \Delta T^3 + \Delta T^4\right] \\ &= 4\varepsilon_c \sigma T_m^3 \Delta T \end{aligned} \tag{4.16a}$$

If the term ΔT is the temperature between the skin and the coat surface, or $\Delta T = T_s - T_c$, then it follows that

$$\mathbf{R}_c = 4\varepsilon_c T_m^3 (T_s - T_c)$$

In such a case, the value $4\varepsilon_c \sigma T_m^3$ can be taken as a coefficient of thermal transfer by radiation:

$$4\varepsilon_c \sigma T_m^3 = \frac{\rho c_p}{r_{Rc}}$$

Therefore, we have

$$r_{Rc} = \frac{\rho c_p}{4\varepsilon_c \sigma T_m^3} \quad \mathrm{s\,m^{-1}}$$

and finally,

$$\mathbf{R}_c = \frac{\rho c_p (T_s - T_c)}{r_{Rc}} \quad \mathrm{W\,m^{-2}} \tag{4.17}$$

It can be easily demonstrated that the error of estimating $\mathbf{R}_c$ by formula 4.17 is small in relation to that obtained by 4.16. That error would be 2.4% only for a 5 K temperature differential, while for a differential of 15 K, it would be 6.9%. Monteith and Unsworth (2008) obtained the same Eq. 4.17 for a similar case.

In the deduction of Eq. 4.16, ΔT^2 and greater terms were neglected by assuming that the temperature difference is generally small. However, it is evident that the amount of radiation transmitted through a given mass of fibres will depend on the number of fibres per unit area and on the fibre dimensions and position (Davis and Birkebak 1974; Cena and Monteith 1975a; Cena and Clark 1978; McArthur 1987).

4.1.4.3 Total Heat Flux Through the Hair Coat

The total flux of sensible heat through the hair coat of an animal will be the combined fluxes by convection, conduction and radiation; in order to estimate this total flux, a combination of the respective thermal resistance coefficients must be available.

In Fig. 4.3, it can be observed that the resistances r_{cK}, r_{cH} and r_{cR} are placed in parallel one to another. Then, an equivalent resistance can be determined by making

$$r_{Eq} = \left(\frac{1}{r_{Kc}} + \frac{1}{r_{Hc}} + \frac{1}{r_{Rc}}\right)^{-1} \quad \mathrm{s\,m^{-1}}$$

However, this equivalent coat resistance is placed in series with the resistance of the tissues and the skin (r_T); it must be added to the parallel resistances in order to obtain the final equivalent resistance:

$$r_{Eq} = r_T + \left(\frac{1}{r_{Kc}} + \frac{1}{r_{Hc}} + \frac{1}{r_{Rc}} \right)^{-1} \text{ s m}^{-1} \tag{4.18}$$

Finally, the total flux of sensible heat from the inner body up to the outer hair coat surface will be given as

$$\Phi_{Sen} = \frac{\rho\, c_p (T_b - T_c)}{r_{Eq}} \text{ W m}^{-2} \tag{4.19}$$

4.1.4.4 Radiation Heat Exchange with Environment

Radiation heat exchange between animals and their environment is among the main factors associated to thermal comfort, especially in the low-latitude regions. Silva et al. (2010) evaluated the thermal radiation received by dairy cows on the range at noon in an equatorial region (Ceará and Rio Grande do Norte, Brazil) and found 742.5 ± 4.6 W m^{-2} on the average, while the mean radiant temperature averaged 65.1 ± 0.5°C.

In order to calculate the radiation exchange of an animal with its environment, the following equation can be used:

$$\mathbf{R} = \varepsilon_s \sigma \left(T_s^4 - T_{rm}^4 \right) \text{ W m}^{-2} \tag{4.20}$$

where ε_s is the emissivity of the outer body surface of the animal, T_s is the body surface temperature (K) and T_{rm} is the mean radiant temperature of the environment (K).

In order to make Eq. 4.20 more compatible with those used for other heat exchange mechanisms, it may be rewritten as

$$\mathbf{R} = \frac{\rho\, c_p (T_s - T_{rm})}{r_R} \text{ W m}^{-2} \tag{4.21}$$

where

$$r_R = \frac{\rho\, c_p}{4 \varepsilon_s \sigma T_m^3} \text{ s m}^{-1} \tag{4.21a}$$

In Eq. 4.21a, the term T_m is the mean temperature $0.5(T_S + T_{rm})$.

Now, two points must be enlightened here. First, in a tropical environment and especially one in the regions closer to the equator, the amount of thermal energy received by animals exposed to sun exceeds that can be usually measured as radiant heat load (RHL), as it was observed by Silva et al. (2010). Those authors suggested that in such cases, the mean radiant temperature T_{rm} in Eqs. 4.20 and 4.21 would be substituted by the temperature T^*_{rm}, which is the *effective mean radiant temperature*, calculated from the *effective radiant heat load* (ERHL) as follows:

$$\mathrm{ERHL} = 0.5\,\mathbf{S}_t + \mathrm{RHL} \quad \mathrm{W\,m^{-2}} \tag{4.22}$$

$$T^*_{rm} = \left(\frac{\mathrm{ERHL}}{\sigma}\right)^{1/4} \quad \mathrm{K} \tag{4.23}$$

where $\mathbf{S}_t$ is the solar short-wave irradiance (W m^{-2}) and σ is the Stefan-Boltzmann constant.

Second, the surface temperature T_s used in Eq. 4.21 must be that of a surface effectively exposed to the radiation, the direct solar one inclusive. In the case the animal has a hair coat, this temperature is that can be given by an infrared radiometer; if the animal has differently coloured body areas, the proportion of each colour area to the total body surface area will be calculated, and T_s must be evaluated separately for each colour. That problem was discussed in detail by Silva (1999).

4.2 The Respiratory Surfaces

4.2.1 Anatomical Aspects

The tracheal tube is not of uniform diameter in cattle and in small ruminant; it has rather a slightly conical form with its widest end at the larynx and the narrowest one opening into the lungs. Its cross section is generally elliptical. See Fig. 4.8 for a schematic representation, while in Table 4.10, there are the measurement values found in several anatomical pieces.

The differences found between the two ends of the tube were not great, especially for the small ruminants; therefore, it is possible to use the mean cross area of both ends for the estimation of the air flux, without much error. On the other hand, as the lungs become full, there has a progressive decrease in the air velocity, which can be neglected because it occurs within a time period that is very short in relation to the duration of the breathing. The same happens during the expiration phase.

4.2.2 Tidal Volume

The amount of air displaced in each respiratory movement is known as *tidal volume*, symbolised here as V_T (m^3 breath^{-1}) and that is associated to the respiratory rate. It can be estimated for sheep by the following equation (Silva et al. 2002):

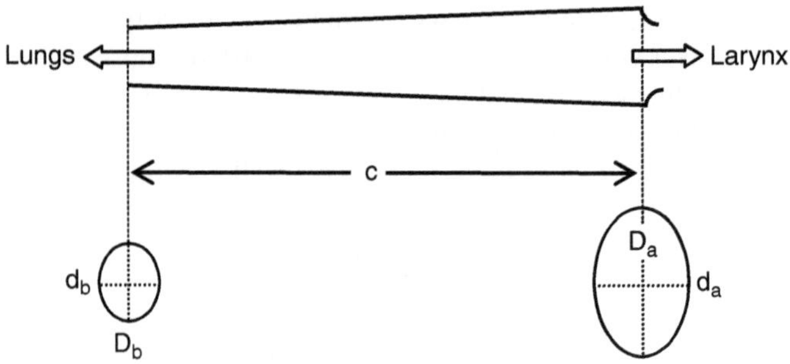

Fig. 4.8 Schematic cut of the tracheal tube of ruminants (cattle, sheep and goats). D_a and D_b are the greatest diameters, d_a and d_b the smallest diameters and c is the length (From Silva 2008)

Table 4.10 Average dimensions of the tracheal tube of cattle, sheep and goats

Dimension	Cattle	Sheep and goat
Lung end		
Greatest diameter (m)	0.035	0.014
Smallest diameter (m)	0.03	0.011
Larynx end		
Greatest diameter (m)	0.047–0.051	0.017
Smallest diameter (m)	0.035–0.037	0.013
Length (m)	0.52–0.65	0.25–0.40
Average diameter (m)	0.037	0.014
Average section area (m^2)	0.00108	0.00015

Silva (2000), non-published report to FAPESP

$$V_T = 0.0496 F_R^{-1.1557} \ \text{m}^3/\text{breath} \tag{4.24}$$

where F_R is the respiratory rate (min^{-1}). That equation was obtained experimentally and has a high determination coefficient, $R^2 = 0.903$.

For dairy cattle, it can be used the formula proposed by Stevens (1981):

$$V_T = 0.0189 F_R^{-0.463} \ \text{m}^3/\text{breath} \tag{4.25}$$

However, Maia et al. (2005b) observed that when Eq. 4.25 was used for cows bred in a tropical environment, there were V_T values significantly lower than those expected ones. Therefore, they proposed a new formula for that case:

$$V_T = \frac{x}{161.93x + 1.97x^2 - 755.32} \ \text{m}^3/\text{breath} \tag{4.26}$$

where $x = F_R - 1.999$. The curve estimated by this equation was very close to the points corresponding to the measurements done in the animals, while the curve given by Eq. 4.25 for the same data resulted in a significant downward deviation. An explanation for such different results would be the possible physiological differences in the respiratory response of Holsteins bred in temperate climates relatively to those in tropical regions.

4.2.3 *Respiratory Convection*

4.2.3.1 General Case

The tidal volume is used for the determination of the air flux into the tracheal tube; the speed of that flux can be estimated by the equation

$$U = \frac{V_T F_R}{60\bar{A}} \ \mathrm{m\,s^{-1}} \tag{4.27}$$

where $\bar{A}$ is the average cross-section area of the tube (m^2), which can be found in Table 4.10.

Now, by taking into account the speed of the air flux within the trachea, the Reynolds number is given as

$$\mathrm{R_e} = U\bar{c}\nu^{-1} \tag{4.28}$$

where $\bar{c}$ is the average tracheal length for a given animal (see Table 4.10) and ν is the kinematic viscosity of air at the temperature within the tracheal tube.

The Prandtl number is calculated by Eq. 2.43, and the respective components must be obtained for the temperature within the trachea. As for the Nusselt number, the better equation to be used will be that proposed by Nusselt (1931) for the convection into short tubes:

$$\mathrm{N_u} = 0.036\,\mathrm{R_e^{0.8}}\,\mathrm{P_r^{1/3}}\left(\frac{\bar{\mathrm{d}}}{\bar{\mathrm{c}}}\right)^{1/18} \tag{4.29}$$

where $\bar{d}$ is the average diameter and $\bar{c}$ the average length of the tracheal tube (see Table 4.10).

Finally, the thermal energy flux by respiratory convection can be given as

$$\mathrm{C_R} = \frac{\rho c_p (T_{exp} - T_a)}{r_{Hr}} \ \mathrm{W\,m^{-2}} \tag{4.30}$$

where

$$r_{Hr} = \frac{\rho c_p \bar{c}}{k N_u} \ \mathrm{s\,m^{-1}} \tag{4.30a}$$

by considering all the properties of the air for the temperature of the expired air, T_{exp}, which can be estimated by the following formulas:

(a) For sheep (Silva et al. 2002),

$$T_{exp} = -142.6193 + 0.291\,T_a + 2.5865\,P_v + 7.3525\,T_r - 0.0016\,T_a^2 - 0.2027P_v^2 - 0.0797T_r^2 \quad (4.31)$$

in degrees °C and e having a determination coefficient of $R^2 = 0.964$.
(b) For cattle (Stevens 1981),

$$T_{exp} = 17 + 0.3\,T_a + exp\{0.01611\,U_R + 0.0387\,T_a\}\ °C \quad (4.32)$$

where U_R is the relative humidity of the atmosphere (%). However, this formula was determined under the conditions of a temperate climate; a more adequate one for tropical conditions would be

$$T_{exp} = 9.47 + 1.18\,T_a - 0.01278\,T_a^2\ °C \quad (4.32a)$$

which is proposed by Maia et al. (2005b), who observed that Eq. 4.32 results into underestimates of T_{exp} when applied to cows in tropical conditions. A new analysis of the same data used by those authors resulted in a more precise equation:

$$T_{exp} = 0.4942\,T_a + 0.7522\,T_r + 2.5228\,P_v - 0.3402\,P_v^2 - 15.0288\ °C \quad (4.32b)$$

with $R^2 = 0.91$ and where T_r is the rectal temperature and P_v is the partial vapour pressure of the atmosphere (kPa).

4.2.3.2 The Special Case of Birds

Thermal exchange at the respiratory surfaces is of particular importance for birds because their plumage cover makes difficult the heat exchange at the body surface. However, their respiratory system is somewhat different from that of other animals.

Birds' lungs are relatively small and are in communication with some bulky structures, the *air sacs*, which are distributed among the internal organs; they have eventually small branches into the skull and in the paw bones. Though such a system has been considered for long time as an adaptation to flight, it is not really needed for that purpose; for example, bats are very good fliers despite they have not air sacs or any similar structure, and their respiratory system is that normal to other mammals.

The air sacs of a given individual are not also associated to any additional need of oxygen nor have any contribution to make its body lighter (Schmidt-Nielsen 1990). Table 4.11 shows that birds have a respiratory volume ($16.08 \times 10^{-5}\,m^3\,kg^{-1}$) about three times greater than that of the mammals ($5.44 \times 10^{-5}\,m^3\,kg^{-1}$) in

Table 4.11 Volumes of the respiratory system in relation to the body weight in birds and mammals ($10^{-5}\ m^3\ kg^{-1}$)

Part	Birds	Mammals
Lung volume	2.96	5.35
Tracheal volume	0.37	0.09
Air sacs	12.75	–
Total	16.08	5.44

Values estimated from data of Lasiewski and Calder (1971)

relation to the body mass. Because of such a great relationship, it is easy to deduce that the respiratory thermal exchange is of greater importance for birds than for other animals. The air sacs are classified into fore or cranial and hind or caudal ones, and their links with other airways can be seen in Fig. 4.9.

The walls of the air sacs are thin, poorly vascularised and inadequate for any gaseous exchange (Schmidt-Nielsen et al. 1969). The more plausible hypothesis for the function of the air sacs is that they act like pumps, moving air in and out throughout the respiratory ways. The studies by Bratz and Schmidt-Nielsen (1972) and Schmidt-Nielsen et al. (1969) support this hypothesis by showing that the respiratory cycle of birds is clearly different from that of the mammals and is characterised by the following sequence of events:

(a) During the inspiration, the caudal sacs expand with the air coming from the outside atmosphere through the great bronchia. At the same time, the lungs deflate into the cranial sacs.
(b) During exhalation, the caudal sacs deflate into the lungs, while the air flows out the caudal ones into the bronchia and then to the ambient atmosphere.
(c) In the following inspiration, the air flows from the lungs into the cranial sacs.
(d) In the next exhalation, the cranial sacs deflate into the great bronchia.

It must be noted that the air always flow through the lungs coming from the caudal sacs and never directly from the atmosphere. Such a standard flux has an important consequence: The blood that leaves the lungs absorbs O_2 and loses CO_2 in proportions higher than that occurs in the mammals. This is very important for birds which fly at great altitude. On the other hand, a respiratory cycle with those characteristics possibly allows a more efficient respiratory loss of both sensible and latent heat. However, at this respect, there has no published data yet.

4.2.4 *Respiratory Evaporation*

4.2.4.1 Importance

There are great differences among species as for the relative importance of cutaneous and respiratory evaporation. For example, the results found by Finch (1985) and McLean (1963a, b) showed that the latent heat loss by respiratory evaporation in cattle at ambient temperatures up to 40°C amounts to about 10–14% of the total

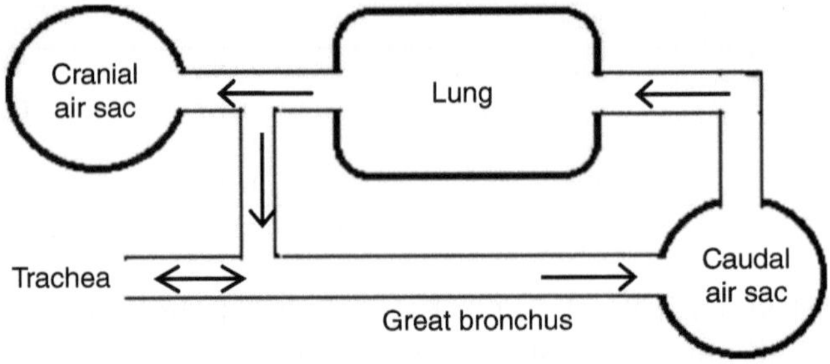

Fig. 4.9 A simplified diagram of the air sacs in birds. The *arrows* point to the air flow direction

evaporation. In birds, the respiratory evaporation is 75% of the total (Richards 1976). As for sheep, those observed into hygrometric tents at high temperatures presented respiratory evaporation rates which represented 10% of the total evaporation (Hopkins et al. 1978) and 17.8% (Silva et al. 2002).

Thompson et al. (1953) evaluated Holstein and Jersey cows exposed to temperatures from 30 to 38°C and observed that the latent heat loss by evaporation was responsible for 70–100% of the losses. The total evaporation rate increased continuously from 80 to 200 g m^{-2} h^{-1} as temperature increased from 12 to 38°C, under 40–50% of air relative humidity; when the humidity increased to 80–85% under a temperature up to 30°C, the evaporation remained at 80–120 g m^{-2} h^{-1}; from then on, the evaporation rate increased up to 160–200 g m^{-2} h^{-1}, decreasing when temperature exceeded 35°C.

McLean (1963a, b) estimated the water loss in Ayrshire calves through cutaneous and respiratory evaporation by using a ventilated capsule and a facial mask, respectively. The animals were exposed to air temperatures from 15 to 40°C and to partial air vapour pressure of 1.07 kPa. The tidal volume decreased and the respiratory rate increased with the temperature. Similar results were obtained by Stevens (1981) in cows and by Silva et al. (2002) in sheep. The amount of water evaporated in the respiratory tract increased slowly with the rise in the air temperature until reaching 59 g h^{-1} at 40°C. At the same time, the rate of water evaporation at the skin surface was close to that in the respiratory tract when the air temperature was <20°C, being about 84% of the total evaporation at 40°C.

A study carried out by Gebremedhin et al. (1981) with Holstein calves exposed to temperatures from zero to 36°C showed that under 35°C, the evaporation rates were 0.83 and 4.3 g kg^{-1} h^{-1} in the respiratory tract and at the skin, respectively. Below 20°C, both the respiratory and the cutaneous evaporation were reduced, but both rates increased under temperatures >20°C. Hofmeyr et al. (1969) observed that in sheep (non-sheared) under high temperatures, the respiratory evaporation was about 65% of the total evaporation.

On the other hand, Silva et al. (2012) evaluated the latent heat loss by Holstein cows in a semiarid equatorial region. There were carried out 486 observations in 178 animals exposed to an average temperature of 33°C, low air humidity and high solar radiation (up to 900–1,100 W m^{-2} at noon). The rate of heat loss by skin evaporation was 116 W m^{-2}, while the respiratory one contributed with 16.2 W m^{-2} only. The dry atmosphere and the frequent, strong wind contributed to the greater importance of the cutaneous evaporation.

Linear increases of evaporation rates with increased air temperature were also recorded by Thompson et al. (1953), McLean (1963a, b), McLean and Calvert (1972), Gebremedhin et al. (1981) and Finch (1985). However, this last author observed a wide difference between the respiratory and cutaneous evaporation rates; cutaneous evaporation was responsible for 90% of the evaporative losses, and the average rate for three steer groups exposed to 44°C was 205 W m^{-2}.

Finally, results found by Maia et al. (2005b) showed that the rate of respiratory evaporation (as that of the cutaneous one) increases exponentially with the temperature in a tropical environment, while the losses of sensible heat decrease almost linearly also.

4.2.4.2 Determination in Mammals

Most studies on respiratory evaporation were carried out under laboratory conditions by using facial masks or hygrometric tents, so a little number of animals has been tested. In order to evaluate the latent heat loss by respiration in mammals, there are used the basic principles already described for sensible heat losses and calculating first the tidal volume (Eqs. 4.24 and 4.26).

The evaporation in the respiratory surfaces had been generally determined by supposing that the air is expired at the inner body temperature and saturated at this temperature (Brockway et al. 1965; McLean 1974; Tenny 1977; McArthur 1987). However, it was experimentally demonstrated by Kibler and Brody (1950), Stevens (1981), Hammarlund et al. (1986) and Silva et al. (2002) that this is not the case and that the ambient temperature and air humidity are important factors to determine temperature of the expired air and consequently the evaporation rate. The temperature of the expired air can be evaluated directly by means of a digital thermometer into the animal nostrils or alternatively by using Eqs. 4.31, 4.32a and 4.32b.

Then, the latent heat loss by respiratory evaporation can be estimated by the equation

$$\mathbf{E}_{\mathrm{R}} = \lambda \dot{m}\, \rho^{-1}(\Psi_{\mathrm{EXP}} - \Psi_{\mathrm{A}})\ \mathrm{W} \qquad (4.33)$$

according to Silva et al. (2002), where λ is the latent heat of vaporisation of water and ρ is the air density, both calculated for temperature T_{EXP}; $\dot{m}$ is the mass flux rate:

$$\dot{m} = V_{\mathrm{T}}\, \rho \left(\frac{F_{\mathrm{R}}}{60}\right)\ \mathrm{kg\,s^{-1}} \qquad (4.34)$$

while Ψ_{EXP} and Ψ_A are the absolute air humidities of the expired air and of the atmosphere, respectively:

$$\Psi_{EXP} = \frac{2,166,869\, P_s(T_{EXP})}{T_{EXP}} \quad \mathrm{g\,m^{-3}} \tag{4.35}$$

$$\Psi_A = \frac{2166,869\, P_v}{T_a} \quad \mathrm{g\,m^{-3}} \tag{4.36}$$

in which temperatures T_a and T_{EXP} are in degrees K.

Maia et al. (2005b) used Eq. 4.33 divided by the body surface area in order to determine the heat loss by respiratory evaporation in dairy cows. McArthur (1987) and Turnpenny et al. (2000) suggested the equation

$$\mathbf{E}_R = \frac{\rho\, c_p\, (P_s(T_r) - P_v)}{P_a\, \gamma\, r_{Vr}} \quad \mathrm{W\,m^{-2}} \tag{4.37}$$

However, by considering the previous discussion about the expired air temperature, Eq. 4.37 can be modified by substituting T_r for T_{EXP} and calculating the saturation vapour pressure at the expired air temperature, $P_s(T_{EXP})$. Then,

$$\mathbf{E}_R = \frac{\rho\, c_p\, (P_s(T_{EXP}) - P_v)}{P_a\, \gamma\, r_{Vr}} \quad \mathrm{W\,m^{-2}} \tag{4.37a}$$

where r_{Vr} is the resistance for vapour transfer at the respiratory surfaces which is a function of the respiratory rate F_R:

$$r_{Vr} = \left(2.7 \times 10^{-4} F_R + 5 \times 10^{-3}\right)^{-1} \quad \mathrm{s\,m^{-1}} \tag{4.38}$$

according to McArthur (1987). By using the data recorded by Maia et al. (2005b) on 332 observations in Holstein cows under several conditions of air temperature and humidity, a new equation can be suggested here:

$$r_{Vr} = 2853.921 - 48.445\, F_R + 0.3414\, F_r^2 - 2.037\, T_a \quad \mathrm{s\,m^{-1}} \tag{4.39}$$

which is more precise than that 4.38, with $R^2 = 0.86$.

From this equation, it can be deduced that the resistance to vapour transfer within the respiratory tract is reduced as the cows increase their respiratory rate, in agreement with the observations by Knapp and Robinson (1954), Hales and Webster (1967) and McArthur (1987). It is also evident that r_{VR} decreases with the increased atmospheric humidity, a fact for which no reference was found in the literature. Despite the contrary would be more logical to be expected, one must think that animals have more difficulties to lose heat by evaporation under high atmospheric humidity levels, by which they increase the respiratory rate; then, as higher the F_R, as lower the r_{VR}.

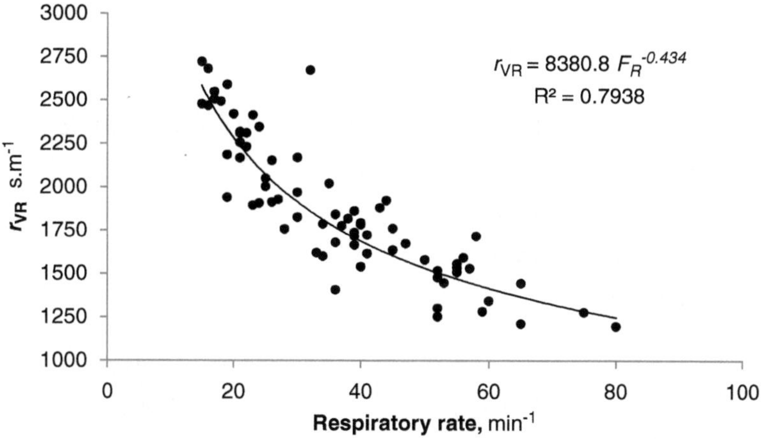

Fig. 4.10 Variation of the resistance to water vapour loss in the respiratory surfaces of Holstein cows, as a function of the respiratory rate

An alternative equation was estimated from the same data, but considering F_R only:

$$r_{Vr} = 8380.8\, F_R^{-0.434} \quad \mathrm{s\,m^{-1}} \tag{4.40}$$

with $R^2 = 0.794$. Despite its lower precision, it is simpler to use than that 4.39 and is to be used also for cattle only. See Fig. 4.10.

4.3 Problems

Problem 4.1. A pig's body can be represented by a horizontal cylinder with dimensions 90 × 40 cm. The animal is placed within a stall in a location at 15°23′ latitude and 195 m altitude and has an average skin surface temperature of 34.5°C. Air temperature in the stall is 30°C, the globe temperature is 35°C and the wind velocity is 0.2 m s^{-1}. Determine the thermal exchange between animal and environment by radiation and convection.

Data:

L_t = latitude = 15°23′ = 15.383333°
z = altitude = 195 m
T_a = air temperature = 30°C = 303.15 K
T_g = globe temperature = 35°C = 308.15 K
T_s = skin surface temperature = 34.5°C = 307.65 K
U = wind velocity = 0.2 m s^{-1}
ε_s = emissivity of the skin surface = 0.98

Air properties at temperature:

$$T_m = 0.5(T_a + T_s) = 32.3°C$$

$$\begin{aligned} g &= 9.78013 + 8.18 \times 10^{-5}L_t + 1.168 \times 10^{-5}L_t^2 - 3.1 \times 10^{-6}z \\ &= 9.78013 + 8.18 \times 10^{-5}(15.83333) \\ &+ 1.168 \times 10^{-5}(15.83333)^2 \\ &- 3.1 \times 10^{-6}(195) \\ &= 9.78375 \text{ m s}^{-2} \end{aligned}$$

$$P_a = 101.325 exp\left\{-\frac{zg}{287.04\ T_a}\right\} = 101.325 exp\left\{-\frac{195(9.78375)}{287.04(303.15)}\right\}$$
$$= 99.13 \text{ kPa}$$

$$\rho = \frac{3484.358 P_a}{T_m} = \frac{3484.358(99.13)}{305.65} = 1.130.065 \text{ g m}^{-2}$$

$$\begin{aligned} c_p &= 1.00522 + 4.577 \times 10^{-4} exp\left\{\frac{T_m}{32.07733}\right\} \\ &= 1.00522 + 4.577 \times 10^{-4} exp\left\{\frac{32.5}{32.07733}\right\} = 1.006481 \text{ J.g}^{-1}.°C^{-1} \end{aligned}$$

$$\begin{aligned} k &= \rho\ c_p\ \left(1.888 \times 10^{-5} + 1.324 \times 10^{-7} T_m\right) \\ &= 1130.065\ (1.006481)\left(1.888 \times 10^{-5} + 1.324 \times 10^{-7} \times 32.5\right) = 0.026368 \end{aligned}$$

$$\begin{aligned} v &= 1.32743 \times 10^{-5} + 9.22286 \times 10^{-8} T_a \\ &= 1.32743 \times 10^{-5} + 9.22286 \times 10^{-8}(32.5) = 1.6272 \times 10^{-5} \text{ m}^2.\text{s}^{-1} \end{aligned}$$

Mean radiant temperature:

$$P_r = \frac{\rho\, c_p\, v}{k} = \frac{(1.130.065)(1.006481)\left(1.6272 \times 10^{-5}\right)}{0.026368} = 0.701896$$

As the wind velocity is too low ($U = 0.2 \text{ m s}^{-1}$), then the convection on the globe surface will be free. As the standard globe was 15 cm in diameter, $d = 0.15$. Then,

$$G_r = \frac{g\, d^3 (T_s - T_a)}{v^2 T_a} = \frac{9.78375(0.15)^3 (35 - 30)}{(1.6272 \times 10^{-5})^2 (303.15)} = 2{,}056{,}883.45$$

$$\begin{aligned} h_c &= k\, d^{-1}\left(2 + 0.43\, G_r^{1/4}\, P_r^{1/4}\right) \\ &= 0.026368\ (0.15)^{-1}\left[2 + 0.43(2, 056, 883.45)^{1/4}(0.701896)^{1/4}\right] \\ &= 2.971716 \end{aligned}$$

$$\text{CTR} = \frac{h_c}{0.95}\left(T_g - T\right)_a + \sigma T_g^4$$
$$= \frac{2.971716}{0.95}(308.15 - 303.15) + \left(5.6697 \times 10^{-8}\right)(308.15)^4 = 526.9\ \text{W m}^{-2}$$

$$T_{rm} = \left(\frac{\text{CTR}}{\sigma}\right)^{1/4} = \left(\frac{526.9}{5.6697 \times 10^{-8}}\right)^{1/4} = 310.5\ \text{K}$$

Thermal exchange by radiation:

$$\mathbf{R} = \varepsilon_s \sigma\left(T_s^4 - T_{rm}^4\right)$$
$$= 0.98\left(5.6697 \times 10^{-8}\right)\left[(307.65)^4 - (310.5)^4\right] = -18.7\ \text{W m}^{-2}$$

The negative signal shows that the animal is gaining heat from the environment.

Thermal exchange by convection:
Characteristic dimension of the cylinder-like animal's body:

$$d = \left[\pi\,(\text{length})\,(\text{radius})^2\right]^{1/3} = 0.77\ \text{m}$$
$$G_r = \frac{g\,d^3(T_s - T_a)}{\nu^2 T_a} = \frac{9.78375\ (0.77)^3(34.5 - 30)}{(1.6272 \times 10^{-5})^2(303.15)} = 250{,}409{,}379.69$$

For a horizontal cylinder with free convection, Eq. 2.39 is used:

$$N_u = \left\{0.6 + \frac{0.387(\text{G}_r\text{P}_r)^{1/6}}{\left[1 + (0.559/\text{P}_r)^{9/16}\right]^{8/27}}\right\}^2$$
$$= \left\{0.6 + \frac{0.387[(250{,}409{,}379.69)(0.701896)]^{1/6}}{\left[1 + (0.559/0.701896)^{9/16}\right]^{8/27}}\right\}^2 = 67.192$$

$$r_H = \frac{\rho\, c_p\, d}{k N_u} = \frac{1{,}130{,}065(1.006481)(0.77)}{0.026368(67.192)} = 494.32$$

$$\mathbf{C} = \frac{\rho\, c_p(T_s - T_a)}{r_H} = \frac{1{,}130{,}065(1.006481)(34.5 - 30)}{494.32} = 10.4\ \text{W m}^{-2}$$

In conclusion, the animal is gaining thermal energy by radiation from the environment; at the same time the heat loss by convection is very low. Then there has a thermal energy excess of $\mathbf{R} - \mathbf{C} = 18.7 - 10.4 = 8.3\ \text{W m}^{-2}$, which must be stored or eliminated through other mechanisms. As a pig does not sweat, its body surface would be sprinkled with water in order to allow heat loss by evaporation.

Problem 4.2. A ewe with 45.5-kg body weight presented a rectal temperature of 39.2°C and an 85 breaths/min respiratory rate when it was in a location at 22°45′ latitude and 560 m altitude; dry bulb and wet bulb temperatures were 31 and 27.5°C, respectively. Determine the respiratory heat loss by the animal.

Data:

L_t = latitude = 22°34′ = 22.75°
z = altitude = 560 m
T_a = dry bulb temperature = 31°C = 304.15 K
T_u = wet bulb temperature = 27.5°C
T_r = rectal temperature = 39.2°C
F_r = respiratory rate = 45 breaths min^{-1}.
P = body weight = 45.5 kg
A = body surface area (Table 3.5)

$$= 0.171\ P^{0.5025} = 0.171\ (45.5)^{0.5025} = 1.165\ \text{m}^2$$

Properties of air at temperature T_a:

$$\begin{aligned} g &= 9.78013 + 8.18 \times 10^{-5} L_t + 1.168 \times 10^{-5} L_t^2 - 3.1 \times 10^{-6} z \\ &= 9.78013 + 8.18 \times 10^{-5}(22.75) + 1.168 \times 10^{-5}(22.75)^2 - 3.1 \times 10^{-6}(560) \\ &= 9.7863\ \ \text{m s}^{-2} \end{aligned}$$

$$P_a = 101.325 exp\left\{-\frac{zg}{287.04\ T_a}\right\} = 101.325 exp\left\{-\frac{560(9.7863)}{287.04(304.15)}\right\}$$

$$= 95.16\ \text{kPa}$$

$$\rho = \frac{3{,}484.358\, P_a}{T_m} = \frac{3{,}484.358\ (95.16)}{304.15} = 1{,}090.1578\ \text{g.m}^{-2}$$

$$\begin{aligned} c_p &= 1.00522 + 4.577 \times 10^{-4} exp\left\{\frac{T_a}{32.07733}\right\} \\ &= 1.00522 + 4.577 \times 10^{-4} exp\left\{\frac{31}{32.07733}\right\} = 1.006423\ \text{J.g}^{-1}.°\text{C}^{-1} \end{aligned}$$

$$\lambda = 2{,}500.788 - 2.37374\, T_a = 2{,}500.788 - 2.37374\ (31) = 2427.2\ \text{J.g}^{-1}$$

$$\gamma = \frac{c_p}{0.6223\ \lambda} = \frac{1.006423}{0.6223(2{,}427.2)} = 6.6631 \times 10^{-4}\ °\text{C}^{-1}$$

$$P_s(T_u) = 0.61078 \times 10^{7.5T_u/(T_u+237.5)}$$
$$= 0.61078 \times 10^{7.5(27.5)/(27.5+237.5)} = 3.666 \text{ kPa}$$

$$P_v = P_s(T_u) - P_a\gamma(T_a - T_u)$$
$$= 3.666 - (95.16)(6.6631 \times 10^{-4})(31 - 27.5) = 3.444 \text{ kPa}$$

Temperature of the expired air (Eq. 4.30):

$$T_{exp} = 0.291\,T_a + 2.5865\,P_v + 7.3525\,T_r - 0.0016\,T_a^2 - 0.2027P_v^2$$
$$- 0.0797\,T_r^2 - 142.6193\ °\text{C}$$
$$= 0.291(31) + 2.5865(3.444) + 7.3525(39.2) - 0.0016(31)^2$$
$$- 0.2027(3.444)^2 - 0.0797(39.2)^2 - 142.6193 = 37.12\ °\text{C}$$

Properties of the expired air at temperature T_{exp}:

$$\rho = \frac{3484.358\,P_a}{T_{exp}} = \frac{3484.358(95.16)}{310.27} = 1{,}068.655 \text{ g.m}^{-3}$$

$$\lambda = 2500.788 - 2.37374\,T_{exp}$$
$$= 2500.788 - 2.37374(37.12) = 2{,}412.68 \text{ J.g}^{-1}$$

$$\gamma = \frac{c_p}{0.6223\,\lambda} = \frac{1.006676}{0.6223(2{,}412.68)} = 6.7049 \times 10^{-4}\ °\text{C}^{-1}$$

$$P_s(T_{exp}) = 0.61078 \times 10^{7.5T_{exp}/(T_{exp}+237.5)}$$
$$= 0.61078 \times 10^{7.5(37.12)/(37.12+237.5)} = 6.30448 \text{ kPa}$$

Tidal volume (Eq. 4.22):

$$V_{RC} = 0.0496\,F_r^{-1.1557} = 0.0496(45)^{-1.1557} = 0.0006094 \text{ m}^3/\text{breath}$$

Mass flux rate (Eq. 4.34):

$$\dot{m} = V_{RC}\,\rho\,\frac{F_r}{60} = (0.0006094)(1.068.655)\frac{45}{60} = 0.488429 \text{ kgs}^{-1}$$

Respiratory evaporation rate (Eq. 4.33):

$$\Psi_A = \frac{2{,}166.869 P_v}{T_a} = \frac{2{,}166.869(3.444)}{304.15} = 24.5362 \text{ g.m}^{-3}$$

$$\Psi_{EXP} = \frac{2{,}166.869 P_s(T_{exp})}{T_{exp}} = \frac{2166.869(6.30448)}{37.12 + 273.15} = 44.0293 \text{ g.m}^{-3}$$

$$\mathbf{E}_R = \frac{\lambda \dot{m} \rho^{-1} (\Psi_{EXP} - \Psi_A)}{A}$$

$$= \frac{(2412.68)(0.488429)(1.068.655)^{-1}(44.0293 - 24.5362)}{1.165} = 18.45 \text{ W.m}^{-2}$$

Problem 4.3. A Holstein cow is on the range exposed to sun in a location at 5° 12′ latitude and 45 m altitude. Data are dry bulb temperature 31.5°C, wet bulb temperature 26°C, globe temperature 43.5°C, atmospheric pressure 100.97 kPa, solar radiation 694 W m^{-2} and wind velocity 1 m s^{-1}. The animal presents a rectal temperature of 39.1°C, a 71 breaths min^{-1} respiratory rate, cutaneous surface temperature 39.6°C and sweating rate 316 g m^{-2} h^{-1}. Determine (a) thermal radiation received by the cow; (b) thermal exchange by convection, by considering that the wind blows parallel to the body axis; (c) heat loss by cutaneous evaporation; and (d) heat loss by respiratory evaporation. Consider the cylindrical model for the body, as follows: The trunk is a horizontal cylinder measuring 1.5 × 0.95 m; each of the four paws is like a vertical cylinder of 0.6 × 0.12 m, and the head is a sphere 0.3 m in diameter.

Data:

Trunk length = 1.5 m
Trunk diameter = 0.95 m
Head diameter = 0.3 m
Paw length = 0.6 m
Paw diameter = 0.12 m
L_t = latitude = 5° 12′ = 5.2°
z = altitude = 45 m
T_a = air temperature = 31.5°C = 304.65 K
T_u = wet bulb temperature = 26°C
T_g = globe temperature = 43.5°C = 316.65 K
P_a = atmospheric pressure = 100.97 kPa
R = solar radiation = 694 W m^{-2}
U = wind velocity = 1 m s^{-1}
T_r = rectal temperature = 39.1°C
F_r = respiratory rate = 71 breaths min^{-1}
T_s = cutaneous surface temperature = 39.6°C = 312.75 K
S = sweating rate = 316 g m^{-2} h^{-1}
ρ_s = reflectance of ground (green grass) = 0.25

Properties of air at temperature Ta:

$$\begin{aligned} g &= 9.78013 + 8.18 \times 10^{-5} L_t + 1.168 \times 10^{-5} L_t^2 - 3.1 \times 10^{-6} z \\ &= 9.78013 + 8.18 \times 10^{-5}(5.2) + 1.168 \times 10^{-5}(5.2)^2 - 3.1 \times 10^{-6}(45) \\ &= 9.780732 \text{ m s}^{-2} \end{aligned}$$

$$\begin{aligned} c_p &= 1.00522 + 4.577 \times 10^{-4} exp\left\{\frac{T_a}{32.07733}\right\} \\ &= 1.00522 + 4.577 \times 10^{-4} exp\left\{\frac{31.5}{32.07733}\right\} = 1.006442 \text{ J g}^{-1} \text{°C}^{-1} \end{aligned}$$

$$\rho = \frac{3484.358 \, P_a}{T_a} = \frac{3484.358\,(100.97)}{304.65} = 1154.82 \text{ g m}^{-3}$$

$$\begin{aligned} k &= \rho \, c_p\left(1.888 \times 10^{-5} + 1.324 \times 10^{-7} T_a\right) \\ &= 1154.82(1.006442)\left[1.888 \times 10^{-5} + 1.324 \times 10^{-7}(31.5)\right] \\ &= 0.026791 \text{ W m}^{-1} \text{°C}^{-1} \end{aligned}$$

$$\begin{aligned} v &= 1.32743 \times 10^{-5} + 9.22286 \times 10^{-8} T_a \\ &= 1.32743 \times 10^{-5} + 9.22286 \times 10^{-8}\,(31.5) = 1.61795 \times 10^{-5} \text{ m}^2\text{s}^{-1} \end{aligned}$$

$$\lambda = 2500.788 - 2.37374\,T_a = 2500.788 - 2.37374(31.5) = 2426.0152 \text{ J g}^{-1}$$

$$\gamma = \frac{c_p}{0.6223\,\lambda} = \frac{1.006442}{0.6223(2.426.0152)} = 0.000667 \text{ °C}^{-1}$$

$$\begin{aligned} D_v &= 2.12138 \times 10^{-5} + 1.4955 \times 10^{-7} T_a \\ &= 2.12138 \times 10^{-5} + 1.4955 \times 10^{-7}(31.5) = 2.59246 \times 10^{-5} \text{ m}^2 \text{ s}^{-1} \end{aligned}$$

$$\begin{aligned} c_{pv} &= 1.86641 - 2.61305 \times 10^{-5} T_a + 1.67249 \times 10^{-5} T_a^2 \\ &= 1.86641 - 2.61305 \times 10^{-5}(31.5) + 1.67249 \times 10^{-5}(31.5)^2 \\ &= 1.882182 \text{ J g}^{-1} \text{°C}^{-1} \end{aligned}$$

$$\begin{aligned} P_s(T_u) &= 0.61078 \times 10^{7.5T_u/(T_u+237.5)} \\ &= 0.61078 \times 10^{7.5(26)/(26+237.5)} = 3.36 \text{ kPa} \end{aligned}$$

$$\begin{aligned} P_v &= P_s(T_u) - P_a\,\gamma\,(T_a - T_u) \\ &= 3.36 - 100.97(0.000667)(31.5 - 26) = 2.99 \text{ kPa} \end{aligned}$$

(a) *Radiation effectively received by the cow:*

$$\mathbf{S}_{\text{ref}} = \rho_s(\mathbf{S}_{\text{dir}} + \mathbf{S}_{\text{dif}}) = \rho_s\mathbf{R} = 0.25\,(694) = 173.5\ \text{W.m}^{-2}$$

$$\mathbf{S}_{\text{t}} = \mathbf{S}_{\text{dir}} + \mathbf{S}_{\text{dif}} + \mathbf{S}_{\text{ref}} = 694 + 173.5 = 867.5\ \text{W.m}^{-2}$$

Calculation of the mean radiant temperature requires the properties of air at temperature $T_m = 0.5(T_a + T_g) = 0.5(31.5 + 43.5) = 37.5°C$; then,

$$\begin{aligned} c_p &= 1.00522 + 4.577 \times 10^{-4} exp\left\{\frac{T_m}{32.07733}\right\} \\ &= 1.00522 + 4.577 \times 10^{-4} exp\left\{\frac{37.5}{32.07733}\right\} = 1.006693\ \text{J g}^{-1}\,°\text{C}^{-1} \end{aligned}$$

$$\rho = \frac{3484.358 P_a}{T_m} = \frac{3484.358(100.97)}{310.65} = 1.132.514\ \text{g m}^{-3}$$

$$\begin{aligned} v &= 1.32743 \times 10^{-5} + 9.22286 \times 10^{-8}\, T_m \\ &= 1.32743 \times 10^{-5} + 9.22286 \times 10^{-8}\,(37.5) = 1.6733 \times 10^{-5}\ \text{m}^2\text{s}^{-1} \end{aligned}$$

$$\begin{aligned} k &= \rho\, c_p \left(1.888 \times 10^{-5} + 1.324 \times 10^{-7} T_m\right) \\ &= \rho\, c_p \left[1.888 \times 10^{-5} + 1.324 \times 10^{-7}(37.5)\right] = 0.027186\ \text{W m}^{-1}\,°\text{C}^{-1} \end{aligned}$$

For a wind velocity of 1 m s^{-1}, free convection is the predominant one; then, considering a standard black globe with diameter $d = 0.15$ m,

$$R_e = U\, d v^{-1} = (1)(0.15)\left(1.6733 \times 10^{-5}\right)^{-1} = 8964.322$$

$$\begin{aligned} P_r &= \rho\, c_p\, v\, k^{-1} \\ &= (1132.514)(1.006693)\left(1.6733 \times 10^{-5}\right)(0.027186)^{-1} = 0.701729 \end{aligned}$$

$$\begin{aligned} h_c &= 0.38\, k d^{-1} R_e^{0.6} P_r^{1/3} \\ &= 0.38(0.027186)(0.15)^{-1}(8964.322)^{0.6}(0.701729)^{1/3} = 14.397 \end{aligned}$$

$$\begin{aligned} \text{RHL} &= \frac{h_c}{0.95}\left(T_g - T_a\right) + \sigma T_g^4 \\ &= \frac{14.397}{0.95}(316.65 - 304.65) + \left(5.6697 \times 10^{-8}\right)(316.65)^4 = 751.9\ \text{W m}^{-2} \end{aligned}$$

$$\text{ERHL} = 0.5\,\mathbf{S}_t + \text{RHL} = 0.5\,(867.5) + 751.9 = 1,185.6\ \text{W.m}^{-2}$$

(b) *Thermal exchange by convection:*
Consider the physical model described in Fig. 4.5a. The characteristic dimension of animal's trunk will be its length, then $d = 1.5$ m. As the convection is forced, we have:

For the trunk ($d = 1.5$):

$$R_e = U\,dv^{-1} = (1)(1.5)\left(1.61795 \times 10^{-5}\right)^{-1} = 92{,}709.911$$

$$\begin{aligned} P_r &= \rho\, c_p\, v\, k^{-1} \\ &= (1154.82)(1.006442)\left(1.6733 \times 10^{-5}\right)(0.026791)^{-1} = 0.725919 \end{aligned}$$

$$\begin{aligned} \mathrm{N_u} &= 0.0296\ \mathrm{R_e^{4/5}}\ \mathrm{P_r^{1/3}} \quad \text{(equation. 2.45)} \\ &= 0.0296(92.709.911)^{4/5}(0.725919)^{1/3} = 250.393 \end{aligned}$$

$$r_{\mathrm{H}} = \frac{\rho\, c_p\, d}{k\,\mathrm{N_u}} = \frac{(1154.82)(1.006442)(1.5)}{0.026791(250.393)} = 259.8863\ \mathrm{s.m^{-1}}$$

$$\mathbf{C}_{\mathrm{trunk}} = \frac{\rho\, c_p(T_s - T_a)}{r_{\mathrm{H}}} = \frac{(1,154.82)(1.006442)(39.6 - 31.5)}{259.8863} = 36.2\ \mathrm{W.m^{-2}}$$

For the head ($d = 0.3$):

$$R_e = Udv^{-1} = (1)(0.3)\left(1.61795 \times 10^{-5}\right)^{-1} = 18,541.981$$

$$\begin{aligned} N_u &= 2 + \left(0.4\,R_e^{1/2} + 0.06\,R_e^{2/3}\right) P_r^{0.4} \quad \text{(Eq. 2.46, simplified)} \\ &= 2 + \left[0.4\,(18{,}541.98)^{1/2} + 0.06\,(18,541.98)^{2/3}\right](0.725919)^{0.4} \\ &= 86.895 \end{aligned}$$

$$r_{\mathrm{H}} = \frac{\rho\, c_p d}{k N_u} = \frac{(1154.82)(1.006442)(0.3)}{0.026791(86.895)} = 149.8\ \mathrm{s\,m^{-1}}$$

$$\mathbf{C}_{\mathrm{head}} = \frac{\rho\, c_p(T_s - T_a)}{r_{\mathrm{II}}} = \frac{(1154.82)(1.006442)(39.6 - 31.5)}{149.8} = 62.8\ \mathrm{W.m^{-2}}$$

For the paws ($d = 0.12$):

$$R_e = Udv^{-1} = (1)(0.12)\left(1.61795 \times 10^{-5}\right)^{-1} = 7{,}416.79$$

$$N_u = 0.3 + \frac{0.62\,R_e^{1/2}\,P_r^{1/3}}{\left[1 + (0.4/P_r)^{2/3}\right]^{1/4}}\left[1 + (R_e/282{,}000)^{5/8}\right]^{4/5} \quad \text{(Eq. 2.44)}$$

$$= 0.3 + \frac{0.62\,(7{,}416.79)^{1/2}(0.725919)^{1/3}}{\left[1 + (0.4/0.725919)^{2/3}\right]^{1/4}}\left[1 + (7416.79/282{,}000)^{5/8}\right]^{4/5}$$

$$= 45.94$$

$$r_{\mathrm{H}} = \frac{\rho\, c_{\mathrm{p}} d}{k N_u} = \frac{(1154.82)(1.006442)(0.12)}{0.026791(45.94)} = 113.32\ \mathrm{s.m^{-1}}$$

$$\mathbf{C}_{\mathrm{paw}} = \frac{\rho\, c_{\mathrm{p}}(T_{\mathrm{s}} - T_{\mathrm{a}})}{r_{\mathrm{H}}} = \frac{(1154.82)(1.006442)(39.6 - 31.5)}{113.32} = 83.08\ \mathrm{W.m^{-2}}$$

$$\mathbf{C}_{\mathrm{total}} = \mathbf{C}_{\mathrm{trunk}} + \mathbf{C}_{\mathrm{head}} + 4\mathbf{C}_{\mathrm{paw}} = 36.2 + 62.8 + 4(83.08) = 431.3\ \mathrm{W.m^{-2}}$$

(c) *Heat loss by cutaneous evaporation:*
The measured sweating rate was $S = 316\ \mathrm{g\ m^{-2}\ h^{-1}} = 0.088\ \mathrm{g\ m^{-2}\ s^{-1}}$. Therefore, by applying Eq. 4.10, we have

$$\begin{aligned}\mathbf{E}_{\mathrm{S}} &= S\left[\lambda - c_{\mathrm{pv}}(T_{\mathrm{s}} - T_{\mathrm{a}})\right] \\ &= 0.088[2426.0152 - 1.882182(39.6 - 31.5)] = 212.15\ \mathrm{W\,m^{-2}}\end{aligned}$$

while Eq. 4.12 gives

$$\begin{aligned}\mathbf{E}_{\mathrm{S}} &= 31.5 + exp\left\{\frac{T_{\mathrm{s}} - 27.9}{2.19115}\right\} \\ &= 31.5 + exp\left\{\frac{39.6 - 27.9}{2.19115}\right\} = 239.9\ \ \mathrm{W\,m^{-2}}\end{aligned}$$

(d) *Heat loss by respiratory evaporation:*

$$\begin{aligned}T_{\mathrm{exp}} &= 0.4942\ T_{\mathrm{a}} + 0.7522\ T_{\mathrm{r}} + 2.5228\ P_{\mathrm{v}} - 0.3402\ P_{\mathrm{v}}^2 - 15.0288\ ^{\circ}\mathrm{C} \\ &= 0.4942(31.5) + 0.7522(39.1) + 2.5228(2.99) \\ &\quad - 0.3402(2.99)^2 - 15.0288 = 34.45\ ^{\circ}\mathrm{C}\end{aligned}$$

$$P_s(T_{exp}) = 0.61078 \times 10^{7.5T_{exp}/(T_{exp} + 237.5)}$$
$$= 0.61078 \times 10^{7.5(34.45)/(34.45 + 237.5)} = 5.445 \text{ kPa}$$

$$r_{Vr} = 2853.921 - 48.445F_R + 0.3414F_r^2 - 2.037T_a$$
$$= 2853.921 - 48.445(71) + 0.3414(71)^2 - 2.037(31.5)$$
$$= 1071.16 \text{ s m}^{-1}$$

$$\mathbf{E}_R = \frac{\rho c_p (P_s(T_{exp}) - P_v)}{P_a \gamma r_{Vr}} = \frac{(1154.82)(1.006442)(5.445 - 2.99)}{100.97(0.000667)(1,071.16)} = 39.6 \text{ W.m}^{-2}$$

References

Alexander, G., Williams, D.: Temperature regulation in the newborn lamb. VI. Heat exchanges in lambs in a hot environment. Aust. J. Agric. Res. **13**, 122–143 (1962)

Allen, T.E., Bligh, J.A.: A comparative study of the temporal patterns of cutaneous water vapour loss from some domesticated mammals with epitrichial sweat glands. Comp. Biochem. Physiol. **31**, 347–363 (1969)

Allen, T.E., Bennett, J.W., Donegan, S.M., Hutchinson, J.C.D.: Moisture, its accumulation and site of evaporation in the coats of sweating cattle. J. Agric. Sci. **74**, 247–258 (1970)

Amakiri, S.F.: Sweat gland measurements in some tropical and temperate breeds of cattle in Nigeria. Anim. Prod. **18**, 285–291 (1974)

Amakiri, S.F.: Melanin and DOPA-positive cells in the skin of tropical cattle. Acta Anat. **103**, 434–444 (1979)

Ames, D.R., Nellor, J.E., Adams, T.: Energy balance during heat stress in sheep. J. Animal Sci. **32**, 784–790 (1971)

Bertipaglia, E.C.A.: Efeito das características do pelame e da taxa de sudação sobre parâmetros reprodutivos em vacas da raça Braford [*Effect of hair coat characteristics and the sweating rate on reproductive parameters of Braford cows*], 163 p. Doctoral Thesis, São Paulo State University (UNESP), Jaboticabal (2007)

Bhume, R.L., Bhokre, A.P., Panchbhai, V.S.: Observations on ocular squamous cell carcinoma in cattle. Indian Vet. J. **69**, 361–362 (1992)

Blaxter, K.L.: Techniques in energy metabolism studies and their limitations. Proc. Nutr. Soc. **26**, 86–96 (1967a)

Blaxter, K.L.: The Energy Metabolism of Ruminants, 2nd edn. Hutchinson, London (1967b)

Blaxter, K.L.: Environmental factors and their influence on the nutrition of farm livestock. In: Haresign, W., Swan, H., Lewis, D. (eds.) Nutrition and the Climatic Environment, pp. 1–16. Butterworths, London (1977)

Blaxter, K.L.: Energy Metabolism in Animals and Man, 336 p. Cambridge University Press, Cambridge (1989)

Bligh, J.: The synchronous discharge of apocrine sweat glands of the Welsh Mountain sheep. Nature **189**, 582–583 (1961)

Bratz, W.L., Schmidt-Nielsen, K.: Movement of gas in the respiratory system of the duck. J. Exp. Biol. **56**, 57–65 (1972)

Brockway, J.M., McDonald, J.D., Pullar, J.D.: Evaporative heat loss mechanisms in sheep. J. Physiol. (Lond.) **179**, 554–568 (1965)

Brook, A.H., Short, B.F.: Regulation of body temperature of sheep in a hot environment. Aust. J. Agric. Res. **11**, 402–407 (1960a)

Brook, A.H., Short, B.F.: Sweating in sheep. Aust. J. Agric. Res. **11**, 557–559 (1960b)

Bullard, R.W., Dill, D.B., Yousef, M.K.: Responses of the burro to desert heat stress. J. Appl. Physiol. **29**, 159–166 (1970)

Campbell, G.S., McArthur, A.J., Monteith, J.L.: Wind dependence of heat and mass transfer through coats and clothing. Boundary-Layer Meteorol. **18**, 485–493 (1980)

Cena, K., Clark, J.A.: Thermal insulation of animal coats and human clothing. Phys. Med. Biol. **23**, 565–591 (1978)

Cena, K., Monteith, J.L.: Transfer processes in animal coats. I. Radiative transfer. Proc. R. Soc. Lond. B **188**, 377–393 (1975a)

Cena, K., Monteith, J.L.: Transfer processes in animal coats. II. Conduction and convection. Proc. R. Soc. Lond. B **188**, 395–411 (1975b)

Chato, J.C.: A method for the measurement of the thermal properties of biological materials. In: Chato, J.C. (ed.) Thermal Problems in Biotechnology. ASME Symposium Series. ASME, New York (1968)

Cohen, M.L.: Measurement of the thermal properties of human skin: a review. J. Invest. Dermatol. **69**, 333–338 (1977)

Costa, C.C.M.: Perda de calor por evaporação cutânea e análise histológica de glândulas sudoríparas de caprinos em um ambiente semi-árido [*Heat loss by cutaneous evaporation and histological analysis of the sweat glands of goats in a semi-arid environment*], 46 p. Monography, Federal Agricultural University of the Semi-Arid Region, Mossoró, RN (2011)

Cowles, R.B.: Black pigmentation: adaptation for concealment or heat conservation? Science **158**, 1340–1341 (1967)

Davis, L.B., Birkebak, R.C.: On the transfer of energy in layers of fur. Biophys. J. **14**, 249–268 (1974)

Dawson, T.J., Brown, G.D.: A comparison of the insulative and reflective properties of the fur of desert kangaroos. Comp. Biochem. Physiol. **37**, 23–38 (1970)

Dill, D.B., Soholt, L.F., Miller, J., Hastings, R.H., Goudy, T.S.: Properties of sweat in burro (*Equus asinus*) and man. Comp. Biochem. Physiol. **62A**, 317–322 (1979)

Dowling, D.F.: The hair follicle and apocrine gland populations of Zebu (*Bos indicus*) and Shorthorn (*Bos taurus*) cattle skin. Aust. J. Agric. Res. **6**, 645–654 (1955)

Dowling, D.F., Nay, T.: Cyclic changes in the follicles and hair coat in cattle. Aust. J. Agric. Res. **11**, 1064–1071 (1960)

Drane, C.R.: The thermal conductivity of the skin of crocodilians. Comp. Biochem. Physiol. **68A**, 107–110 (1981)

Dumas, A., Barozzi, G.S.: Laminar heat transfer of blood flowing in a circular duct. Int. J Heat Mass Transf. **27**, 391–398 (1984)

Eckert, R., Randall, D., Augustine, G.: Animal Physiology: Mechanisms and Adaptations, 3rd edn. Freeman, New York (1988)

Ellenburger, W.: Anatomie der Haustiere, vol. 1. Paul Parey, Berlin (1906)

Ferguson, K.A., Dowling, D.F.: The function of cattle sweat gland. Aust. J. Agric. Res. **6**, 640–644 (1955)

Finch, V.A.: Comparison of non-evaporative heat transfer in different cattle breeds. Aust. J. Agric. Res. **36**, 497–508 (1985)

Finch, V.A., Dmi'el, R., Boxman, R., Shkolnik, A., Taylor, C.R.: Why black goats in hot deserts? Effects of coat color on heat exchanges of wild and domestic goats. Physiol. Zool. **53**, 19–25 (1980)

Finch, V.A., Bennett, I.L., Holmes, C.R.: Coat colour in cattle: effect on thermal balance, behavior and growth, and relationship with coat type. J. Agric. Sci. **102**, 141–147 (1984)

Findlay, J.D., Yang, S.H.: The sweat glands of Ayrshire cattle. J. Agric. Sci. **40**, 126–133 (1950)

Gardiner, M.R., Anderson, J.L., Robertson, D.E.: Cancer eye of cattle. J. Agric. West. Aust. **13**, 53–59 (1972)

Gatenby, R.M.: Evaporimeter measuring sweat rate of cattle outdoors. J. Therm. Biol. **5**, 21–27 (1980)

Gatenby, R.M.: Exponential relation between sweat rate and skin temperature in hot climates. J. Agric. Sci. **106**, 175–183 (1986)

Gates, D.M.: Biophysical Ecology. Springer, New York (1980)

Gebremedhin, K.G., Cramer, C.O., Porter, W.P.: Predictions and measurements of heat production and food and water requirements of Holstein calves in different environments. Trans. ASABE **3**, 715–720 (1981)

Gebremedhin, K.G., Porter, W.P., Cramer, C.O.: Quantitative analysis of the heat exchange through the fur layer of Holstein calves. Trans. ASABE **26**, 188–193 (1983)

Gebremedhin, K.G., Ni, H., Hillmann, P.E.: Modelling temperature profile and heat flux through irradiated fur layer. Trans. ASABE **40**, 1441–1447 (1997)

Goodwin, P., Josey, M.J., Cowan, M.: Coat color and its effect on production in Holstein-Friesian in South-West Queensland. In: Proceedings of the 11th Conference of the Australian Association of Animal Breeding and Genetics, pp. 295–298. University of Adelaide, Roseworthy Campus, South Australia (1995)

Goodwin, P.J., Gaughan, J., Skele, P., Josey, M., Hall, A., Young, B.: Coat color and alleviation of heat load in Holstein-Friesian cows. In: Proceedings of the International Livestock Environment Symposium, 5th, Minneapolis, pp. 923–927. American Society of Agricultural Engineers, St. Joseph (1997)

Grojean, R.E., Souza, J.A., Henry, M.C.: Utilization of solar radiation by polar animals: an optical model for pelts. Appl. Opt. **19**, 339–346 (1980)

Gruijl, F.R., Van der Leun, J.C.: The wavelength dependence of ultraviolet carcinogenesis in humans and its relationship to risk assessment of stratospheric ozone depletion. Health Phys. **67**, 317–329 (1995)

Gurlt, E.F.: Vergleichende Untersuchungen über di Haut der Menschen und der aussangetiere besonders in Beziehund auf die Absondesungsorgane des Haustalges und des Schwisses. Arch. Anat. w. Physiol., (Berlin), (1835)

Hafez, E.S.E., Badreldin, A.L., Shafei, M.M.: The hair coat in bovine. Emp. J. Exp. Agric. **23**, 34–39 (1955)

Hales, J.R.S.: Physiological responses to heat. In: Robertshaw, D. (ed.) Environmental Physiology, pp. 107–162. Butterworths, London (1974)

Hales, J.R.S., Webster, M.E.D.: Respiratory function during thermal tachypnoea in sheep. J. Physiol. (Lond.) **190**, 241–260 (1967)

Hamilton, W.J.: Life's Color Code. McGraw-Hill, New York (1973)

Hamilton, W.J., Heppner, F.: Radiant solar energy and the function of black homeotherm pigmentation: an hypothesis. Science **155**, 196–197 (1967)

Hammarlund, K., Riesenfeld, T., Sedin, G.: Measurement of respiratory water loss in newborn lambs. Acta Physiol. Scand. **127**, 61–65 (1986)

Hammel, H.T.: Thermal properties of fur. Am. J. Physiol. **182**, 369–376 (1955)

Hansen, P.J.: Effects of coat color on physiological responses to solar radiation in Holsteins. Vet. Rec. **127**, 133–334 (1990)

Hansen, P.J., Lander, M.F.: Coat color and heat stress in cattle. In: International Conference on Livestock in the Tropics, Gainesville, FL (1988) (mimeo)

Hayman, R.H., Nay, T.: Observation on hair growth and shedding in cattle. Aust. J. Biol. Sci. **12**, 513–527 (1961)

Hillman, P.E., Lee, C.N., Carpenter, J.R., Back, K.S., Parkhurst, A.: Impact of Hair Color on Thermoregulation of Dairy Cows to Direct Sunlight. American Society of Agricultural Engineers, Paper no. 014031. St. Joseph, MI (2001)

Hofmeyr, M.D., Louw, G.N.: Thermoregulation, pelage conductance and renal function in the desert-adapted springbok, *Antidorcas marsupialis*. J. Arid Environ. **13**, 137–151 (1987)

Hofmeyr, H.S., Guidry, A.J., Waltz, F.A.: Effects of temperature and wool length on surface and respiratory evaporative losses of sheep. J. Appl. Physiol. **26**, 517–523 (1969)

Holmes, K.R.: Thermal conductivity of selected tissues. Ann. N. Y. Acad. Sci. **858**, 18–20 (1998)
Hopkins, P.S., Knights, G.I., Lefevre, A.S.: Studies on the environmental physiology of tropical Merinos. Aust. J. Agric. Res. **29**, 161–171 (1978)
Hutchinson, J.C.D., Brown, G.D.: Penetrance of cattle coats by radiation. J. Appl. Physiol. **26**, 454–464 (1969)
Hutchinson, J.C.D., Allen, T.E., Spence, F.B.: Measurements of the reflectances for solar radiation of the coats of live animals. Comp. Biochem. Physiol. **52A**, 343–349 (1975)
Jenkinson, D.M.E.: The distribution of nerves, monoamine oxidase and cholinesterase in the skin of the guinea pig, hamster, mouse, rabbit and rat. Res. Vet. Sci. **11**, 60–67 (1970)
Jenkinson, D.M.E., Nay, T.: The sweat glands and hair follicles of Asian, African, and South American cattle. Austr. J. Biol. Sci. **26**, 261–275 (1973)
Johnson, E.: Inherent rhythms of activity in the hair follicle and their control. In: Lyne, A.G., Short, B.F. (eds.) Biology of Skin and Hair Growth. Angus and Robertson, Sydney (1965)
Johnson, K.G.: Sweat storage as a factor influencing sweat discharge in sheep. J. Physiol. (Lond.) **235**, 523–534 (1973)
Joyce, J.P., Blaxter, K.L.: The effect of air movement, air temperature and infrared radiation on the energy requirements of sheep. Br. J. Nutr. **18**, 5–27 (1964)
Joyce, J.P., Blaxter, K.L., Park, C.: The effect of air movement, air temperature and infrared radiation on the energy requirements of sheep. Res. Vet. Sci. **7**, 342–359 (1966)
Kibler, H.H., Brody, S.: Influence of temperature, 50 to 95°F, to evaporative cooling from the respiratory and exterior body surfaces in Jersey and Holstein cow. Missouri Agric. Exp. Sta. Res. Bull. **461**, 1–19 (1950)
Kimura, T., Doi, K.: Responses of the skin over the dorsum to sunlight in hairless descendants of Mexican hairless dogs. Am. J. Vet. Res. **55**, 199–203 (1994)
Knapp, B.J., Robinson, K.W.: The role of water for heat dissipation by a Jersey cow and a Corriedale ewe. Aust. J. Agric. Res. **5**, 568–577 (1954)
Kovarik, M.: Flow of heat in an irradiated protective cover. Nature **201**, 1087 (1964)
Kovarik, M.: Radiation penetrance of protective covers. J. Appl. Physiol. **35**, 562–563 (1973)
Kvadscheim, P.H., Folkow, L.P., Blix, A.S.: A new device for measurement of thermal conductivity of fur and blubber. J. Therm. Biol. **19**, 431–435 (1994)
Kvadscheim, P.H., Folkow, L.P., Blix, A.S.: Thermal conductivity of Minke whale blubber. J. Therm. Biol. **21**, 123–128 (1996)
Lasiewski, R.C., Calder Jr., W.A.: A preliminary allometric analysis of respiratory variables in resting birds. Resp. Physiol. **11**, 152–166 (1971)
Lee, D.H.K., Robinson, K.W.: Reactions of the sheep to hot atmospheres. Proc. R. Soc. Qld **53**, 189–200 (1941)
Ligeiro, E.C., Maia, A.S.C., Silva, R.G., Loureiro, C.M.B.: Perda de calor por evaporação cutânea associada às características morfológicas do pelame de cabras leiteiras criadas em ambiente tropical. Rev. Bras. Zootec. **35**, 544–549 (2006)
Lyne, A.G., Heideman, M.J.: The pre-natal development of skin and hair in cattle (*Bos taurus*). Aust. J. Biol. Sci. **12**, 72–95 (1959)
Maia, A.S.C., Silva, R.G., Loureiro, C.M.B.: Respiratory heat loss of Holstein cows in a tropical environment. Int. J. Biometeorol. **49**, 332–336 (2005a)
Maia, A.S.C., Silva, R.G., Loureiro, C.M.B.: Sensible and latent heat loss from the body surface of Holstein cows in a tropical environment. Int. J. Biometeorol. **50**, 17–22 (2005b)
Marston, H.R.: Wool growth. In: Hammond, J. (ed.) Progress in the Physiology of Farm Animals, vol. 2. Butterworths, London (1955)
Matolstsy, A.G.: The chemistry of keratinization. In: Montagna, W., Ellis, R.A. (eds.) Biology of Hair Growth. Academic, New York (1958)
McArthur, A.J.: Thermal interaction between animal and microclimate: a comprehensive model. J. Theor. Biol. **126**, 203–238 (1987)
McArthur, A.J., Monteith, J.L.: Air movement and heat loss from sheep. 2. Thermal insulation of fleece in wind. Proc. R. Soc. Lond. B **209**, 209–217 (1980)

McDowell, R.E.: Improvement of Livestock Production in Warm Climates. Freeman, San Francisco (1972)
McLean, J.A.: Measurement of cutaneous moisture vaporization from cattle by ventilated capsules. J. Physiol. (Lond.) **167**, 417–426 (1963a)
McLean, J.A.: The partition of insensible losses of body weight and heat from cattle under various climatic conditions. J. Physiol. (Lond.) **167**, 427–447 (1963b)
McLean, J.A.: Loss of heat by evaporation. In: Monteith, J.L., Mount, L.E. (eds.) Heat Loss from Animals and Man, pp. 19–31. Butterworths, London (1974)
McLean, J.A., Calvert, D.J.: Influence of air humidity on the partition of heat exchanges of cattle. J. Agric. Sci. **78**, 303–307 (1972)
McMannus, C., Paludo, G.R., Louvandini, H., Garcia, J.A.S., Egito, A.A., Mariante, A.S.: Heat tolerance in naturalized cattle in Brazil: physical factors. Archivos de Zootecnia (Cordoba) **54**, 453–458 (2005)
Monteith, J.L.: Latent heat of vaporization of sweat in thermal physiology. Nature **236**, 96 (1972)
Monteith, J.L., Unsworth, M.H.: Principles of Environmental Physics, 3rd edn. Academic, New York (2008)
Nay, T.: Sweat glands in cattle: histology, morphology, and evolutionary trends. Aust. J. Agric. Res. **10**, 121–128 (1959)
Nay, T., Hayman, R.H.: Sweat glands in Zebu (*Bos indicus*) and European (*Bos taurus*) cattle. I. Size of individual glands, the denseness of their population and their depth below the skin surface. Aust. J. Agric. Res. **7**, 482–495 (1956)
Nicolau, C.V.J.: Variação das características da epiderme e do pelame em bovinos da raça Caracu [Variation of the skin and hair coat of Caracu cattle]. Monograph, São Paulo State University, Jaboticabal, SP (1993)
Nikula, K.J., Benjamin, S.A., Angleton, G.M., Saunders, W.J., Lee, A.C.: Ultraviolet radiation, solar dermatosis, and cutaneous neoplasia in Beagle dogs. Radiat. Res. **129**, 11–18 (1992)
Nishimura, H., Frisch, J.E.: Eye cancer and circumocular pigmentation in *Bos taurus, Bos indicus* and crossbred cattle. Aust. J. Exp. Agric. Anim. Husb. **17**, 709–711 (1977)
Nogueira, O.R., Fares, M.A., Morão, F., Costa, M.J.R.P., Silva, R.G.: Sudação e temperatura retal em ovinos deslanados no Estado de São Paulo [Sweating and rectal temperature of woolless sheep in the State of São Paulo]. In: Proceedings of the 6º Congresso Brasileiro de Iniciação em Ciências Agrárias, Areias, PB (1987)
Nusselt, W.D.: Wärmeraustausch zwischen Wand and Wasser im Roher. Forsch. Geb. Ingenieurwes. **2**, 309–314 (1931)
Oliveira, A.L., Silva, R.G., Aiura, F.S., Maia, A.S.C.: Evaporação total em cabras Saanen sob ambiente de sol e sombra [Total evaporation rates in Saanen goats under the sun and shade]. In: Proceedings of the 43th Annual Meeting of the Brazilian Society of Animal Science, João Pessoa, PB, pp. 137–139 (2006)
Pan, Y.S., Donegan, S.M., Hayman, R.H.: Sweating rate at different body regions in cattle and its correlation with some quantitative components of sweat gland for a given area of skin. Aust. J. Agric. Res. **20**, 395–403 (1969)
Pinheiro, M.G., Silva, R.G.: Pelame e produção de vacas da raça Holandesa em ambiente tropical. I. Características do pelame [Hair coat and milk yield of Holstein cows in a tropical environment. I. Hair coat characteristics]. Boletim de Indústria Animal (São Paulo) **55**, 1–6 (1998)
Priestley, C.H.B.: The heat balance of sheep standing in the sun. Aust. J. Agric. Res. **8**, 271–280 (1957)
Rai, A.K., Mehta, B.S., Gour, D., Singh, M.: Sweating in sheep and goats. Indian J. Anim. Sci. **49**, 546–548 (1979)
Richards, S.A.: Evaporative water loss in domestic fowls and its partition in relation to ambient temperature. J. Agric. Sci. **87**, 527–532 (1976)
Rieck, R.F., Hardy, M.H., Lee, D.H.K., Carter, H.B.: The effect of the dietary plane upon the reactions of two breeds of sheep during short exposures to hot environment. Aust. J. Agric. Res. **1**, 217–230 (1950)

Riemerschmidt, G.: Some aspects of solar radiation in its relation to cattle in South Africa and Europe. Onderstepoort J. Vet. Sci. **18**, 327–353 (1943)
Riemerschmidt, G., Elder, J.S.: The absorptivity for solar radiation of different coloured hairy coats of cattle. Onderstepoort J. Vet. Anim. Ind. **20**, 223–234 (1945)
Riley, P.A.: Molecules in focus: melanin. Int. J. Biochem. **29**, 1235–1239 (1997)
Robertshaw, D.: Sweat gland functioning of the domestic ruminants. In: Proceedings of the 4th International Congress of Biometeorology, New Brunswick, NJ (1966)
Ryder, M.L.: Nutritional factors in hair and wool growth. In: Montagna, W., Ellis, R.A. (eds.) Biology of Hair Growth. Academic, New York (1958)
Ryder, M.L.: Coat structure and seasonal shedding in goats. Anim. Prod. **8**, 289–302 (1966)
Ryder, M.L.: Cycles of wool follicle activity in some Shetland sheep. Anim. Prod. **13**, 511–521 (1971)
Salimos, E.P.: Alguns fatores que afetam a função sudorípara em vacas das raças Jersey e Holandesa [Some factors affecting sweating function of Jersey and Holstein cows]. M.Sc. dissertation, São Paulo State University, Jaboticabal, SP (1980)
Schleger, A.V.: Physiological attributes of coat color in beef cattle. Aust. J. Agric. Res. **13**, 943–959 (1962)
Schleger, A.V., Bean, K.G.: The melanocyte system of cattle skin. II. Melanotic melanocytes of epidermis and dermis. Aust. J. Biol. Sci. **26**, 985–997 (1973)
Schleger, A.V., Turner, H.G.: Sweating rates of cattle in the field and their reaction to diurnal and seasonal changes. Aust. J. Agric. Res. **16**, 92–106 (1965)
Schmidt-Nielsen, K.: Animal Physiology: Adaptation and Environment, 4th edn, 602 p. Cambridge University Press, Cambridge (1990)
Schmidt-Nielsen, K., Kanwischer, J., Lsiewski, R.C., Cohn, J.E., Bratz, W.L.: Temperature regulation and respiration in the ostrich. Condor **71**, 341–352 (1969)
Scholander, P.F., Walters, V., Hock, R., Irving, L.: Body insulation of some artic and tropical mammals and birds. Biol. Bull. **99**, 225–234 (1950)
Seiji, M.: Melanogenesis. In: Zelickson, A.S. (ed.) Ultrastructure of Normal and Abnormal Skin, pp. 183–201. Lea and Febiger, Philadelphia (1967)
Shkolnik, A., Borut, A., Choshniak, I.: Water economy and the Bedouin goat. Symp. Zool. Soc. Lond. **31**, 229–242 (1972)
Silva, R.G.: Estimativa do balanço térmico por radiação em vacas Holandesas expostas ao sol e à sombra em ambiente tropical [Estimate of the heat balance by radiation in Holstein cows at the sun and under shade in tropical environment]. Braz. J. Anim. Sci. **28**, 1403–1411 (1999)
Silva, R.G.: Um modelo para a determinação do equilíbrio térmico de bovinos em ambientes tropicais [A model for the determination of thermal balance of cattle in tropical environments]. Braz. J. Anim. Sci. **29**, 1244–1252 (2000)
Silva, R.G.: Assessment of body surface temperature in cetaceans: an iterative approach. Braz. J. Biol. B **64**, 719–724 (2004)
Silva, R.G.: Biofísica Ambiental: Os Animais e seu Ambiente. FUNEP/FAPESP, Jaboticabal, SP, 393 p (2008)
Silva, R.G., Maia, A.S.C.: Evaporative cooling and skin surface temperature of Holstein cows in tropical conditions. Braz. J. Anim. Sci. **40**, 1143–1147 (2011)
Silva, R.G., Starling, J.M.C.: Evaporação cutânea e respiratória em ovinos sob altas temperaturas ambientes [Cutaneous and respiratory evaporation of sheep at high ambient temperatures]. Braz. J. Anim. Sci. **32**(suppl.2), 1956–1961 (2003)
Silva, R.G., Arantes-Neto, J.G., Holtz-Filho, S.V.: Genetic aspects of the variation of the sweating rate and coat characteristics of Jersey cattle. Braz. J. Genetics **11**, 335–347 (1988)
Silva, R.G., Costa, M.J.R.P., Nogueira, O.R.: Sweating rate of heat stressed sheep. In: Proceedings of the 12th International Congress of Biometeorology, Vienna, Austria, August 26–September 3, 1990, p. 64
Silva, R.G., LaScala Jr., N., Lima-Filho, A.E., Catharin, M.C.: Respiratory heat loss in the sheep: a comprehensive model. Int. J. Biometeorol. **46**, 136–140 (2002)

Silva, R.G., LaScala Jr., N., Tonhati, H.: Radiative properties of the skin and haircoat of cattle and other animals. Trans. ASABE **46**, 913–918 (2003)

Silva, R.G., Guilhermino, M.M., Morais, D.A.E.F.: Thermal radiation absorbed by dairy cows in pasture. Int. J. Biometeorol. **54**, 5–11 (2010)

Silva, R.G., Maia, A.S.C., Costa, L.L.M., Queiroz, J.P.A.F.: Latent heat loss of dairy cows in an equatorial semi-arid environment. Int. J. Biometeorol. **56**, 927–932 (2012). doi:10.1007/s00484-011-0501-y

Soholt, L.F., Dill, D.B., Oddershede, I.: Evaporative cooling in desert heat: sun and shade; rest and exercise. Comp. Biochem. Physiol. **57A**, 369–375 (1977)

Stevens, D.G.: A model of respiratory vapor loss in Holstein dairy cattle. Trans. ASAE **24**, 151–158 (1981)

Stewart, R.E.: Absorption of solar radiation by the hair of cattle. Agric. Eng. **34**, 235–238 (1953)

Taneja, G.C.: Sweating in cattle. V. Sweat prints. J. Agric. Sci. **52**, 168–169 (1959)

Tenny, S.M.: Respiration in mammals. In: Swenson, M.H. (ed.) Duke's Physiology of Domestic Animals, pp. 175–202. Comstock Publishing, Ithaca (1977)

Thompson, M.L.: Melanins. In: Florkin, M., Mason, H.S. (eds.) Comparative Biochemistry, vol. 3. Academic, New York (1962)

Thompson, H.J.; Worstell, D.M.; Brody, S.: The effect of humidity on the insensible weight loss, total vaporized moisture, and surface temperature in cattle. Missouri Agric. Exp. Sta. Res. Bulletin nr. **531** (1953)

Titto, E.A.L., Baccari, F., Jr., Toledo, L.R.A., Bombarda, A.F., Nogueira-Filho, J.L.M.: Taxa de sudação e composição mineral do suor de eqüinos das raças Bretão, Anglo-Árabe e Mangalarga [Sweating rate and mineral composition of the sweat in Breton, Anglo-Arabian and Mangalarga horses]. In: Proceedings of the First Congress on Equine Medicine, São Paulo, p. 201 (1994)

Treggear, R.T.: Hair density, wind speed, and heat loss in mammals. J. Appl. Physiol. **20**, 796–801 (1965)

Turner, H.G., Nay, T., French, G.T.: The hair follicle population of cattle in relation to breed and body weight. Aust. J. Agric. Res. **13**, 960–973 (1962)

Turnpenny, J.R., McArthur, A.J., Clark, J.A., Whates, C.M.: Thermal balance of livestock. I. A parsimonious model. Agric. For. Meteorol. **101**, 15–27 (2000)

Urbach, F.: Ultraviolet radiation and skin cancer of humans. J. Photochem. Photobiol. B **40**, 3–7 (1997)

Urbach, F.: Biologic effects of increased ultraviolet radiation: plankton, plants and people. In: Proceedings of the 13th International Congress of Biometeorology, Calgary, Alberta, Canada, September 12–18. Part 2, 1, pp. 88–98 (1993)

Villares, J.B., Berthet, L.A.: Contribuição para o estudo do aparelho pilo-sebáceo-musculo-sudoríparo no *Bos taurus* e *Bos indicus* [Contribution for the study of the sweating apparatus of Bos taurus and Bos indicus]. Boletim de Indústria Animal (São Paulo) **12**, 3–14 (1951)

Walsberg, G.E.: Coat color and solar heat gain in animals. Bioscience **33**, 88–91 (1978)

Wathes, C., Clark, J.A.: Sensible heat transfer from the fowl. II. Thermal resistance of the pelt. Brit. Poultry Sci. **22**, 175–183 (1981)

Webster, A.J.F.: Physiological effects of cold exposure. In: Robertshaw, D. (ed.) Environmental Physiology, pp. 33–69. Butterworths, London (1974)

Worthy, G.A., Edwards, E.F.: Morphometric and biochemical factors affecting heat loss in a small temperate cetacean (*Phocoena phocoena*) and a small tropical cetacean (*Stenella attenuate*). Physiol. Zool. **63**, 432–442 (1990)

Yamane, J., Ono, Y.: Rassenanatomische Untersuchungen der Hautstruktur von Buffet, Zebu, Formosarind und Friesisch-Hollander: ein Hinblick auf das Problem der Tropenanpassung. Memories Fac. Sci. Agric. Taihoku (Japan) **19**, 3 (1936)

Yang, S.H.: A method of assessing cutaneous pigmentation in bovine skin. Aust. J. Agric. Res. **6**, 891–902 (1952)

Yeates, N.T.M.: Modern Aspects of Animal Production. Butterworths, London (1965)

Yousef, M.K.: Physiological adaptation of less well known types of livestock in arid zones: donkeys. In: Yousef, M.K. (ed.) Stress Physiology of Livestock, vol. 2. CRC Press, Boca Raton (1985). cap.6

Yousef, M.K., Dill, D.B., Mayes, M.G.: Shifts in body fluids during dehydration in the burro, *Equus asinus*. J. Appl. Physiol. **29**, 345–350 (1970)

Chapter 5
Heat Exchange Between Animals and Environment: Aquatic Mammals

Abstract Introduction to the thermoregulatory problems of cetaceans and other marine mammals is provided. Heat transfer mechanisms are discussed, together with some practical methods for the determination of heat exchange by radiation, conduction, convection and heat loss by evaporation. Assessment of skin temperature in cetaceans at large, metabolic rate and body mass and practical problems are explained.

Keywords Cetaceans • Conduction • Convection • Evaporation • Evaporation Metabolism • Radiation • Seals • Thermoregulation

5.1 Cetaceans

5.1.1 Introduction

Cetaceans – whales, dolphins or porpoises – are the only homeothermic animals that live in the open sea without leaving water, from the borders of the polar ice caps (where water temperature is often equal or close to −1.86°C, the freezing point of sea water) to the equatorial zone (sea surface temperature often above 20°C and eventually reaching 36°C, according to Heath and Ridgway 1999). Because the high thermal conductance of water, which results into extremely fast thermal exchanges at the body surface (10–100 times that that occur in the atmosphere), the cutaneous surface temperature of aquatic animals has been often assumed as practically equal to that of the surrounding water (Irving and Hart 1957). However, the internal temperature of cetaceans is almost constant in the range of 35–38°C that is usual among mammalians.

Until recent times, there had few published papers about cetacean thermoregulation, as it was recognised by Kanwischer and Sundnes (1966); for example, it was generally supposed that a high metabolic rate would be a logical consequence of the need to maintain the body heated into an aquatic environment (Irving and Hart 1957;

R. Gomes da Silva and A.S. Campos Maia, *Principles of Animal Biometeorology*, Biometeorology 2, DOI 10.1007/978-94-007-5733-2_5,

Hart and Irving 1959; Kanwischer and Sundnes 1965). More recent studies as those of Lavigne et al. (1986a, b) showed that such an assumption was a misunderstanding and does not attain the conditions required for the determination of resting metabolism (Kleiber 1975); by the way, the metabolic level of an aquatic animal does not differ substantially from that of terrestrial animals of the same size.

Some authors as Kanwischer and Ridgway (1983) say that marine mammals have critical temperatures greater than those of their environment, a fact that would imply very high metabolic rates or the need to migrate to less cold zones (Kshatriya and Blake 1988); however, according to other authors (Hokkanen 1990; Lavigne et al. 1990), those animals would not need to increase its metabolic rate even under the coldest ambient temperatures. There has yet much controversy on this matter, as was recognised by Watts et al. (1993).

Sure, marine mammals – as endotherms in general – were able to minimise heat loss in cold environments through a reduction of their body surface temperature in order to sustain a minimal temperature differential with environment. Such a reduction is attained by the thick insulating blubber layers that encase most of the body surface just below the skin; the blubber permits a deep body temperature of 39°C, much higher than that of the surrounding water. For example, Noren et al. (1999) observed that dolphins in tropical waters at 28–31°C presented cutaneous temperatures 1°C only above that of the water.

On the other hand, whales and dolphins have countercurrent arrangements of blood vessels in their flukes and fins that permit the control the transfer of heat from the body core to the body surface (Scholander and Schevill 1955).

The study of Williams et al. (1999) with bottlenose dolphins showed that the heat flow from the body ranged from 42.9 W m^{-2} for the flank during diving to 126.2 W m^{-2} for the dorsal fin immediately after a dive. It is interesting to note that the greatest heat flow values were observed just after diving, for both flank and dorsal fin, while the lowest values were measured during dive; the authors interpreted this as a result of alterations in the heart rate. One consequence of the delayed heat transfer during diving would be the increase in core temperature in this period. The authors concluded that in this species of dolphin, the heat transfer is reduced by the dive response, perhaps due to the need for conservation of a limited oxygen reserve.

5.1.2 Heat Transfer Mechanisms

The thermal balance Eq. 3.1 given previously in Chap. 3 cannot be applied to aquatic animals as the cetaceans, because evaporative and radiative heat exchanges cannot occur in such an environment.[1] Therefore, thermal balance of those animals

---

[1] Solar short-wave radiation can affect animals remaining just below water surface, but long-wave thermal radiation is absorbed rapidly into the water.

involves mainly heat exchange by cutaneous convection ($\mathbf{C_S}$) and conduction (**K**). When swimming at the sea surface, the animals can also exchange heat with atmosphere by respiratory convection ($\mathbf{C_R}$) and evaporation ($\mathbf{E_R}$). Then, their thermal balance can be described by the equation:

$$\mathbf{M} + \mathbf{C_S} + \mathbf{C_R} + \mathbf{E_R} + \mathbf{K} = 0 \tag{5.1}$$

where **M** is metabolic rate.

Hampton et al. (1971) carried out several observations during 2 years on a dolphin (*Tursiops truncatus*), measuring skin surface and rectal temperatures, respiratory rate and oxygen consumption; water temperature varied from 25 to 26.5°C. Average rectal temperature was 37.27 ± 0.043°C at rest and increased for about 0.14°C after exercise, a nonsignificant difference; respiratory rate was 2.6 ± 0.25 breaths $\min^{-1}$ at rest and 4 breaths $\min^{-1}$ after exercise, and O_2 consumption varied from 0.2 to 2.6 L $\min^{-1}$ and was higher than that expected from terrestrial animals of the same size. It was observed that at respiratory rate of 2–6 breaths $\min^{-1}$, the average O_2 consumption was 0.969 L $\min^{-1}$. As for the heat flow, it varied from 102 kcal $(m^2\,h)^{-1}$ in the flank to 3 kcal $(m^2\,h)^{-1}$ in the tip of the flipper.

Equation 5.1 can be used to describe the thermal balance in aquatic mammals under water. However, their basic problem is how to conserve body heat and reduce the intense losses by conduction to the skin surface and then to the environment by convection. At this respect, the thermal insulation given by the blubber is the main control mechanism, as it is recognised since the studies by Scholander et al. (1950). In extreme cases, a dolphin with 2-cm-thick blubber is able to maintain its deep body temperature as higher as 40°C than that of the water.

5.1.2.1 Conduction and Convection

Differently from terrestrial mammals, aquatic species have a relatively constant temperature between the body core and the blubber. Several studies have shown that that temperature is essentially constant in the range from 35 to 38°C common to mammals in general (Kanwischer and Leivestad 1957; Morrison 1962; Kanwischer and Sundnes 1966). The metabolic rate needed to maintain such levels of internal temperature is similar to that of other mammals: The metabolic activity and subsequent heat production increase about ten times from resting to the maximum activity.

Assuming that the needed effective thermal insulation is inversely related to metabolic rate, then its variation is of the same magnitude; then, considering the variation of the water temperature, a cetacean resting into cold water would need a thermal insulation 25–50 times greater than that required by the same animal swimming at full speed in tropical waters (Kanwischer and Sundnes 1966).

A simple electrical-equivalent model can explain the thermal flux from deep body tissues to the environment in aquatic animals. See Fig. 5.1.

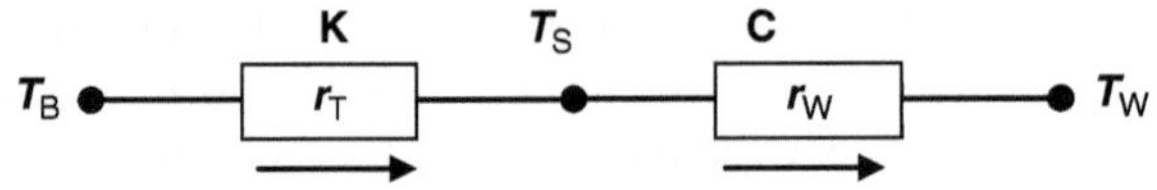

Fig. 5.1 Electrical-equivalent model of heat transfer from the deep body of an aquatic mammal to the ambient water. Body temperature (T_B) is much higher, and skin temperature (T_S) is close to than that of the water (T_W). Thermal resistance of the blubber (r_T) is very high, and heat loss by conduction (**K**) and convection (**C**) is substantially reduced; r_W is the thermal resistance of the boundary layer of water at the skin surface

The body of a cetacean may be assumed, *grosso modo*, as a cylinder covered by an insulation layer of uniform thickness (Scholander et al. 1950; Lavigne et al. 1990; Watts et al. 1993). Thermal balance is achieved when the heat produced by metabolic activities equals the losses through the skin surface; some heat can also be lost by respiratory evaporation when the animal is swimming or resting at the water surface.

According to Tomilin (1951) and Williams et al. (1999), flippers and fins can act as efficient heat exchangers due to the absence of thermal insulation in those appendages; excessive heat losses from them are controlled by a countercurrent mechanism characterised by a special arrangement of the blood vessels (Scholander and Schevill 1955). Each artery is surrounded by veins, and as warm arterial blood flows into the appendage, it is cooled by the cold venous blood that comes through the veins at temperatures close to that of the ambient water. On the other hand, the cold venous blood is warmed before entering the deep body sites.

With respect to the trunk, the problem of heat transfer by conduction and convection can be solved by applying the methods found in engineering textbooks (e.g., Chapman 1987) for calculating heat transfer from insulated pipes. The heat is transferred by conduction from the body core to the skin through the blubber, at a rate given by

$$\mathbf{K} = \frac{k_T(T_r - T_s)}{R \ln(R/r)} \text{ W m}^{-2} \tag{5.2}$$

(Silva 2004) where k_T is the thermal conductivity of the blubber (W m^{-1} K^{-1}), T_r is the deep body temperature (estimated by the rectal one, K), T_s is the skin temperature (K) and the other values are body dimensions defined in Fig. 5.2. Blubber thickness is given by the difference between radii lengths ($R-r$) and has average values of 0.02 m in dolphins and porpoises, and 0.2 m in whales (Kanwischer and Sundnes 1966).

Values found in the literature for the thermal conductivity of blubber of some cetaceans are given in Table 5.1.

In order to determine the convection heat exchange in dolphins and porpoises, the ellipsoid model can be assumed (Fig. 5.3).

According to measurements done in Antarctic (*Arctocephalus gazelle*) and sub-Antarctic seals (*Arctocephalus tropicalis*) by Luque et al. (2007), average body

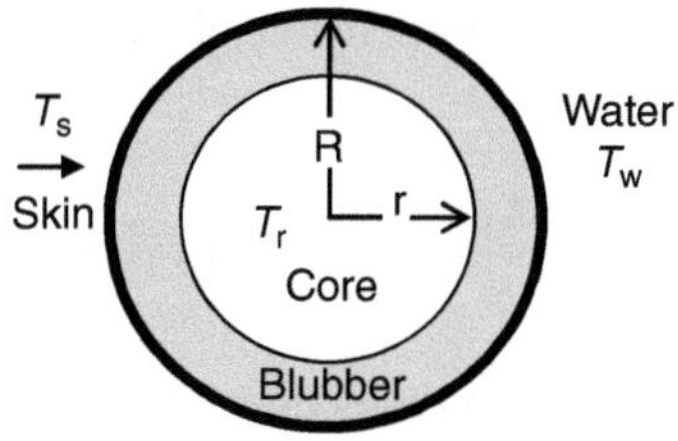

Fig. 5.2 Schematic transverse cut of a cetacean's body. The core has a temperature T_r and is surrounded by the insulating blubber layer with thermal conductivity k_T. Temperature of water is T_w and that of the skin surface is T_s

Table 5.1 Thermal conductivity (k_T) for the bubble in cetaceans, as given in the literature

Animal	k_T (W m^{-1} K^{-1})	References
Porpoise	0.1	Worthy and Edwards (1990)
Bottlenose dolphin	0.15	Hampton et al. (1971)
	0.18	Dunkin et al. (2005)
	0.1025	Silva (2004)
Whale	0.2	Innes (1986), Lavigne et al. (1990) and Watts et al. (1993)
Minke whale	0.2–0.28	Kvadsheim et al. (1996)

Fig. 5.3 Ellipsoidal model for dolphins and porpoises, where L is the body length and D is the maximum diameter, in metres

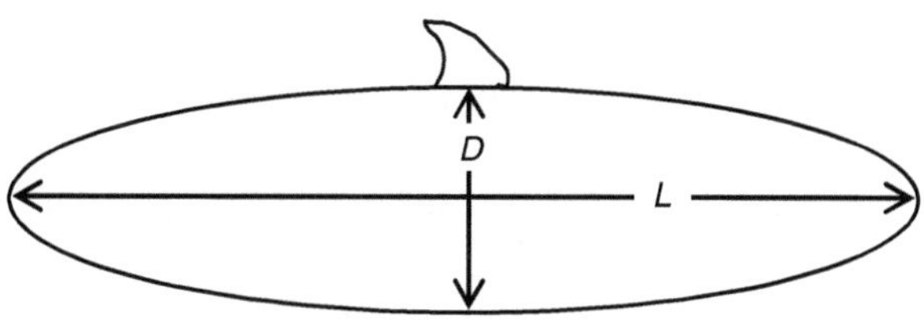

length of adult animals ranges 1.8–2.7 m and for body weights from 135 to 225 kg. In general, the greatest body diameter (just below the dorsal flipper) is about 44% of the total body length, or $D = 0.44\ L$.

Body surface area can be calculated with reasonable accuracy by considering ellipsoidal model and using Knud Thomsen's approximation (Eq. 2.21). The volume is given by

$$V = \frac{4}{3}\pi L D^2 \text{ m}^3 \tag{5.3}$$

where L (m) is the length and D (m) is the maximum diameter of the body.

Once stored at the body surface, the heat is transferred by convection to the surrounding water at a rate given by the equation:

$$\mathbf{C} = h_c(T_s - T_w) \text{ W m}^{-2} \tag{5.4}$$

where d (m) is the characteristic dimension of the body, h_c is the coefficient of convection,

$$h_c = \frac{k_T}{d} N_u \qquad \mathrm{W\,m^{-2}\,K^{-1}} \tag{5.5}$$

and N_u is the Nusselt number. Two cases may be considered:

(a) When the animal is moving through water, the forced convection must be considered, and in this case, d is the body length (L, m). The Reynolds number and the Nusselt number are calculated by Eqs. 2.42 and 2.45, respectively.
(b) If the animal is resting and remains motionless in the water, then d is the body diameter (D, m). The Grashof number is calculated as

$$G_r = \frac{9.8 d^3 (T_s - T_w)}{v^2 (T_w + 273.15)} \tag{5.6}$$

and the Nusselt number is calculated by Eq. 2.39. In those calculations, the water temperature is taken instead of that of the air.

In both cases, the Prandtl number is calculated by Eq. 2.35, and all the thermo-physical properties of the sea water are obtained for the temperature $T_m = 0.5$ $(T_s + T_w)$ by using the following formulas:

$$k = 0.56631 + 0.0018\,T_m - 7.0039 \times 10^{-6} T_m^2 \quad \mathrm{W\,m^{-1}\,°C^{-1}} \tag{5.6a}$$

$$\rho = 1028.22 + 0.1046 T_m - 0.00402 T_m^2 \ \mathrm{kg\,m^{-3}} \tag{5.6b}$$

$$c_p = 3.9919 - 5.0995 \times 10^{-5} T_m + 6.065 \times 10^{-6} T_m^2 \ \mathrm{J\,kg^{-1}\,°C^{-1}} \tag{5.6c}$$

$$v = 0.00181 - 5.014 \times 10^{-5} T_m + 5.867 \times 10^{-7} T_m^2 \ \mathrm{m^2\,s^{-1}} \tag{5.6d}$$

5.1.2.2 Respiratory Heat Transfer

There has a lack of information about sensible and latent heat exchange in the respiratory system of cetaceans. Kanwischer and Sundnes (1966) mentioned that in the bottlenose dolphin (*Tursiops truncatus*), less than 5% of the heat is lost by evaporation in the lungs, without more details about the process.

According to the study carried out by Hampton et al. (1971) on the bottlenose dolphin, this animal has an average respiratory rate of 2.6 breaths min^{-1} when resting and 4 breaths min^{-1} after exercise; the respective rectal temperatures were 37.3 and 37.4°C. The oxygen consumption increased with the respiratory rate according to the function:

$$O_{2\,cons} = 0.6033 + 0.0317F_r + 0.0007F_r^2 \text{ L/min} \tag{5.7}$$

that is higher than that expected for terrestrial mammals of the same size.

As there has no information in the literature on the dimensions of the air passages of cetaceans, together with respective air fluxes, it is not possible to calculate the sensible heat exchange through respiration for those animals. However, the respective rate is probably negligible as compared to the evaporative heat loss.

In order to determine the evaporative heat exchange through respiration, it is needed the *tidal volume* (V_T) about which there has no information in the literature. By using the data of Hampton et al. (1971), the following equation can be established in order to estimate V_T from the respiratory rate:

$$V_T = 0.009125F_r^{-0.451} \text{ m}^3/\text{breath} \tag{5.8}$$

This value is then used to calculate the mass flux rate ($\dot{m}$) by using Eq. 4.34. Now, assuming that the air is expired by a cetacean at its deep body temperature, the absolute humidity of the expired air would be given by

$$\Psi_{EXP} = \frac{2,166.869P_s(T_r)}{T_r + 273.15} \text{ g m}^{-3} \tag{5.9}$$

where $P_S(T_r)$ is the saturation pressure of air at temperature T_r (°C) which is calculated by Eq. 1.50. In the same way, it is obtained the saturation vapour pressure at the atmospheric temperature, Ψ_{ATM}.

The rate of latent heat loss from the respiratory system is given by

$$\mathbf{E}_R = \frac{\lambda\rho^{-1}\dot{m}\,(\Psi_{EXP} - \Psi_{ATM})}{A} \quad \text{W m}^{-2} \tag{5.10}$$

where λ and ρ are atmospheric properties obtained by Eqs. 1.46 and 1.41, respectively, and A is the body surface area, determined as

$$A = 0.0725\,W^{2/3} \quad \text{m}^2 \tag{5.11}$$

where W is the body weight (kg). This formula was based on the data of Hampton et al. (1971) for dolphins; an alternative is Eq. 5.3.

Finally, the thermal balance of a cetacean would be given as

$$\mathbf{M} + \mathbf{K} + \mathbf{C} + \mathbf{E}_R = 0 \tag{5.12}$$

and the metabolic rate can be estimated as $\mathbf{M} = \mathbf{K} + \mathbf{C} + \mathbf{E}_R$.

5.1.3 Estimation of Skin Temperature

The hydrodynamic and thermodynamic aspects of thermal convection require knowledge of the surface temperature, T_s, together with the convection coefficient h. However, in real problems involving cetaceans, neither T_s nor h is known; this last parameter depends on T_s, and in turn, this temperature depends on h, such a dependence involving also the temperature differential among the skin and the surrounding water. Besides, in several circumstances, it is not possible to measure the skin surface directly.

Silva (2004) proposed an indirect iterative approach for the prediction of T_s in cetaceans from deep body temperature, swimming speed and temperature and thermodynamic properties of the water. First, a trial T_s value is assumed; as an example, bottlenose dolphins have skin temperature about 1°C above that of the water, T_w (Noren et al. 1999). The mean temperature $T_m = 0.5(T_s + T_w)$ is used to calculate the thermal properties of the sea water by using Eqs. 5.6a, 5.6b, 5.6c, and 5.6d. Then, the overall transfer coefficient is calculated by the equation:

$$\Phi = \left[\frac{R\ln(R/r)}{k_T} + \frac{1}{h}\right]^{-1} \text{ W m}^{-2}\text{ K}^{-1} \tag{5.13}$$

where $h = k N_u d^{-1}$ is the coefficient of convection (W m^{-2} °C^{-1}), k is the thermal conductivity of sea water (W m^{-1} °C^{-1}), d is the body length (m) and k_T the thermal conductivity of blubber (W m^{-1} °C^{-1}). Then the skin temperature will be

$$T_s = \frac{\Phi T_r + T_w(h - \Phi)}{h} \text{ °C} \tag{5.14}$$

If this result is different from the previously assumed T_s, it can be substituted in the equations, and all the calculations are carried out again. The process continues until the results converge and a satisfactory agreement between two successive T_s values is achieved. In general, less than ten iterations may be needed until convergence occurs.

5.2 Seals and Sea Lions

5.2.1 Introduction

Sea lions, seals and other similar aquatic mammals are in general covered by a coat of thin, smooth hairs that provide thermal insulation. In fact, seals have skin temperatures often 20°C above that of the surrounding water (Boyd 2000).

However, thermal properties of the hair coats of those animals were not yet adequately studied. For the polar bear (*Ursus maritimus*) – which is just the main predator of the Arctic seals – there were carried out some studies at that respect. Hurst and Øristland (1982) observed that the presence of oil in the hair coat of polar bears has an important effect on its thermal insulation properties; in a still atmosphere, the coefficient of convection (h) was two to five times greater than that of a normal hair coat; but under windy conditions, this coefficient increased about 290%, and the transmission of solar thermal energy through the coat was 55% higher. As for the thermal conductance of the oiled hair coat, it remained high under air temperatures as low as 0.6°C and decreased under temperatures of 24.7°C. In general, the greater the oil viscosity, the lower the thermal insulation of the hair coat. It was suggested that the same effect would be observed in the hair coat of seals.

When seals remain under much exercise on land, they are not able to transfer the excess heat to the atmosphere with efficiency; then they may have hyperthermia and may die if they cannot enter the water immediately. As a consequence, those animals avoid unnecessary physical exercise when landed, unless there has prompt, unobstructed way to the water.

An electrical-equivalent model for the heat exchanges in seals and sea lions in land is presented in Fig. 5.4. This diagram can be used to establish practical thermoregulation models for seals and sea lions, if adequate values for the thermal resistances are available. Methods for the calculation of some of those resistances are given in this chapter, in Chap. 4 and in the paper of Silva et al. (2002). For thermal exchange by respiratory convection and respiratory evaporation, methods developed for other mammals can be used with some adaptations.

5.2.2 *Heat Gains and Losses*

5.2.2.1 Metabolic Rate

The metabolic rate of harp seals (*Phoca groenlandica*) measured when they were on land was 2.23 W kg^{-1} at an air temperature of −30°C; when they were immersed into 2.3°C water, the rate was 2.31 W kg^{-1}, regardless of T_a (Folkow and Blix 1989).

For elephant seal (*Mirounga angustirostris*) pups with body weights from 62 to 108 kg, Noren (2002) observed O_2 consumption rates from 293.6 to 512.7 mL min^{-1}. By using the methods explained in Chap. 3, those values can be transformed into metabolic rates, giving estimates from 53.2 to 55.7 W m^2. For the body surface area (m^2), there was used the formula $A = 0.09\ W^{2/3}$, while for the metabolic rate, the formula of Kleiber (1961) for mammals in general; this last formula was considered as appropriate for marine mammals by Kanwischer and Sundnes (1966).

Liwanag (2010) studied fur seal (*Callorhinus ursinus*) pups weighing 11.8–12.8 kg and resting in water at a temperature close to the neutral zone

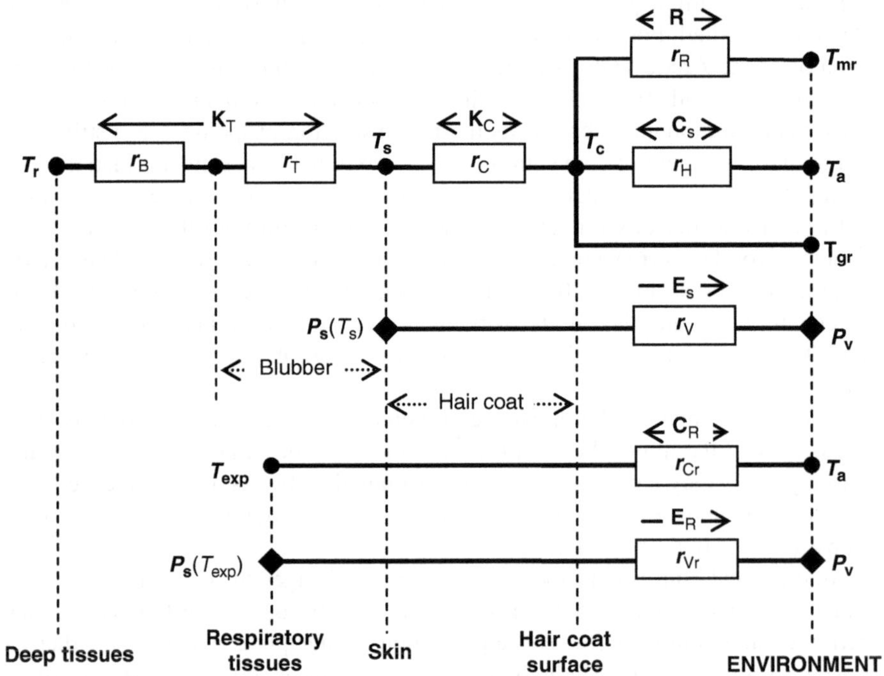

Fig. 5.4 Electrical-equivalent model for the heat fluxes between environment and the body of seals and sea lions on land. Temperatures: T_r = deep body, T_s = skin surface, T_c = hair coat surface, T_{exp} = expired air, T_a = atmosphere, T_{mr} = mean radiant and T_{gr} = ground surface. Vapour pressures: P_v = partial, atmospheric; $P_s(T_{exp})$ = saturated, expired air and $P_s(T_s)$ = saturated, skin surface. Thermal fluxes: $\mathbf{K_T}$ = condution, tissues; $\mathbf{K_C}$ = conduction, hair coat; $\mathbf{R}$ = radiation; $\mathbf{C_S}$ = convection, body surface; $\mathbf{E_S}$ = evaporation, body surface; $\mathbf{C_R}$ = convection, respiratory and $\mathbf{E_R}$ = evaporation, respiratory. The values r_B, r_T, r_C, r_H, r_V, r_{Cr} and r_{Vr} are thermal resistances

(8.3–24.3°C); oxygen consumption was determined as $10.03 \pm 2.26\,\text{mL kg}^{-1}\,\text{min}^{-1}$ on the average. The average metabolic rate can be estimated as 46.5 W m^{-2}.

Donohue et al. (2000) determined the oxygen consumption in the air by fur seal pups (*Callorhinus ursinus*) in the pre-moult (7.4 kg on the average) and post-moult (13.3 kg on the average) stages, with averages 113 (74–149) and 160 (128–200) mL min^{-1}. The O_2 consumption rate was considered as proportional to the body mass (W, kg) according to the equation:

$$O_2 = 26.24\,W^{0.7}\ \text{mL}_{O_2}\ \text{min}^{-1} \quad (5.15)$$

Again, from the above formula and considering that 1 L of O_2 is equivalent to 20.1 kJ, we estimate the metabolic rate of seals as

$$\mathbf{M} = 8.7904\,W^{0.7}\ \text{W} \quad (5.16)$$

In terms of body area,

$$\mathbf{M} = \frac{8.7904\, W^{0.7}}{A}\ \mathrm{W\ m^{-2}} \tag{5.16a}$$

where A can be determined by using the ellipsoid model and Eq. 2.21. By using this last equation, the average metabolic rate values obtained by Donohue et al. (2000) would be transformed into 104.4 and 106.5 W m^{-2} for pre-moult and post-moult stages, respectively.

5.2.2.2 Radiation

Ohata and Miller (1977a) observed the occurrence of hyperthermia in Arctic seals on land because forced physical activity; solar radiation contributed to that overheating, while rain caused larger heat loss and prevented the overheating. Therefore, radiation seems to be an important avenue of heat exchange for those animals.

The variation of the body surface of grey seal (*Halichoerus grypus*) pups exposed to sun was studied by McCafferty et al. (2005) in Scotland, determining that solar radiation accounted for 43–48% of the variation of the skin surface during the day, together with air temperature and humidity. These authors found an average surface temperature of 15.8°C (7.7–29.7°C) at air temperatures of 10.2°C. By applying the Planck's law (Eq. 2.7) to their data, we can obtain heat losses by long-wave thermal radiation from 345.7 to 467.4 W m^{-2}. Such losses were to be compensated with heat gain from environmental thermal radiation; because there has no information about the radiant heat load or the mean radiant temperature, it is not possible to calculate the radiant heat exchange between animals and environment (formula 4.20).

5.2.2.3 Conduction

When they are on land, both seals and sea lions exchange thermal energy with the atmosphere in the same way the terrestrial mammals do, through the same physical mechanisms of radiation, convection, conduction and evaporation. There are differences, however.

One of those differences is the greater importance of the conduction, due to the large surface area of those animals in direct contact with the ground. A seal, for example, has about 2/5 of its total body surface area in contact with the ground and exchanging with it great amounts of thermal energy by conduction.

The thermal conductivity of the blubber was given as an average of 0.18 W m^{-1} K^{-1} by Worthy (1991) as valid for several species (harp seal, grey seal).

Noren (2002) determined the combined thermal conductance values of the blubber in elephant seal pups in the range from 3.1 (within cold water) to 15.2 W K^{-1} (in the air); however, the measurement unit used by this author differs from that of the SI, which is W m^{-2} K^{-1}. In another way, our interest is just on the thermal conductivity, k (W m^{-1} K^{-1}). Therefore, the estimates of Noren (2002) can be transformed into conductivity ones by dividing them by the blubber thickness (m).

The data given by Scholander et al. (1950) for the insulation of the seal fur in terms of W^{-1} m^2 K can be transformed into thermal conductance (W m^{-2} K^{-1}) and then in thermal conductivity, considering fur thickness from 0.06 to 0.07 m. Estimated conductivity was 2.69–4.49 $Wm^{-1}K^{-1}$ in the air and 4.24–5.15 $Wm^{-1}K^{-1}$ in the water.

5.2.2.4 Convection

The large fins of a seal greatly contribute to the heat loss. For example, it was concluded that because the limited ability of fur seals to lose heat when landed, their main response to the heat stress is thermal storage; they are tolerant to body temperatures as high as 43°C.

Swimming in very cold waters, seals need a thick subcutaneous fat layer which is the result of a balance between feeding and the energy expenses; if this balance is affected, the ability of the animals to sustain homeothermy is reduced, and death may be the result. However, there has no information about the thermal exchange by convection of those animals when under water.

Ohata and Miller (1977b) measured the temperature of 11 different body regions in Arctic seals (*Callorhinus ursinus*) immersed into water at 7°C, determining average values of 37.4 ± 0.6°C within the thorax and lower rectal temperatures (35.8 ± 1.1°C); the temperatures just below the subcutaneous fat layer were 32.9 ± 1.1°C on the average, while the flippers were uniformly cold (10.2 ± 1.1°C).

Sea lions (*Eumetopias jubatus*) were observed by Willis et al. (2005), who found that the heat flux from shoulders and rumps was higher than that from the trunk and armpits, both in animals swimming and in those resting into 8°C water; it is possible that the fat layer is thinner in those regions, by which the heat exchange is higher there.

In order to estimate the amount of heat exchange by convection in seals, best prediction would arose from a model which describes the animal as a cylinder, as it was suggested by Watts et al. (1993) and that is similar to the approach of Silva (2004) for cetaceans.

5.2.3 Evaporation

Latent heat loss at the skin surface is possible only when this surface is moistened with water from the environment; the skin of aquatic animals has no sweating glands and is barely permeable to water; at least, there has no published data at this respect.

The heat loss by cutaneous evaporation would be estimated by using the cylindrical model and considering that great part of the body surface of a seal is in direct contact with the ground – thus preventing evaporation from those parts.

As for the respiratory evaporation, estimates of respiratory rate and tidal volume and temperature of the expired air must be previously obtained. Folkow and Blix (1989) determined the mean temperature of the expired air in harp seals as 9.5, 13.0 and 25.0°C at air temperatures of −30, −10 and 10°C, respectively; when the animals were immersed in water at 2.3°C, the values were 8.0, 9.5 and 15.5°C, respectively. The authors suggested that the low T_{exp} recorded at air temperatures of −30 and −10°C in both air and water was due to the conservation of heat by thermal exchange in the respiratory ways.

5.3 Problems

Problem 5.1. A dolphin having 2.5 m in length, 0.39 m as maximum body diameter and 1.5-cm-thick blubber is swimming at a speed of 10 knots in 17°C water. Its deep body temperature is supposed to be 37.5°C. What would be the heat loss by convection?

Data:

T_w = water temperature = 17°C
T_r = deep body temperature = 37.5°C
L = body length = 2.5 m
b = blubber thickness = 0.015 m
D = body diameter = $2R$ = 0.39 m
$r = R - b = 0.195 - 0.015 = 0.18$ m
k_T = blubber thermal conductivity = 0.18 W m^{-1} $°C^{-1}$
U = swimming speed = 10 knots = 5.1 m s^{-1}

Solution:

In order to obtain the heat loss by convection, there is needed the skin temperature (T_s) of the animal; it can be estimated by using the iterative method presented in Sect. 5.1.3 (Eqs. 5.12 and 5.13), as follows:

Initial T_s value = $T_W + 1 = 18$°C

Sea water properties at $T_m = 0.5\ (T_s + T_w) = 17.5$°C (Eqs. 5.6a, 5.6b, 5.6c, and 5.6d):

$$k = 0.56631 + 0.0018T_m - 7.0039 \times 10^{-6}T_m^2$$
$$= 0.56631 + 0.0018(17.5) - 7.0039 \times 10^{-6}(17.5)^2 = 0.5957\ \mathrm{W\ m^{-1}\ ^\circ C^{-1}}$$

$$\rho = 1028.22 + 0.1046T_m - 0.00402T_m^2$$
$$= 1028.22 + 0.1046(17.5) - 0.00402(17.5)^2 = 1028.82\ \mathrm{kg\ m^{-3}}$$

$$c_p = 3.9919 - 5.0995 \times 10^{-5}T_m + 6.065 \times 10^{-6}T_m^2$$
$$= 3.9919 - 5.0995 \times 10^{-5}(17.5) + 6.065 \times 10^{-6}(17.5)^2 = 3.9929\ \mathrm{J\ kg^{-1}\ ^\circ C^{-1}}$$

$$v = 0.00181 - 5.014 \times 10^{-5}T_m + 5.867 \times 10^{-7}T_m^2$$
$$= 0.00181 - 5.014 \times 10^{-5}(17.5) + 5.867 \times 10^{-7}(17.5)^2 = 0.001112\ \mathrm{m^2\ s^{-1}}$$

Body volume (characteristic dimension):

$$V = d = \frac{4}{3}\pi LD^2 = \frac{4}{3}\pi(2.5)(0.39)^2 = 1.59\,\mathrm{m^3}$$

Prandtl number (Eq. 2.35):

$$P_r = \frac{\rho\, c_p v}{k} = \frac{1028.82(3.9929)(0.001112)}{0.5957} = 7.668405$$

Reynolds number (Eq. 2.42):

$$R_e = \frac{Ud}{v} = \frac{5.1(1.59)}{0.001112} = 7,292.2662$$

Nusselt number (Eq. 2.45):

$$N_u = 0.0296R_e^{4/5}P_r^{1/3}$$
$$= 0.0296(7,292.2662)^{4/5}(7.668405)^{1/3} = 71.86$$

Coefficient of convection:

$$h_c = kN_u d^{-1} = \frac{0.5957(71.86)}{1.59} = 26.9226\ \mathrm{W\ m^{-2}\ ^\circ C^{-1}}$$

Overall heat transfer (Eq. 5.12):

$$\mathbf{\Phi} = \left[\frac{R\ln(R/r)}{k_{\mathrm{T}}} + \frac{1}{h_{\mathrm{c}}}\right]^{-1}$$

$$= \left[\frac{0.195\ln(0.195/0.18)}{0.18} + \frac{1}{26.9226}\right]^{-1} = 8.0739\ \mathrm{W\ m^{-2}\ ^{\circ}C^{-1}}$$

Skin surface temperature (Eq. 5.13):

$$T_{\mathrm{s}} = \frac{\mathbf{\Phi} T_{\mathrm{r}} + T_{\mathrm{w}}(h_{\mathrm{c}} - \mathbf{\Phi})}{h_{\mathrm{c}}} = \frac{8.0739(37.5) + 17(26.9226 - 8.0739)}{26.9226} = 23.15\ ^{\circ}\mathrm{C}$$

The final estimates after 5 iterations were $T_{\mathrm{s}} = 23.0°\mathrm{C}$ and $h_{\mathrm{c}} = 27.8618$.
Rate of heat loss by convection:

$$\mathbf{C} = h_{\mathrm{c}}(T_{\mathrm{s}} - T_{\mathrm{w}}) = 27.8618(23 - 17) = 167.2\ \mathrm{W\ m^{-2}}$$

Problem 5.2. Let us have the same animal described in Problem 5.1 but resting at the water surface in another place. Its rectal temperature is $T_{\mathrm{r}} = 37.3°\mathrm{C}$ and the respiratory rate $F_{\mathrm{r}} = 3$ breaths min^{-1}. Environmental data: water temperature $T_{\mathrm{w}} = 26°\mathrm{C}$, dry bulb temperature $T_{\mathrm{a}} = 30°\mathrm{C}$, wet bulb temperature $T_{\mathrm{u}} = 28°\mathrm{C}$ and atmospheric pressure $P_{\mathrm{a}} = 101.325$ kPa. Calculate the respiratory latent heat loss.

Data:

T_{r} = deep body temperature = 37.3°C
F_{r} = respiratory rate = 3 breaths min^{-1}
$2R$ = body length = 2.5 m $\therefore R = 1.25$
$2r$ = maximum body diameter = 0.39 m $\therefore r = 0.195$
T_{w} = water temperature = 26°C
T_{a} = dry bulb temperature = 30°C
T_{u} = wet bulb temperature = 28°C
P_{a} = atmospheric pressure = 101.325 kPa

Solution:

By using equations given in Chap. 1 (1.40–1.51), the thermo-physical properties of the atmosphere can be obtained as

$\rho = 1{,}164.6\ \mathrm{gm^{-3}}$, $\lambda = 2{,}429.576\ \mathrm{Jg^{-1}}$, $c_{\mathrm{p}} = 1.006388\ \mathrm{Jg^{-1}\ ^{\circ}C^{-1}}$, $\gamma = 0.000666°\mathrm{C}^{-1}$

Then, the following values are calculated:

Saturation vapour pressure at air temperature:

$$\begin{aligned} P_s(T_a) &= 0.61078 \times 10^{7.5T_a/(T_a+237.5)} \\ &= 0.61078 \times 10^{7.5(30)/(30+237.5)} = 4.236 \text{ kPa} \end{aligned}$$

Saturation vapour pressure at wet bulb temperature:

$$\begin{aligned} P_s(T_u) &= 0.61078 \times 10^{7.5T_u/(T_u+237.5)} \\ &= 0.61078 \times 10^{7.5(28)/(28+237.5)} = 3.774 \text{ kPa} \end{aligned}$$

Atmospheric partial vapour pressure:

$$\begin{aligned} P_v &= P_s(T_u) - P_a\gamma(T_a - T_u) \\ &= 3.774 - 101.325(0.000666)(30 - 28) = 3.639 \text{ kPa} \end{aligned}$$

Saturation vapour pressure at the temperature of expired air:

$$\begin{aligned} P_s(T_r) &= 0.61078 \times 10^{7.5T_r/(T_r+237.5)} \\ &= 0.61078 \times 10^{7.5(37.3)/(37.3+237.5)} = 6.367 \text{ kPa} \end{aligned}$$

Absolute humidity of atmosphere:

$$\Psi_{\text{ATM}} = \frac{2166.869P_v}{T_a} = \frac{2166.869(3.639)}{30 + 273.15} = 26.011 \text{ g m}^{-3}$$

Absolute humidity of the expired air:

$$\Psi_{\text{EXP}} = \frac{2166.869P_s(T_r)}{T_r} = \frac{2166.869(6.367)}{37.3 + 273.15} = 44.436 \text{ g m}^{-3}$$

Tidal volume:

$$V_T = 0.009125F_r^{-0.451} = 0.009125(3)^{-0.451} = 0.00556 \text{ m}^3/\text{breath}$$

Respiratory mass flux:

$$\dot{m} = V_T\,\rho\left(\frac{F_r}{60}\right) = 0.00556(1{,}164.6)\left(\frac{3}{60}\right) = 0.32376 \text{ kg s}^{-1}$$

Body surface area (Eq. 2.21):

$$A = 4\pi\left[\frac{2(Rr)^{1.6075} + r^{3.125}}{3}\right]^{0.6221}$$

$$= 4\pi\left[\frac{2(1.25)^{1.6075}(0.195)^{1.6075} + (0.195)^{3.125}}{3}\right]^{0.6221} = 2.4\ \mathrm{m}^2$$

Rate of respiratory latent heat loss:

$$\mathbf{E}_{\mathrm{R}} = \frac{\lambda\rho^{-1}\dot{m}(\Psi_{\mathrm{EXP}} - \Psi_{\mathrm{ATM}})}{A}$$

$$= \frac{2429.576(1164.6)^{-1}(0.32376)(44.436 - 26.011)}{2.4} = 5.2\ \mathrm{W\ m^{-2}}$$

Problem 5.3. A seal has 33 kg of body weight, 1.2 m body length, 3-cm fur thickness, 2-cm blubber thickness, deep body temperature 38°C and skin surface temperature 18°C. It is resting on land at air temperature 10°C, ground surface temperature 15°C and wind speed 2.5 m s^{-1}. Animal's body is positioned with its axis parallel to the wind. Determine the heat exchange by conduction and convection.

Data:

T_{w} = deep body temperature = 38°C
T_{s} = skin surface temperature = 18°C
f = fur thickness = 0.03 m
b = blubber thickness = 0.02 m
W = body weight = 33 kg
L = body length = 1.2 m
k_{T} = blubber thermal conductivity = 0.18 W m^{-1} °C^{-1}
k_{F} = fur thermal conductivity = 2.7 W m^{-1} °C^{-1}
T_{a} = air temperature = 10°C
T_{g} = ground surface temperature = 15°C
U = wind velocity = 2.5 m s^{-1}

Solution:

Atmospheric properties at temperature T_{a} =10°C:

$\rho = 1{,}246.9$ gm^{-3}, $k = 0.02487$ Wm^{-2} °C^{-1}, $c_{\mathrm{p}} = 1.005847$ Jg^{-1} °C^{-1}, $\nu = 1.42 \times 10^{-5}$ m^2 s^{-1}.

The seal's body is assumed as a horizontal cylinder with axis parallel to the wind. Therefore, the characteristic dimension is the body length $d = 1.2$, and convection is calculated by Eq. 2.45.

Prandtl number (Eq. 2.35):

$$P_r = \frac{\rho c_p \nu}{k} = \frac{1{,}246.9(1.005847)\left(1.42 \times 10^{-5}\right)}{0.02487} = 0.716104$$

Reynolds number (Eq. 2.42):.

$$R_e = \frac{Ud}{\nu} = \frac{2.5(1.2)}{1.42 \times 10^{-5}} = 211,267.6$$

Nusselt number (Eq. 2.45):

$$\begin{aligned} N_u &= 0.0296\, R_e^{4/5} P_r^{1/3} \\ &= 0.0296(211,267.6)^{4/5}(0.716104)^{1/3} = 481.74 \end{aligned}$$

Coefficient of convection:

$$h_c = k N_u d^{-1} = \frac{0.02487(481.74)}{1.2} = 9.9841 \text{ W m}^{-2}\,{}^{\circ}\text{C}^{-1}$$

Thermal resistance of the boundary layer at body surface:

$$r_H = \frac{\rho c_p d}{k N_u} = \frac{1{,}246.9(1.005847)(1.2)}{0.02487(481.74)} = 125.6 \text{ s m}^{-1}$$

Rate of heat exchange by convection:

$$\mathbf{C} = \frac{\rho c_p (T_s - T_a)}{r_H} = \frac{1246.9(1.005847)(15 - 10)}{125.6} = 49.9 \text{ W m}^{-2}$$

Rate of heat exchange by conduction to the ground:
Thermal resistances (Eq. 2.29):

$$r_T = \frac{\rho c_p \mathrm{b}}{k_T} = \frac{1246.9(1.005847)(0.02)}{0.18} = 139.4 \text{ s m}^{-1}$$

$$r_F = \frac{\rho c_p \mathrm{f}}{k_F} = \frac{1246.9(1.005847)(0.03)}{2.7} = 13.9 \text{ s m}^{-1}$$

$$\mathbf{K} = \frac{\rho c_p (T_r - T_s)}{r_T + r_F} = \frac{1246.9(1.005847)(38 - 15)}{139.4 + 13.9} = 188.1 \text{ W m}^{-2}$$

References

Boyd, I.L.: Skin temperatures during free-ranging swimming and diving in Antarctic fur seals. J. Exp. Biol. **203**, 1907–1914 (2000)

Chapman, A.J.: Fundamentals of Heat Transfer. Macmillan, New York (1987)

Donohue, M.J., Costa, D.P., Goebel, M.E., Baker, J.D.: The ontogeny of metabolic rate and thermoregulatory capabilities of northern fur seal, *Callorhinus ursinus*, pups in air and water. J. Exp. Biol. **203**, 1003–1016 (2000)

Dunkin, R.C., McLellan, W.A., Blum, J.E., Pabst, D.A.: The ontogenetic changes in the thermal properties of blubber from Atlantic bottlenose dolphin *Tursiops truncatus*. J. Exp. Biol. **208**, 1469–1480 (2005)

Folkow, L.P., Blix, A.S.: Thermoregulatory control of expired air temperature in diving harp seals. Am. J. Physiol. Reg. Int. Comp. **257**, R306–R310 (1989)

Hampton, I.F.G., Whittow, G.C., Szekerczes, J., Rutherford, S.: Heat transfer and body temperature in the Atlantic bottlenose dolphin, *Tursiops truncatus*. Int. J. Biometeorol. **15**, 247–253 (1971)

Hart, J.S., Irving, L.: The energetic of harbour seals in air and in water with special consideration of seasonal changes. Can. J. Zool. **37**, 447–457 (1959)

Heath, M.E., Ridgway, S.H.: How dolphins use their blubber to avoid heat stress during encounters with warm water. Am. J. Physiol. **276**, R1188–R1194 (1999)

Hokkanen, J.E.J.: Temperature regulation of marine mammals. J. Theor. Biol. **145**, 465–485 (1990)

Hurst, R.J., Øristland, N.A.: Polar bear thermoregulation: effect of oil on the insulative properties of fur. J. Therm. Biol. **7**, 201–208 (1982)

Innes, S.: How fast should be a dead whale cool? Can. J. Zool. **64**, 2064–2065 (1986)

Irving, L., Hart, J.S.: The metabolism and insulation of seals as bare-skinned mammals in cold water. Can. J. Zool. **35**, 497–510 (1957)

Kanwischer, J., Leivestad, H.: Thermal regulation in whales. Norsk Hvalfangs-tid. **1**, 1–5 (1957)

Kanwischer, J., Ridgway, S.H.: The physiological ecology of whales and porpoises. Sci. Am. **248**, 110–120 (1983)

Kanwischer, J., Sundnes, G.: Physiology of a small cetacean. Hval. Skrifter **48**, 46–53 (1965)

Kanwischer, J., Sundnes, G.: Thermal regulation in cetaceans. In: K.S. Norris, (ed.) Whales, Dolphins and Porpoises, pp. 397–407. University of California Press, Berkeley (1966)

Kleiber, M.: The Fire of Life: An Introduction to Animal Energetics. Wiley, New York (1961)

Kleiber, M.: The Fire of Life: An Introduction to Animal Energetics. Krieger, New York (1975)

Kshatriya, M., Blake, R.W.: Theoretical model of migration energetic in the blue whale, *Balaenoptera musculus*. J. Theor. Biol. **133**, 474–498 (1988)

Kvadsheim, P.H., Folkow, L.P., Blix, A.S.: Thermal conductivity of Minke whale blubber. J. Therm. Biol. **21**, 123–128 (1996)

Lavigne, D.M., Innes, S., Worthy, G.A.J., Kovacs, K., Schmitz, O., Hickie, J.P.: Metabolic rates of seals and whales. Can. J. Zool. **64**, 279–284 (1986a)

Lavigne, D.M., Innes, S., Worthy, G.A.J., Kovacs, K.: Metabolic rate-body size relations in marine mammals. J. Theor. Biol. **122**, 123–124 (1986b)

Lavigne, D.M., Innes, S., Worthy, G.A.J., Edwards, E.F.: Lower critical temperatures of blue whales, *Balaenoptera musculus*. J. Theor. Biol. **144**, 249–257 (1990)

Liwanag, H.E.M.: Energetic costs and thermoregulation in northern fur seal (*Callorhinus ursinus*) pups: the importance of behavioral strategies for thermal balance in furred marine mammals. Physiol. Biochem. Zool. **83**, 898–910 (2010)

Luque, S.P., Miller, E.H., Arnould, J.P.Y., Chambellant, M., Guinet, C.: Ontogeny of body size and shape of Antarctic and Subantarctic fur seals. Can. J. Zool. **85**, 1275–1285 (2007)

McCafferty, D.J., Moss, S., Bennett, K., Pomeroy, P.P.: Factors influencing the radiative surface temperature of grey seal (*Halichoerus grypus*) pups during early and late lactation. J. Comp. Physiol. B **175**, 423–431 (2005)

Morrison, P.: Body temperatures in some Australian mammals. III. Cetacea (*Megaptera*). Biol. Bull. **123**, 154–169 (1962)
Noren, D.P.: Thermoregulation of weaned northern elephant seal (*Mirounga angustirostris*) pups in air and water. Physiol. Biochem. Zool. **75**, 513–523 (2002)
Noren, D.P., Williams, T.M., Berry, P., Butler, E.: Thermoregulation during swimming and diving in bottlenose dolphins, *Tursiops truncates*. J. Comp. Physiol. B **169**, 93–99 (1999)
Ohata, C.A., Miller, I.K.: Northern fur seal thermoregulation: thermal responses to forced activity on land. J. Therm. Biol. **2**, 135–140 (1977a)
Ohata, C.A., Miller, I.K.: Northern fur seal thermoregulation: thermal responses to pelagic conditions. J. Therm. Biol. **2**, 141–143 (1977b)
Scholander, P.F., Schevill, W.E.: Counter-current vascular heat exchange in the fins of whales. J. Appl. Physiol. **8**, 279–282 (1955)
Scholander, P.F., Walters, V., Hock, R., Irving, L.: Body insulation of some arctic and tropical mammals and birds. Biol. Bull. **99**, 225–236 (1950)
Silva, R.G.: Assessment of body surface temperature in cetaceans: an iterative approach. Braz. J. Biol. **64**, 719–724 (2004)
Silva, R.G., LaScala Jr., N., Lima-Filho, A.E., Catharin, M.C.: Respiratory heat loss in the sheep: a comprehensive model. Int. J. Biometeorol. **46**, 136–140 (2002)
Tomilin, A.G.: On the thermal regulation in cetaceans. Priroda **6**, 55–58 (1951)
Watts, P., Hansen, S., Lavigne, D.M.: Models of heat loss by marine mammals: thermoregulation below the zone of irrelevance. J. Theor. Biol. **163**, 505–525 (1993)
Williams, T.M., Noren, D., Berry, P., Estes, J.A., Allison, C., Kirtland, J.: The diving physiology of bottlenose dolphins (*Tursiops truncatus*). III. Thermoregulation at depth. J. Exp. Biol. **202**, 2763–2769 (1999)
Willis, K., Horning, M., Rosen, D.A.S., Trites, A.W.: Spatial variation of heat flux in Steller sea lions: evidence for consistent avenues of heat exchange along the body trunk. J. Exp. Marine Biol. Ecol. **315**, 163–175 (2005)
Worthy, G.A.J.: Insulation and thermal balance of fasting harp and grey seal pups. Comp. Biochem. Physiol. A **100**, 845–851 (1991)
Worthy, G.A.J., Edwards, E.F.: Morphometric and biochemical factors affecting leat loss in a small temperate cetacean (*Phocoena phocoena*) and a small tropical cetacean (*Stenella attenuata*). Physiol. Zool. **63**, 432–442 (1990)

Chapter 6
Shade and Shelter

Abstract Importance of shade for the animals is explained. Factors affecting shade efficiency are described: ground cover and temperature, exposure to the sky and sun elevation and its calculation. Protection is given by a shelter and factors affecting it: position of shelter, height, walls, ground cover and building materials. Calculation and prediction of the shade are given by a shelter. Tree shade and thermal comfort are also discussed. Calculation and prediction of the shade are also given by a tree, according to the shape of its crown.

Keywords Shade • Shelters • Tree shade • Tree type

6.1 Shade and Radiation

6.1.1 Importance of Shade

Shade is the absence of direct solar radiation over a given surface, for example, that of an animal's body. Despite it not affecting the air temperature and not being considered as a protection of that surface against the heat as itself, a shade is a protective measure of utmost importance in tropical regions; it is generally known that animals' performance can be favourably changed if there are shelters or tree shades in the pasture. For example, Hahn (1985) reported an incident occurred in Chino Valley, California, 1977, when more than 700 dairy cows got dead due to an intense heat wave; it was observed that where shelter and shade were available production, losses were 50% and mortality was 1/3 only of those that occurred in locations without shelter or shade.

The first paper on the subject was probably that of Kelly et al. (1950), and since then, several studies had been reported that shading is particularly beneficial for the production of cows of European breeds in hot environments. Roman-Ponce et al. (1977) observed a 10.7% increase in the milk yield of Holstein, Guernsey and

R. Gomes da Silva and A.S. Campos Maia, *Principles of Animal Biometeorology*, Biometeorology 2, DOI 10.1007/978-94-007-5733-2_6,

Jersey cows with access to shade. Ingraham et al. (1979) and Collier et al. (1981) reported similar results with Jersey and Holstein cows.

On the other hand, Baccari et al. (1982) observed 28 dairy cows during summer in Botucatu, SP, Brazil, and did not found significant differences between animals under shade and those exposed to sun; however, those animals were Zebu crossbred ones, which have generally good tolerance to sun. Similar were reported by Buffington et al. (1983), Igono (1986), Silver (1987) and Davison et al. (1988) for Holstein cows. Even in a subtropical region as Florida, animals chronically exposed to direct solar radiation under high air temperatures show decreases of 1.5–3.3 kg per cow per day in its milk yield (Hansen 1990).

Those problems are considered as more serious for the beef cattle, which is generally bred on the range, while dairy cows have access to milk parlours and shelters during at least part of the daytime. McDaniel and Roark (1956) observed in Louisiana that the use of shade in the pastures during summer resulted into increases of the weight gain of Hereford and Angus steers. An experiment by McIlvain and Shoop (1970) showed that Hereford steers with access to shade gained 8.6 kg more than those without shade.

The effect of shade is related to the animal's ability to tolerate intense solar radiation. A study carried out in Australia by Bennett et al. (1985) showed that Shorthorn steers actively seek shade (3.48 h day^{-1} on the average) more than do the Zebu ones (1.64 h day^{-1}), under mean radiant temperatures of 73°C at the sun and 53°C under shade.

The pattern of grazing behaviour can be influenced by the availability of shade. According to Daly (1984), beef cattle on the range in Northern Australia spend the hottest hours of the day under shade and only leave it to seek water; the animals grazed mainly in the latter afternoon or at night. In the rangelands of Oklahoma, USA, where air temperatures can be as high as 46°C in the summer, beef cattle sought shade in the hottest hours (McIlvain and Shoop 1970). Similar results were reported for dairy cattle in Brazil by Paranhos da Costa (1985) in the subtropical region of São Paulo.

In the equatorial region of Ceará, Brazil, the average daily air temperature on the rangeland can be up to 32.2°C, with little change along the year; but the average solar radiation is 640 ± 3.1 W.m^{-2} or even more, and the mean radiant temperature is 65°C or even higher (Silva et al. 2010); therefore, cattle is often maintained there under shelter during the day, while those animals grazing on the range seek after any available shade. See Fig. 6.1.

Schütz et al. (2010) observed the behaviour of Holstein cows in a location near Hamilton, New Zealand (37°47′S, 175°19E), under three treatments: no shade, small shade (2.4 m^2 per cow) and large shade (9.6 m^2 per cow). The results showed that cows with access to large shade spent more than twice as much time under shade, as compared to cows with access to small shade. The cows spent more time in shade and less time lying with increasing heat load; in addition, respiration rate was higher when there had less shade available. These results were obtained under the following average values for the period from 1,000 to 1,550 h: solar radiation 704 W m^{-2}, air temperature 22°C and black globe temperature 30°C.

Fig. 6.1 Crossbred Holstein cows seek after shade during the hottest hours of the day in the equatorial region of Ceará, Brazil (Photo by R.G.S.)

Blackshaw and Blackshaw (1994) carried out a comprehensive review about the effects of shade on production and behaviour of cattle, which is recommended for the interested reader.

6.1.2 Factors Affecting Shade Efficiency

6.1.2.1 Effect of the Ground Cover

A plant-covered ground surface is insulated against an excessive heating by the direct solar radiation and has generally a lower reflectance for this a radiation. As a result, the radiant heat load over a grass field was about 270 W m^{-2} lower than that of a bare ground surface (Kelly et al. 1950); therefore, the animals within a shelter or shade can receive less thermal radiation if the surrounding ground is covered with grass. However, if there has a building wall near the shelter or tree, depending on its reflectance, such wall will add more to the heat load of animals under shade.

The fractions of incident radiation which are transmitted and reflected by plant leaves depend on the zenith angle of sun (θ), especially that from 50° to 90°. As in the plant leaves, the absorption of the green light (0.492- to 0.535-μm wavelength) is lower than that of the blue light (0.424–0.492 μm) or the red one (0.647–0.700 μm); then the radiation reflected by healthy leaves is basically that of the green band, constituting 10–20% of the total incident solar radiation (Monteith and Unsworth 2008). Leaves are excellent emitters of the long-wave infrared radiation, with emissivity from 0.94 to 0.98 in that band. In another way, several studies referred by Jones (1983) showed that greater reflectance values are

Table 6.1 Area of shade per animal, as recorded in some studies with cattle

m^2 per animal	Animal	References
2.3–5.6	Beef cattle	Ittner et al. (1954)
5.6	Cattle	Bond et al. (1958)
2.8	Beef cattle	McIlvain and Shoop (1970)
3.5–5.6	Dairy cows	Buffington et al. (1983)
1.8–2.5	Cattle	Hahn (1985)
3.5	Beef cattle	Mader et al. (1997)
2.4–9.6	Dairy cows	Schütz et al. (2010)

presented by those leaves of light colour that are highly pubescent and have low humidity content or a wax layer on the surface. Pubescence makes the leaf surface reflectance greater.

6.1.2.2 Amount of Available Shade

There were published few studies on the effects of the amount of shade in controlled experiments, and most of them focused on cattle production traits, as milk yield or body weight gain. A recent study by Schütz et al. (2010) revealed that even in the conditions of a temperate climate as that of New Zealand, cows spend more time in the shade when they have access to shade greater enough to provide thermal comfort. These authors suggested a shade area of 9.6 m^2 per cow.

Table 6.1 shows the values for the area of shade recorded in some studies and presented as desirable figures.

However, in tropical regions, solar radiation can be very strong, and the mean radiant temperature of the environment is often as high as 65°C or more (Silva et al. 2010); animals exposed to such conditions need access to shade areas wide enough to shelter all the animals without crowding.

Hahn (1985), for example, suggested values from 1.8 to 2.5 m^2 per animal. In our personal observations in Brazil, such values resulted into an excessive crowding of the animals and caused increased thermal discomfort because the amount of heat exchange by radiation among animals and problems with heat loss by convection. In fact, assuming that a cow lying down on the ground occupies an approximate area of 1.8 m^2, the numbers given by Hahn (1985) and McIlvain and Shoop (1970) are inadequate for a tropical environment because animals would be very close one to another.

On the other hand, a shade area of 5.6 m^2 (about three times greater) would suffice because it results in an approximate distance of 0.5 m between animals, while a shade area of 9.6 m^2 per cow will give a space of about 1 m around each one.

Therefore, when planning a shelter for cattle in a tropical region, it is important to take into consideration a shade area great enough to be used simultaneously for all or at least most of the animals, with an occupation density of about 5–6 m^2 per animal. Tree planting for shade must be done with the same caution in order to provide enough shade area for all the animals in a pen.

6.1.2.3 Other Factors

The effective shading provided by a shelter depends on the solar radiation, the open sky, the ground surface, the shelter structure and the several surfaces around the shelter. As it was previously discussed in other chapters, those real or virtual surfaces around an animal emit thermal energy by radiation whose intensity depends on their temperature and emissivity. At the same time, those surfaces receive thermal radiation. The amount of thermal radiation exchanged among those surfaces and an animal is related to the difference between the energies intercepted by the animal and the surrounding surfaces. Therefore, the position of animal's body in relation to those surfaces must be taken into account.

In short, the radiative environment of an animal under a shelter can be divided into five parts:

(a) Cold ground under shade
(b) Hot ground out of the shade
(c) Lower surface of the shelter roof
(d) The sky up to 10° above horizon
(e) The rest of the open sky that is not blocked by the roof

6.1.3 Sun Elevation and Shade

In order to quantify the shade given by a tree or shelter, it is necessary to know the incidence angle of solar radiation. This angle is evaluated with respect to the horizon (elevation angle of the sun, β) or with respect to a perpendicular line from the zenith point (90° in relation to the horizon) to the ground surface (zenith angle of the sun, θ), as it is shown in Fig. 6.2.

As for the calculations, the following data are needed: date, time of day, latitude (L_t) and longitude (L_g); these two last values are in decimal angles. Zenith angle of sun is calculated by using Eqs. 1.2, 1.3 and 1.4. Elevation angle is given as $\beta = 90 - \theta$, according to Fig. 6.2.

The azimuth angle is calculated with respect to the south by using the formula:

$$\omega = 180 - \arcsin\left[-\frac{\cos(\delta)\sin(\eta)}{\sin(\theta)}\right] \text{ degrees} \qquad (6.1)$$

This angle shows the position of the sun as it is displaced counterclockwise over a horizontal plane in relation to a given point. With those data in hands, one can calculate the area of the shade given by a tree or shelter in a given place, a given date and a given time of the day.

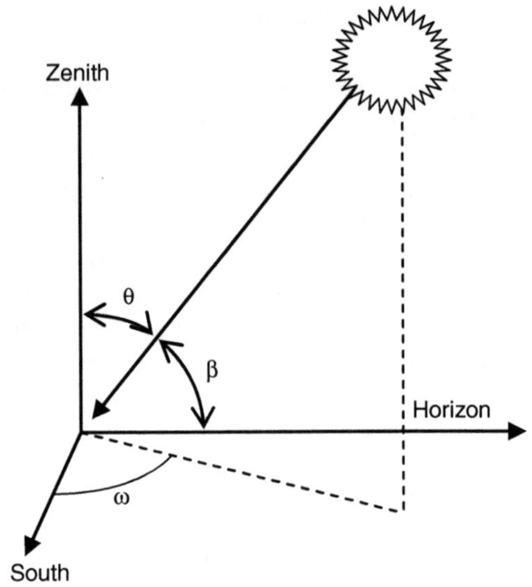

Fig. 6.2 *Angles of the sun: zenith (θ), elevation (β) and azimuth (ω)*

6.2 Shelters

6.2.1 Factors to Be Considered

The utility of a shelter, as well as the protection it gives to the animals, is primarily related to the shade area that can be effectively occupied by a given number of animals without lack or excess. The use of shelters can reduce the radiant heat load on the animals as much as 30%, thus favouring the thermal comfort of animals (Bond and Laster 1975; Blackshaw and Blackshaw 1994; Eigenberg et al. 2005).

However, the effect of a shelter is barely measurable indirectly through the response of the animals in terms of rate of weight gain, milk yield or food conversion, for animal organisms present great functional diversity in order to face environmental challenges. Therefore, the environmental conditions given by a shelter must be directly evaluated, and the factors associated to such an evaluation are the following, according to Kelly et al. (1950) and Bond and Kelly (1955):

(a) Azimuth orientation of the building axis
(b) Roof height
(c) Size of the animals
(d) Walls
(e) Vegetation around the shelter
(f) Building materials

Those factors will be discussed as follows.

6.2.1.1 Shelter Orientation

Shelters with the longitudinal axis oriented in the north–south direction are generally advantageous because they can be maintained dry more easily; however, they have higher RHL values because the direct sun received by the one sidewall the day along; heating will be much lower if there have no walls at all. This is a very old problem, as it can be easily deduced from the words of the Roman architect Marcus Vitruvius Pollio (first century BC): *"Quod spectat ad occidentam sole exorto tepexit meridie calet vespere fervet"* (buildings facing west are warm in the morning, hot at noon and burning by evening).

Therefore, an east–west orientation of the building axis would be the most advantageous choice in low-latitude regions.

6.2.1.2 Shelter Height

As the roof height increases, the animals within a non-walled shelter are exposed to a wider area of open sky while in the shade. The open sky can be assumed as a virtual surface whose temperature is lower (and often below 0°C, even in tropical regions) than that of other portions of the environment; in such a case, there has an increased heat loss by radiation from the skin surface of the animals to the open sky.

Kelly et al. (1950) tested hay-covered shelters with 2.1 and 3.5 m height and observed that all the cows remained under the higher shelter. It was concluded that most of the benefit, as far as animal comfort is concerned, should be ascribed to the reduced radiation heat load because of more exposure to the open sky.

6.2.1.3 Size of the Animals

The RHL measured within a shelter varies in relation to the height above ground. This is an important fact to remember because a pig or a chicken placed in the shade at the centre of the building will receive an amount of thermal energy which is lower than that received by a cow or a horse in the same place and position; this is because of the relative distance of the body with respect to the roof. It has been observed that chickens maintained into cages high above the floor presented greater thermal discomfort than the ones into cages close to the floor in the same shed.

6.2.1.4 Wall Effect

Walls preclude entering to the shelter of outside thermal radiation, but that advantage is reduced by the decrease in the ventilation, by the increased heat coming from internal radiation sources (the animals themselves) and also by the possibility the wall material can absorb thermal energy from the outside and transmit it to the inside environment.

Walls whose internal surface is covered with some reflecting material tend to give back the thermal radiation emitted by the animals; this is the explanation for the fact that most climatic chambers are internally covered with polished aluminium. The same effect is observed with respect to an animal standing close to the external wall of a building (Bond et al. 1969). Those authors observed that a building wall does present an important source of thermal radiation on a sunny day that can add significantly to the RHL of an animal placed nearby; an animal standing 15 m away from a 3 × 12-m wall could receive as much as 44.2 $W.m^{-2}$ of thermal energy additional from the wall. Although there has no data at this respect from tropical regions, such a reflection effect would be probably more intense in them.

As a rule for tropical climates, shelters must have no walls at all in order to permit free ventilation. In that case, there has the mentioned problem of thermal radiation entering the shelter, as it was above discussed. However, this problem may be attenuated by planting vegetation around the building.

6.2.1.5 Vegetation Around the Shelter

It is evident that the RHL within a shelter will be greater if it is built over a bare ground than if it is on the grass. In the case of a bare ground, the shelter must be ventilated inside and have walls to preclude the thermal radiation coming from outside, but they need an adequate thermal insulation. When it is possible, grass would be planted around any shelter or animal facility in order to reduce reflection of solar radiation into it.

6.2.1.6 Building Materials

The materials used for the shelter roof must have high reflectance coefficients; in some ways, it is quite a good idea to paint the upper surface of the roof with an acrylic white. In addition, the roof material must be of low thermal conductivity in order to reduce the transmission of heat into the shelter. The structure must be favourable to the loss of heat by convection. Bond and Kelly (1955), Kelly and Bond (1958) and Bond et al. (1961) tested several materials for cattle shelters, and some of the results are presented in Table 6.2.

In many tropical regions, the thatch is traditionally used as roof material; its main advantages are the low cost and the excellent thermal insulation, but its life is short, and it may act as a refuge for insects, rats and even snakes. A study carried out in California by Kelly and Bond (1958) showed thatch as the best roof material as for the thermal comfort.

Those classical studies showed that a thick layer of hay (or dried grass or palm leaves) would be the best roof material as for the thermal comfort. Undulated new,

Table 6.2 Radiant heat load (RHL) measured within shelters with different roof materials

Roof material	Conditions	RHL (W m^{-2})
Thatch	15 cm thick	473.2
Aluminium	Outside white, inside black	504.7
Aluminium	Outside polished, inside black	507.9
Aluminium	New, polished	536.3
Aluminium	Oxidised, non-painted	545.7
Aluminium	Outside white, inside polished	520.5
Wood chipboard	Natural	526.8
Wood chipboard	Outside white, inside natural	526.8
Asbestos cement	Natural	548.9
Galvanised iron sheet	New, non-painted	539.4
Galvanised iron sheet	Oxidised, non-painted	542.6
Galvanised iron sheet	Outside white, inside black	514.2
Galvanised iron sheet	Outside white, inside natural	520.5

Adapted from Kelly and Bond (1958)

polished aluminium sheet gives a result inferior to that of the same material but painted white in its outside and black in its inside face. Similar results were observed for galvanised iron sheet. On the other hand, the use of both aluminium and galvanised iron sheets results in quite higher RHL values when oxidised.

Asbestos cement was considered among the worst roof materials in its natural condition, that is, non-painted. However, Panelli and Silva (1989) observed in Jaboticabal, SP, Brazil, that shelters with a roof made of this material but painted with acrylic white in the outside and matt black in the inside face presented RHL values significantly lower than the shelters covered by non-painted asbestos. The advantage of a black inside face is the absence of reflection; then, the upward thermal radiation is not reflected upon the animals inside the shelter. In addition, the matt-black surface absorbs the thermal energy emitted by the animals, favouring their heat losses. On the other hand, if the roof internal surface temperature is as high as that of the external surface – which is likely to happen if the roof material has a high thermal conductivity – then the matt-black painting would leave to increased thermal radiation upon the animals. In order to avoid such a problem, an additional layer of insulating material can be applied on the inside face of the roof; for example, there is easily available an undulated polished aluminium sheet with an additional polystyrene layer.

Polished aluminium with either natural or white-painted surface has the same reflectance value (0.74–0.75, see Table 6.3), but emissivity of the acrylic white paint that is often used is much greater (0.90) than that of the polished aluminium (0.05). Therefore, the outside roof surface that is white painted reflects almost the same amount of incident radiation as the polished one; however, while this last surface can be heated without emitting most of the stored thermal energy, the white-painted aluminium can emit much of that energy, so reducing the inside surface temperature.

Table 6.3 Emissivity (ε) and solar radiation absorptance (α_s) for some surfaces

	Material	α_s	ε	α_s/ε
Metals	Aluminium, polished	0.26	0.05	5.20
	Aluminium, commercial new	0.32	0.10	3.20
	Chrome	0.49	0.08	6.13
	Inoxidable steel type 410	0.76	0.16	4.75
	Copper, polished	0.18	0.04	4.50
	Galvanised iron, new	0.65	0.13	5.00
	Galvanised iron, oxidised	0.80	0.28	2.86
Paints	Epoxy or acrylic, white	0.25	0.90	0.28
	Epoxy or acrylic, matt black	0.85	0.91	0.93
	Aluminium	0.20	0.43	0.47
	Graphite	0.78	0.41	1.90
	Carbon black	0.98	0.95	1.03
Plants	Dried grass	0.68	0.90	0.76
	Green grass	0.67	0.98	0.68
	Tree leaves	0.75	0.93	0.81
Ground	Dry, ploughed soil	0.78	0.90	0.87
	Black, humid soil	0.90	0.95	0.95
	Brown, dry soil	0.68	0.95	0.72
	Brown, humid soil	0.84	0.90	0.93
	White sand	0.60	0.90	0.67
	Concrete	0.60	0.88	0.68
	Asphalt, clean	0.93	0.94	0.99
Diverse	Asbestos cement, plate	0.59	0.90	0.66
	Glass	0.93	0.89	1.04
	Common brick	0.55	0.93	0.59

Adapted from Chapman (1987) and Esmay (1969)

After a long time without significant progress about this matter, the new evidences on climatic changes have induced more studies on shelter for livestock (Blackshaw and Blackshaw 1994; Brown-Brandl et al. 2005; Eigenberg et al. 2005, 2007, 2010). For example, Eigenberg et al. (2010) evaluated during the summer in Nebraska, USA, the effect of using polyethylene cloth in shade structures for feedlot cattle, taking into account three protection levels: (a) 100%, (b) 60% with a silver reflective coating and (c) 60% black without coating. Type (a) presented the best results, with black globe temperatures lower (maximum 36°C) than those for the other types, 38 and 39°C for (b) and (c), respectively. The globe temperature recorded under the open sky was up to 43°C, quite lower than those generally found in tropical regions; Silva et al. (2010) observed black globe temperatures of 65°C or higher under the sun in Ceará, Brazil.

The results of the above-cited studies show that the relation between absorptance and emissivity is an important property of the surfaces exposed to direct solar radiation. A high α_s/ε_s value is desirable for a surface that must be able to absorb thermal energy, while the inner surface of the roof of a shelter must have an α_s/ε_s value as low as possible under tropical conditions.

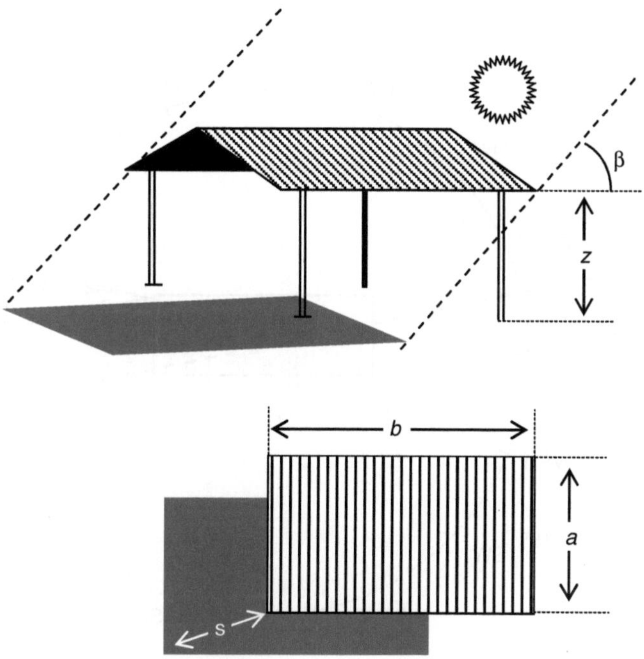

Fig. 6.3 Schematic design of the shade projected by a shelter without walls. a = roof width, b = roof length, z = height and β = angle of de elevation of the sun

6.2.2 *Shade Prediction of a Shelter*

Figure 6.3 illustrates the problem. According to the day of the year, the geographic location and day time, the position of the sun is defined by the azimuth ω

$$\omega = 180 - \arcsin\left[-\frac{\cos(\delta)\sin(\eta)}{\sin(\theta)}\right] \text{ degrees} \tag{6.1}$$

the zenith angle θ, the elevation angle above horizon $\beta = 90 - \theta$, the sun declination δ, and the hour angle of the sun η (Eqs. 1.2, 1.3 and 1.4a). According to a given position of the sun, the shade produced by the shelter roof is displaced to a distance s relatively to the roof position, in a direction given by the sun azimuth.

For a shelter without walls whose roof has a length b, a width a and a height z (all dimensions in metres), shade displacement will be given as

$$s = \frac{z}{\tan\beta} \text{ m} \tag{6.2}$$

and the shade area as:

$$A_s = a \cdot b \qquad \text{m}^2 \tag{6.3}$$

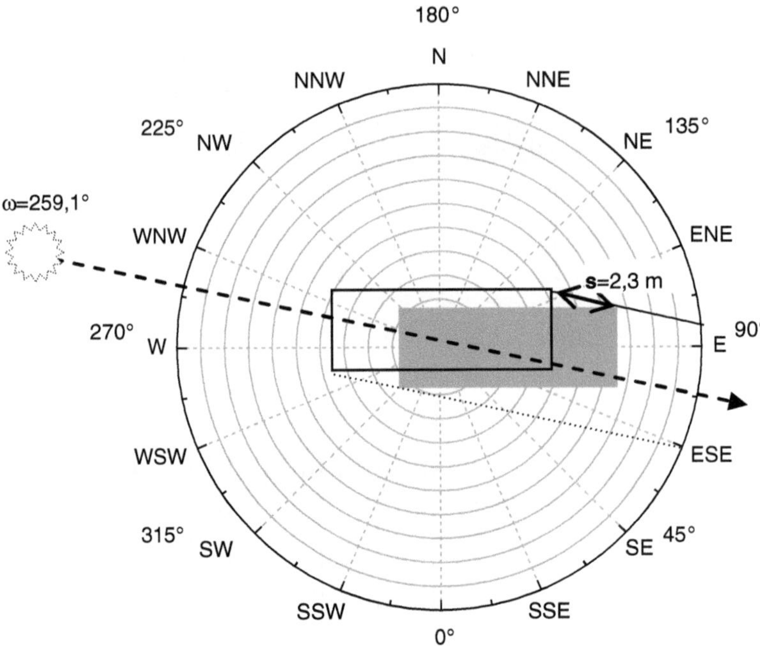

Fig. 6.4 Shade position of a shelter with a roof 8 m length, 3.5 m wide and 3 m high, in a location at 21°35′ south latitude and 48°54′ west longitude in December 22 at 15:00 h

Once obtained the azimuth ω, it is plotted on a polar graph divided into sections of 45° corresponding to the cardinal points; the degrees must be counted counter-clockwise from the south, as it is illustrated by the example in Fig. 6.4.

The next step is to plot the roof contours on the graph, according to the orientation of its axis. The shade is then plotted according to its displacement s and direction, remembering that its area is equal to that of the roof. The graph given by Fig. 6.4 can be designed for several situations of a shelter along the year and the time of day in order to take a notion about the shade behaviour.

***Example*:**

Suppose a shelter without walls with a roof 8 m length, 3.5 m wide and 3 m high, whose greater axis is east–west oriented. Considering that it is in a location at 21°35′ south latitude and 48°54′ west longitude, calculate the shade that is projected in December 22 (southern summer peak) at 15:00 h.

Data:

$$L_t = -21°35' = -21.583333° \quad L_g = 48°54' = 48.9°$$

$$a = 3.5\ \text{m} \quad b = 8\ \text{m} \quad z = 3\ \text{m}$$

$$d = \text{December } 22 = 356\text{th day of the year}$$

$$\begin{aligned}\delta &= 23.45 \text{ sen}\left[\left(\frac{360}{365.242}\right)(284+d)\right]\\ &= 23.45 \text{ sin}\left[\left(\frac{360}{365.242}\right)(284+356)\right] = -23.44763^\circ\end{aligned}$$

$$\eta = 15\left(h - 12 + \frac{\mathrm{M_h} - L_\mathrm{g}}{15}\right) = 15\left(15 - 12 + \frac{45 - 48.9}{15}\right) = 41.1^\circ$$

$$\cos\,\theta = (\cos L_\mathrm{t})(\cos\delta)(\cos\eta) + (\sin L_\mathrm{t})(\sin\delta) = 0.789236$$

$$\theta = \arccos(0.789236) = 37.88583^\circ$$

$$\beta = 90 - \theta = 90 - 37.88583 = 52.11417^\circ$$

$$\begin{aligned}\omega &= 180 - \arcsin\left[-\frac{\cos(\delta)\sin(\eta)}{\sin(\theta)}\right]\\ &= 180 - \arcsin\left[-\frac{\cos(-23.44763)\text{sen}(41.1)}{\sin(37.88583)}\right] = 259.1^\circ\end{aligned}$$

$$s = \frac{z}{\tan\beta} = \frac{3}{\tan(52.11417)} = 2.3 \text{ m}$$

$$A_\mathrm{s} = a \cdot b = (3.5)(8) = 28 \text{ m}^2$$

6.3 Tree Shades

6.3.1 *Vegetation and Thermal Comfort*

An important effect of the plants is the thermal insulation they give to the soil, thus reducing its heating by the solar radiation; in addition, the reflectance of the ground surface is reduced. In another way, ground heating is a significant factor for the plant physiology and for the support of microbial flora and fauna equilibrium which is of so great importance for the vegetation.

On the other hand, the low reflectance of the plant surfaces is an important effect upon the thermal comfort of animals. For example, a grass-covered ground around a shelter results into a RHL that is 270 W m^{-2} lower into the shelter, in relation to the figure observed in the case of a bare ground (Kelly et al. 1950). The fractions of the

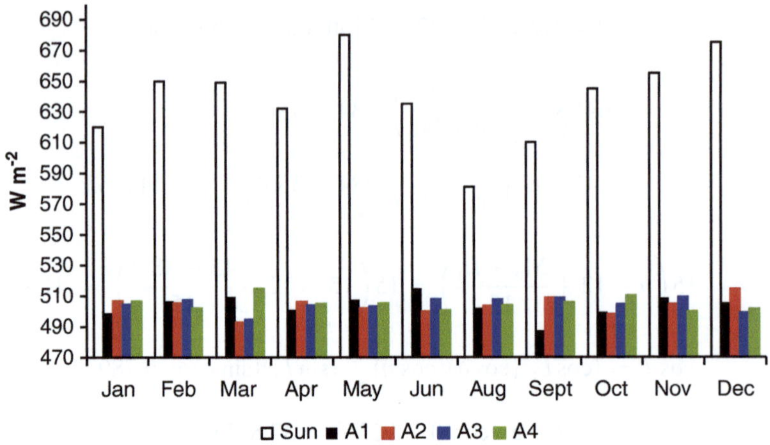

Fig. 6.5 Radiant heat load measured at the sun and under the shade of four tree types: *A1* = mango tree (*Mangifera indica*), high, wide crown and dense shade; *A2* = mango tree, low-sized, dense shade; *A3* = *Casuarina* sp., *Aleurite mollucana* and *Caesalpinia peltophoroides*, all of them with high, wide, sparse crown and sparse shade; and *A4* = *Pinus* sp., high, narrow crown and long, low-density shade (Graph based on the data from Waldige 1994)

radiation either transmitted or reflected by the plant leaves depend on the angle of incidence of solar radiation, especially the zenith angles from 50° to 90°.

Despite the protection against the direct solar radiation, the RHL received by an animal under the shade of a tree can be of significance. Figure 6.5 summarises the results of a study carried out along a year in Jaboticabal, SP (Waldige 1994). In this study, the differences among the four tree types were statistically significant ($P < 0.01$), and the respective average values were quite lower than the corresponding RHL values recorded under the sun. As the ground was grass covered in all the cases, the upward thermal radiation from the ground was quite lower than the downward radiation coming from the sky. The grass was of the bermuda variety (*Cynodon dactylon*). See Figs. 6.6, 6.7, and 6.8.

In a study carried out by Alcântara et al. (1978) in Nova Odessa, SP, Brazil, there was observed that Napier grass (*Pennisetum purpureum*, cv. Taiwan A-144) produced a significantly higher amount of thermal energy than other forage species observed at the same time in the same place. Despite the differences among species were not statistically significant, the amount of radiation measured at 1 m above the ground was directly associated with plant height, but depending on leaves' colour. A similar study was done by Reis (1984), who compared five different forage types during 12 months; the results showed significant differences among the forages with respect to the RHL measured 1 m above the ground. See Table 6.4.

6.3.2 Tree Shade Prediction

Though there were several papers in the literature about shade prediction for man-built shelters, two references only were found as for the tree shade prediction; the

Fig. 6.6 The dense shade projected by a mango tree, *Mangifera indica*

Fig. 6.7 The wide, sparse shade of a *Caesalpinia peltophoroides*. Note the *black globe* above ground in the centre of the *shade*

first was the mention in the book of Monteith and Unsworth (2008) about the shade projected by conic-shaped trees (*Taxus baccata*), and the second is the paper by Silva (2006).

Aiming to the determination of the form, position and area of a tree shade, there were needed the tree dimensions and the approximate form of its crown. Silva (2006) presented formulas for trees with spherical, lentil-shaped, cylindrical, conic, inverse conic and ellipsoid crown. The following notes will take into consideration the lentil-shaped, cylindrical and inverse conic types, which can be applied to most part of the tree species found in tropical regions.

Fig. 6.8 The long, narrow, sparse shade of a *Pinus* sp.

Table 6.4 Radiant heat load (*RHL*) measured 1 m above the ground covered by five different forage types during 1 year (Reis 1984)

Grass type	Radiant heat load $W.m^{-2}$
Brachiaria humidicola	570.02
Brachiaria decumbens	572.05
Cynodon dactylon, cv. coast cross	565.71
Cynodon dactylon, cv. coastal bermuda	562.32
Melinis minutiflora	563.31

6.3.2.1 Lentil-Shaped Crown

The crown is round but flat like a lentil grain as it is shown in Fig. 6.9. The respective equations were the following:

Shade area:

$$A_s = \frac{\pi r R}{\sin\beta}\ \text{m}^2 \tag{6.4}$$

after Silva (2006) and where z is the height of the tree, r is the smaller radius (vertical) and R the greater radius (horizontal) of the crown, y is the trunk height and β is the angle of the sun elevation above horizon.

Shade length:

$$c = 0.6(z-y)\left(\frac{1}{\sin\beta} - 1\right) + 2R\quad \text{m} \tag{6.5}$$

Distance of shade from the trunk:

$$s = \frac{z-r}{\tan\beta} + 0.3(z-y)\left(\frac{1}{\sin\beta} - 1\right) + R\quad \text{m} \tag{6.6}$$

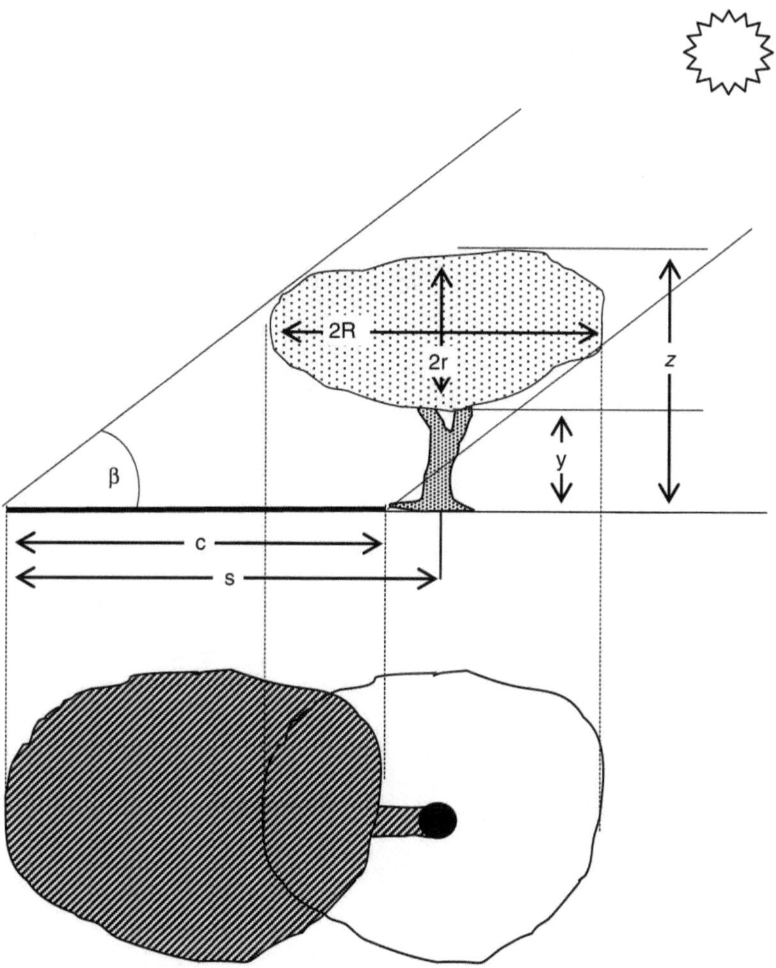

Fig. 6.9 Schematic design of the shade projected by a tree with lentil-shaped crown. Symbols: R = greater radius, r = smaller radius, c = shade length, s = distance of shade from the trunk, z = total tree height, y = trunk height and β = angle of sun elevation above horizon

6.3.2.2 Inverse Conic Crown

There has a quite great number of tree species in tropical rangelands, especially those of the *Leguminosae* family, with high, wide crowns that are inverse-conic shaped (Fig. 6.10).

Shade length:

$$c = \frac{z - y}{\tan \beta} + r \text{ m} \tag{6.7}$$

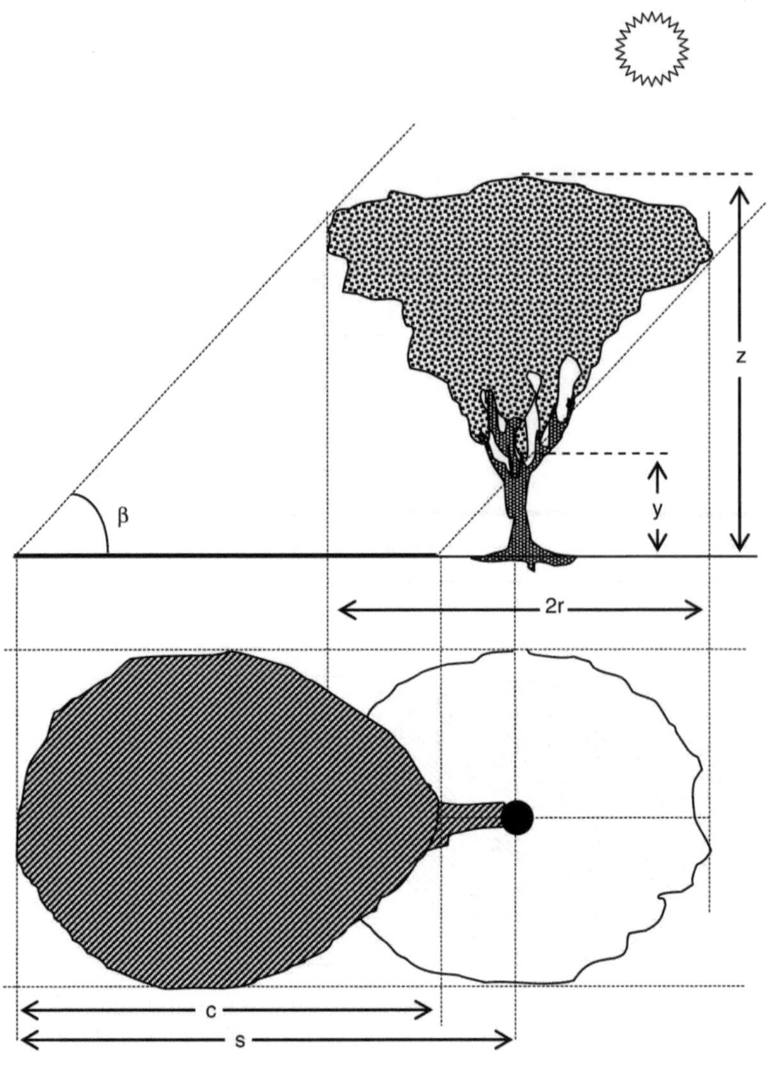

Fig. 6.10 Schematic design of the shade projected by a tree with inverse conic crown. Symbols: r = radius of the crown c = shade length, s = distance of shade from the trunk, z = total tree height, y = trunk height and β = angle of sun elevation above horizon

Distance of shade from the trunk:

$$s = \frac{z}{\tan\beta} \quad \text{m} \tag{6.8}$$

Shade area:

$$A_s = 2\sqrt{x(x-a)(x-b)(x-e)} + \pi r^2 - \beta r^2 - d^2 \tan\beta \quad \text{m}^2 \tag{6.9}$$

where:

$d = r \cos \beta$

$a = c - r - d$

$b = d \ \tan \beta$

$e = \sqrt{b^2 + r^2}$

$x = 0.5(a + b + e)$

6.3.2.3 Cylindrical Crown

Several species of Eucalyptus and conifers have narrow, long crowns that are sometimes like cylinders with rounded ends, as it is shown in Fig. 6.11.

Shade length:

$$c = \frac{2r}{\sin \beta} + \frac{z - y - 2r}{\tan \beta} \quad \text{m} \tag{6.10}$$

Distance of shade from the trunk:

$$s = \frac{z - r}{\tan \beta} + \frac{r}{\sin \beta} \quad \text{m} \tag{6.11}$$

Shade area:

$$A_s = \frac{\pi r^2}{\sin \beta} + \frac{2r(z - y - 2r)}{\tan \beta} \quad \text{m}^2 \tag{6.12}$$

6.3.3 Qualitative Aspects of the Shade

6.3.3.1 Problems Related to Fruits

Trees with very dense foliage offer denser shades and better protection against solar radiation; however, such trees are not always the best choice for pasture shading because their fruits. It is the case of the mango tree (*Mangifera indica*), which presents one of the better shades but that is not adequate for cattle shade; its fruits are easily ingested by a cow and block up its oesophagus, causing acute meteorism or tympanites which leaves often to death by cardiac arrest if the animal is not

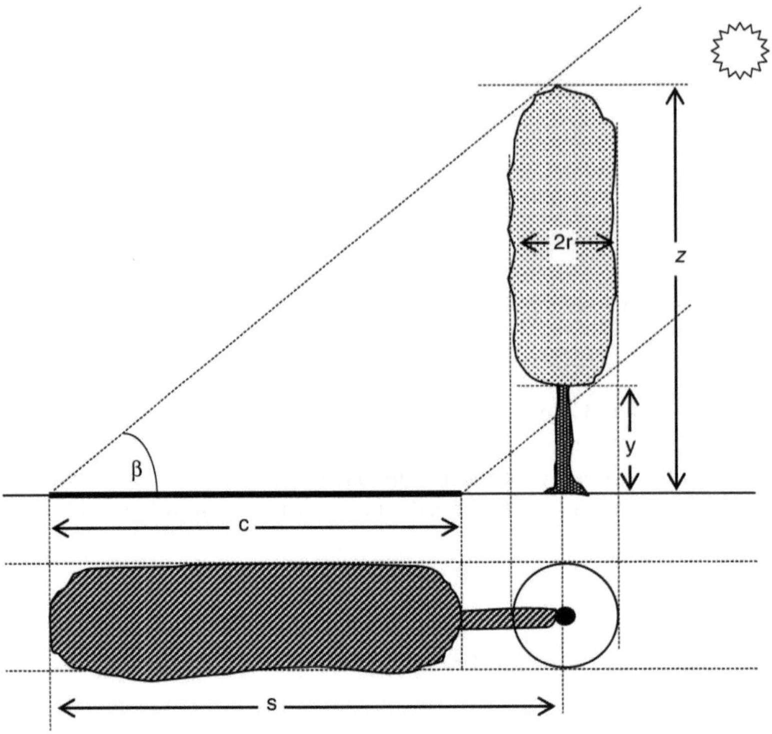

Fig. 6.11 Schematic design of the shade projected by a tree with cylindrical crown. Symbols: r = radius of the crown, y = trunk height, z = total tree height, c = shade length, s = distance of shade from the trunk and β = angle of sun elevation above horizon

promptly aided. Orange trees present the same problem, but they are more frequent around the farmhouses and rarely found in pastures.

6.3.3.2 Effective Protection Given by the Shade

Beyond the dimension and position aspects of the shade, it is important to take into consideration the effective protection given by any shade. The radiant heat load must be evaluated always in the centre of the shade and compared to that RHL measured near the tree in the sun. It is clear that as lower the RHL under the shade in relation to that in the sun, as better will be the tree as a shelter.

Finally, with respect to the shade area needed for adult cattle, several different values had been found in the literature. Buffington et al. (1983) proposed that at least 4.2 m^2 per animal be used; however, they are also in accordance with Bond et al. (1958) that 5.6 m^2 per animal would be a more desirable figure.

In some ways, it would be interesting to evaluate the trees available in a given pasture with respect to the total area of shade they give in order to be sure that all the animals can use the shade without problems. If the actual shade area is not enough, then it is the case to plant additional trees of an adequate type, which can give enough shade when fully grown.

6.4 Problems

Problem 6.1. A shelter without walls was built in a pasture at 25°30′ south latitude and 48°53′ west longitude with the following dimensions: 3.5 m height, 8.6-m roof length and 4.2-m roof width. The building axis runs in the northeast-southwest direction. Determine the shade projected by the shelter in January 10 at 14:00 h.

Data:

L_t = latitude = 25°30′ south = −25.5°
L_g = longitude = 48°53′ west = 48.883333°
d = January 10 = 10th day of the year
a = roof width = 4.2 m
b = roof length = 8.6 m
z = total height = 3.5 m

Sun angle of elevation above horizon:

$$\delta = 23.45 \sin\left[\left(\frac{360}{365.242}\right)(284 + d)\right]$$
$$= 23.45 \sin\left[\left(\frac{360}{365.242}\right)(284 + 10)\right] = -22.06636^\circ$$

$$\eta = 15\left(h - 12 + \frac{M_h - L_g}{15}\right) = 15\left(14 - 12 + \frac{45 - 48.88333}{15}\right) = 26.116667^\circ$$

$$\cos\theta = (\cos L_t)(\cos\delta)(\cos\eta) + (\sin L_t)(\sin\delta)$$
$$= \cos(-25.5)\cos(-22.06636)\cos(26.116667) + \sin(-25.5)\sin(-22.06636)$$
$$= 0.912801$$

$$\theta = \arccos(0.912801) = 24.1^\circ$$

$$\beta = 90 - \theta = 90 - 24.1 = 65.9^\circ$$

Calculation of azimuth with respect to south:

$$\omega = 180 - \arcsin\left[-\frac{\cos(\delta)\sin(\eta)}{\sin(\theta)}\right]$$
$$= 180 - \arcsin\left[-\frac{\cos(-22.06636)\sin(26.116667)}{\sin(24.1)}\right] = 267.5^\circ$$

Distance of the shade from the roof line:

$$s = \frac{z}{\tan\beta} = \frac{3.5}{\tan(65.9)} = 1.6\ \text{m}$$

$$A_s = a \cdot b = (4.2)(8.6) = 36.1\ \text{m}^2$$

Shade positioning:

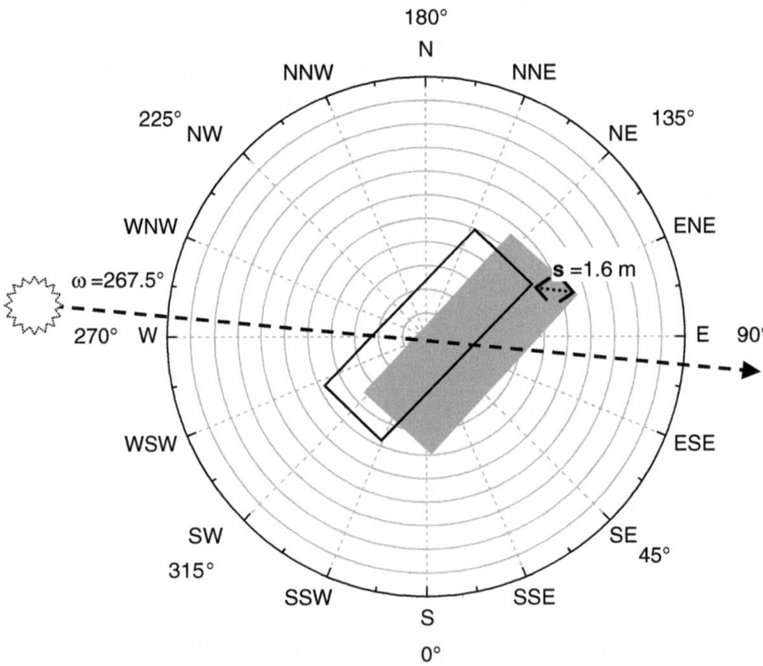

Problem 6.2. Let us have a tree with 6.7 m total height and a lentil-shaped crown 8 m in diameter, placed in a location at 5°31′ south latitude and 38°13′ west longitude. Its trunk is 2 m high. Calculate the shade projected at 15:30 h in June 21 of a non-leap year.

Data:

L_t = latitude = $-5°31' = -5.516667°$
L_g = longitude = $38°13' = 38.216667°$
d = June 21 = 172th day of the year
h = local day time = 15:30 h = 15.5 h
z = total height = 6.7 m
y = trunk height = 2 m
$2R$ = greater crown diameter (horizontal) = 8 m
R = 4 m
$2r$ = smaller crown diameter (vertical) = $z - y = 6.7 - 2.0 = 4.7$ m
r = 2.35 m

Angle of sun elevation above horizon:

$$\begin{aligned}\delta &= 23.45\sin\left[\left(\frac{360}{365.242}\right)(284+d)\right]\\ &= 23.45\sin\left[\left(\frac{360}{365.242}\right)(284+172)\right] = 23.448941^\circ\end{aligned}$$

$$\eta = 15\left(h - 12 + \frac{M_{\mathrm{h}} - L_{\mathrm{g}}}{15}\right) = 15\left(15.5 - 12 + \frac{45 - 38.216667}{15}\right) = 59.283333^\circ$$

$$\begin{aligned}\cos\theta &= (\cos L_{\mathrm{t}})(\cos\delta)(\cos\eta) + (\sin L_{\mathrm{t}})(\sin\delta)\\ &= \cos(-5.516667)\cos(23.448941)\cos(59.283333)\\ &\quad + \sin(-5.516667)\sin(23.448941)\\ &= 0.428183\end{aligned}$$

$$\theta = \arccos(0.428183) = 64.65^\circ$$

$$\beta = 90 - \theta = 90 - 24.1 = 25.35^\circ$$

Azimuth:

$$\begin{aligned}\omega &= 180 - \arcsin\left[-\frac{\cos(\delta)\sin(\eta)}{\sin(\theta)}\right]\\ &= 180 - \arcsin\left[-\frac{\cos(23.448941)\sin(59.283333)}{\sin(64.65)}\right] = 241^\circ\end{aligned}$$

Shade area:

$$A_{\mathrm{s}} = \frac{\pi r R}{\sin\beta} = \frac{\pi(2.35)(4)}{\sin(25.35)} = 68.9\ \mathrm{m}^2$$

Shade length:

$$\begin{aligned}\mathrm{c} &= 0.6(z-y)\left(\frac{1}{\sin\beta} - 1\right) + 2R\\ &= 0.6(6.7-2)\left(\frac{1}{\sin(25.35)} - 1\right) + 2(4) = 11.8\ \mathrm{m}\end{aligned}$$

Distance of shade from the trunk:

$$s = \frac{z-r}{\tan\beta} + 0.3(z-y)\left(\frac{1}{\sin\beta} - 1\right) + R$$
$$= \frac{6.7-2.35}{\tan(25.35)} + 0.3(6.7-2)\left(\frac{1}{\mathrm{sen}(25.35)} - 1\right) + 4 = 15.1\ \mathrm{m}$$

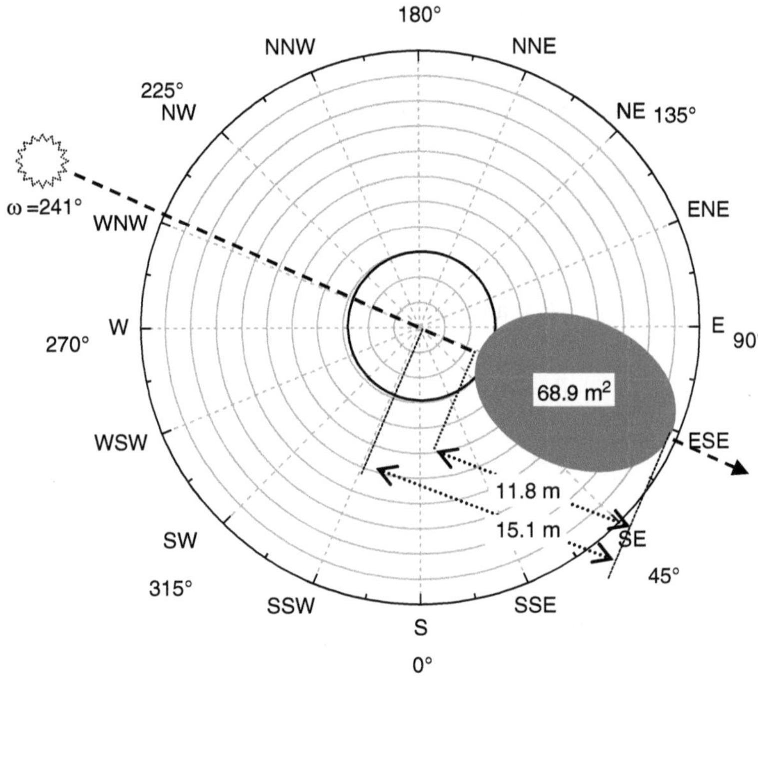

References

Alcântara, P.B., Ghisi, O.M.A.A., Kuhn-Neto, J., Alcântara, V.B.G., França, A.E.S.: Influência da espécie forrageira no incremento de calor por radiação em pastagens. Non-published report to Post-Graduate Course on Animal Science. São Paulo State University, Jaboticabal, SP (1978)

Baccari Jr., F., Assis, P.S., Polastre, R., Fré, C.A.: Shade management in tropical environment for milk production in crossbred cows. Proc. Am. Soc. Anim. Sci. **33**, 209–210 (1982)

Bennett, I.L., Finch, V.A., Holmes, C.R.: Time spent in shade and its relationship with physiological factors of thermoregulation in three breeds of cattle. Appl. Anim. Behav. Sci. **13**, 227–236 (1985)

Blackshaw, J.K., Blackshaw, A.W.: Heat stress in cattle and the effect of shade on production and behaviour: a review. Aust. J. Exp. Agric. **34**, 285–295 (1994)

Bond, T.E., Kelly, C.F.: The globe thermometer in agricultural research. Agric. Eng. **36**, 251–260 (1955)

Bond, T.E., Laster, D.: Influence of shading on production of Midwest feedlot cattle. Trans. ASAE **18**, 957–959 (1975)

Bond, T.E., Kelly, C.F., Heitman Jr., H.: Improving livestock environment in high-temperature areas. J. Hered. (Washington, DC) **49**, 75–79 (1958)

Bond, T.E., Garret, W.N., Hahn, G.L.: Livestock shades. Calif. Agric. **15**, 7–10 (1961)

Bond, T.E., Morrison, S.R., Givens, R.L.: Influence of surroundings on radiant heat load of animals. Trans. ASAE **12**, 246–248 (1969)

Brown-Brandl, T.M., Eigenberg, R.A., Nienaber, J.A., Hahn, G.L.: Dynamic response indicators of heat stress in shaded and non-shaded feedlot cattle, Part 1: Analyses of indicators. Biosyst. Eng. **91**, 451–462 (2005)

Buffington, D.E., Collier, R.J., Canton, G.H.: Shade management systems to reduce heat stress for dairy cows in hot, humid climates. Trans. ASAE **26**, 798–802 (1983)

Chapman, A.J.: Fundamentals of heat transfer. Macmillan, New York (1987)

Collier, R.J., Elley, A.K., Sharma, A.K., Pereira, R.M., Buffington, D.E.: Shade management in subtropical environment for milk yield and composition in Holstein and Jersey cows. J. Dairy Sci. **64**, 844–847 (1981)

Paranhos da Costa, M.J.R.P.: Alguns aspectos do comportamento de vacas leiteiras em pastagens neotropicais [Some aspects of the behaviour of dairy cows in neotropical grasslands]. In: Proc Encontro Paulista de Etologia, Ribeirão Preto, 1985, vol. 3, pp. 199–217. USP, Ribeirão Preto (1985)

Daly, J.J.: Cattle need shade trees. Qld. Agric. J. **109**, 21–24 (1984)

Davison, T.M., Silver, B.A., Lisle, A.T., Orr, W.N.: The influence of shade on milk production of Holstein-Friesian cows in a tropical upland environment. Aust. J. Exp. Agric. **28**, 149–154 (1988)

Eigenberg, R.A., Brown-Brandl, T.M., Nienaber, J.A., Hahn, G.L.: Dynamic response indicators of heat stress in shaded and non-shaded feedlot cattle, Part 2: Predictive relationships. Biosyst. Eng. **91**, 111–118 (2005)

Eigenberg, R.A., Brown-Brandl, T.M., Nienaber, J.A.: Shade material evaluation using a cattle response model. Paper nº 074082. American Society of Agricultural and Biological Engineering. St. Joseph, ASABE (2007)

Eigenberg, R.A., Brown-Brandl, T.M., Nienaber, J.A.: Shade material evaluation using a cattle response model and meteorological instrumentation. Int. J. Biometeorol. **54**, 601–607 (2010)

Esmay, M.L.: Principles of Animal Environment. AVI Publications, Westport (1969)

Hahn, G.L.: Management and housing of animals in hot environments. In: Yousef, M.K. (ed.) Stress Physiology in Livestock, vol. 2. CRC Press, Boca Raton (1985)

Hansen, P.J.: Effects of coat colour on physiological responses to solar radiation in Holsteins. Vet. Rec. (Lond.) **127**, 333–334 (1990)

Igono, M.O.: Effect of a humid temperate climate and environmental modifications with shade, spray and fan, on milk production, thermal balance and hormone function of dairy cows. Diss. Abstr. Int. Sect. B: Sci. Eng. **46**, 3645 (1986)

Ingraham, R.H., Stanley, R.W., Wagner, W.C.: Seasonal effects of tropical climate on shaded and non-shaded cows as measured by rectal temperature, adrenal cortex hormones, thyroid hormone, and milk production. Am. J. Vet. Res. **40**, 1792–1796 (1979)

Ittner, N.R., Bond, T.E., Kelly, C.F.: Increasing summer gains of livestock: with cool water, concentrate roughage, wire corrals, and adequate shades. J. Anim. Sci. **13**, 867–877 (1954)

Jones, H.G.: Plants and Microclimates. Cambridge University Press, London (1983)

Kelly, C.F., Bond, T.E.: Effectiveness of artificial shade materials. Agric. Eng. **39**, 758–759 (1958)

Kelly, C.F., Bond, T.E., Ittner, N.R.: Thermal design of livestock shades. Agric. Eng. **31**, 601–606 (1950)

Mader, T.L., Fell, L.R., McPhee, M.J.: Behavior response of non-Brahman cattle to shade in commercial feedlots. Livest. Environ. **5**, 795–802 (1997)

McDaniel, A.H., Roark, C.B.: Performance and grazing habits of Hereford and Aberdeen-Angus cows and calves on improved pastures as related to types of shade. J. Anim. Sci. **15**, 59–63 (1956)

McIlvain, E.H., Shoop, M.C.: Shade for improving cattle gains and rangeland use. J. Range Manage. (Denver) **24**, 181–184 (1970)

Monteith, J.L., Unsworth, M.H.: Principles of Environmental Physics, 3rd edn. Academic Press, New York, (2008) 418 p

Panelli, M.A., Silva, R.G.: Conforto térmico e tipos de cobertura em abrigos para animais [Thermal comfort and shade for livestock]. In: Workshop Brasileiro de Bioclimatologia Animal, 2°, Jaboticabal, SP, November 1989

Reis, F.J.: Efeitos da espécie forrageira sobre a carga térmica radiante em pastagem [Effects of the Forage Type on the Radiant Heat Load in Pastures]. Monograph. São Paulo State University, Jaboticabal (1984)

Schütz, K.E., Rogers, A.R., Poulouin, Y.A., Cox, N.R., Tucker, C.B.: The amount of shade influences the behavior and physiology of dairy cattle. J Dairy Sci **93**, 125–133 (2010)

Silva, R.G.: Predição da configuração de sombras de árvores em pastagens para bovinos [Prediction of shades in pastures for cattle]. Engenharia Agrícola (Jaboticabal) **26**, 268–281 (2006)

Silva, R.G., Guilhermino, M.M., Morais, D.A.E.F.: Thermal radiation absorbed by dairy cows in pasture. Int. J. Biometeorol. **54**, 5–11 (2010)

Silver, B.A.: Shade is important for milk production. Qld. Agric. J. (Brisbane, Australia) **113**, 95–96 (1987)

Waldige, V.: Avaliação do sombreamento proporcionado por alguns tipos de árvores em pastagens [Evaluation of the shade projected by some tree types in pastures]. Monograph, São Paulo State University, Jaboticabal (1994)

Chapter 7
Thermal Stress Indexes

Abstract There is an introduction to the concept of indexes used to evaluate environments as for the thermal stress they cause upon animals. Several indexes in use are described and discussed. Practical methods of index construction are presented.

Keywords Environment evaluation • Index • Thermal stress • Thermal comfort

7.1 Thermal Stress Indexes

7.1.1 Index Concepts

Since it was recognised that animals were different among themselves as for their ability to face environmental challenges, there were many attempts to establish some type of criterion to select the best individuals for specified environments.

In the previous chapters, there were discussed several factors involved in the mechanisms of thermal exchange between organisms and their environments. The relative importance of such factors is quite variable according to the organism and the circumstances in which it lives, but it is possible to establish particular criteria to choose a specific environment based on the thermal comfort it allows, in order to evaluate its adequacy for a given organism.

The main factors governing thermal exchange between organisms and environments can be summarised as follows:

External environment:

Mean radiant temperature
Air temperature
Solar radiation
Partial vapour pressure
Atmospheric displacements (wind)
Atmospheric pressure

R. Gomes da Silva and A.S. Campos Maia, *Principles of Animal Biometeorology*,
Biometeorology 2, DOI 10.1007/978-94-007-5733-2_7,

Body surface:

Surface temperature
Surface humidity and permeability to water vapour
Radiative properties

Body surface covering:

Morphological characteristics
Thermal insulation
Permeability to wind
Permeability to water vapour
Radiative properties

Other factors:

Respiratory functions
Behavioural aspects

An index with such an objective must be linear, with adequate weights in order to define the relative importance of the chosen variables (Yamamoto 1983). Physiological performance, behaviour or any apparent manifestation of an animal depends on the status of its energetic balance – a fundamental part of which is the thermal component.

Most research and information about the effects of thermal environment until present times have been those associated to thermoregulatory physiological reactions. Such indexes can be divided into three groups (Epstein and Moran 2006):

(a) *Rational indexes* based on the thermal energy balance
(b) *Empirical indexes* based on the evaluation of the strain
(c) *Direct indexes* based on measurements of the environment

The first two types involve many environmental and physiological variables and are more difficult to assess, while the third one is the most widely used.

Yamamoto (1983) suggested that an index aimed to specify the thermal environment would be linear, as, for example,

$$I = aE + bT_a + cU_R + dU + e\mathbf{R} \tag{7.1}$$

where E is the energy input to the organism, T_a is the dry bulb temperature, U_R the relative humidity, U is the wind velocity and $\mathbf{R}$ is the incident radiation. Such a concept is coherent with the actual formal definition of thermal stress as it was discussed in Chap. 3.

The concept of index in the field of Animal Science was developed perhaps from the paper of Hazel (1943) about multivariate selection indexes for livestock genetic improvement. However, with respect to the aspects associated to the thermal environment, it must be mentioned the pioneering work of Albert O. Rhoad based on his observations of cattle behaviour and productive performance in the Federal University of Viçosa, MG, Brazil, where he worked for many years. This work resulted into the first thermal comfort index, the Iberia index or heat tolerance coefficient (Rhoad 1944):

$$\mathrm{HTC} = 100 - 10(\bar{T}_R - 38.33) \tag{7.2}$$

where $\bar{T}_{\mathrm{R}}$ is the average of six successive rectal temperature (°C) measurements, always done in the hottest hours of the day after 5 h of exposure to sun. As higher its HTC value, the more adequate for tropical conditions an animal was considered to be. This index, together with similar later appeared others, is now a matter of historical interest only.

Several indexes were later proposed with different objectives, as the human thermal comfort, as those of Thom (1959) and of Bianca (1962) which were based on air temperature and humidity only. However, air temperature is a measure of the sensible heat content of atmosphere only, while the humidity estimates the latent heat content. Despite both these factors are important as for the heat exchange processes among animals and environment, it is actually recognised that thermal radiation has a fundamental role as a thermal stress factor either for the man (Matzarakis et al. 2007) or the animals (Buffington et al. 1981; Yamamoto 1983; Shioya et al. 1997). As a matter of fact, solar radiation is of great magnitude in an intertropical region, where the mean radiant temperature is usually higher than the air temperature (Silva et al. 2010). Therefore, thermal radiation has been incorporated into the more recently developed indexes for sunny, hot regions, together with the wind speed – which is of great importance as for the convective heat exchange. However, it must be stressed that an animal's ability to face the effects of solar radiation is highly dependent on several characteristics of the skin and the hair coat (Riemerschmidt 1943; Silva 2000; Hillman et al. 2001; Silva et al. 2003).

Many indexes have been estimated from primary meteorological measurements, as the *temperature-humidity index*, THI (Thom 1959); *effective temperature*, ET (Bianca 1962); *black-globe humidity index,* BGHI (Buffington et al. 1981); *equivalent temperature index,* ETI (Baeta et al. 1987) and several others. Some of them that were originally developed for human use have been applied to animals also, as the THI.

In the case of humans, the indexes are used with focus on the thermal comfort, while in the case of animals, there has interest on indexes that are able to help a rational environmental management and allow to decisions related to their performance, health and well-being (Hahn et al. 2003).

All the way, such indexes have been developed under the climatic conditions of temperate regions; they were based on data from animals bred in those regions and observed mainly under the artificial conditions of climatic chambers. It is now recognised that animals confined into such chambers have reactions often substantially different from those presented by the same individuals under the natural conditions of the open field.

7.1.2 Indexes for Animals

7.1.2.1 Operative Temperature (T_o)

If it is neglected the heat transfer by conduction between an animal and its environment, sensible heat exchange takes place by convection and radiation only. Three factors are then concerned with (air temperature, mean radiant temperature and wind velocity), but they can be reduced to two by the operative temperature approach (Herrington et al. 1937), as follows.

Let be **C** the rate of heat exchange by convection and **R** that exchange by radiation. Then,

$$\mathbf{C} + \mathbf{R} = h_c(T_s - T_a) + h_r(T_s - T_{rm}) \tag{7.3}$$

where the temperatures are in °C and h_c and h_r are the coefficients of thermal exchange by convection and radiation, respectively:

$$h_r = F_c \varepsilon \sigma \left(T_s^3 + T_s^2 T_{rm} + T_{rm}^2 T_s + T_{rm}^3\right) \tag{7.4}$$

where F_c is the shape factor of a given body, ε is the emissivity of the body surface, σ is the Stefan-Boltzmann constant, T_{rm} is the mean radiant temperature of the environment (K), T_a the dry bulb temperature (K) and T_s the body surface temperature (K).

Now, we can redefine

$$\mathbf{C} + \mathbf{R} = h_o(T_s - T_o) \tag{7.5}$$

where T_s and T_o (operative temperature) are in degrees °C. By equating expressions (7.3) and (7.5), it is obtained:

$$h_o(T_s - T_o) = h_c(T_s - T_a) + h_r(T_s - T_{rm})$$

$$T_s - T_o = \frac{h_c T_s - h_c T_a + h_r T_s - h_r T_{rm}}{h_c + h_r}$$

Finally, the equation to calculate the operative temperature is

$$T_o = \frac{h_c T_a + h_r T_{rm}}{h_c + h_r}$$

A most practical version of the above equation would be

$$T_o = \frac{r_R T_a + r_H T_{rm}}{r_R + r_H} \ ^\circ\mathrm{C} \tag{7.6}$$

where r_R and r_H are the resistances (s m^{-1}) for heat exchange by radiation and convection at the body surface, respectively.

According to Herrington et al. (1937), the values T_o and h_o would be independent of the skin surface temperature (T_s), in such a way that the operative temperature would be a function of environmental factors only. However, as it was pointed by Kerslake (1972), this is not strictly true because the coefficients h_c and h_r depend on T_s and on other factors that are directly linked to the skin surface.

The above-described concept of operative temperature takes into account the sensible heat exchange only. Considering that for cattle in tropical environments the thermal comfort depends mainly on a combination of four environmental factors (mean radiant temperature, wind speed, air temperature and air humidity), the flux of thermal energy between animals' body and the environment can be represented by

$$\mathbf{\Phi} = \mathbf{R} + \mathbf{C} + \mathbf{E}$$

where **R**, **C** and **E** are the thermal fluxes by radiation, convection and evaporation, respectively. Then, according to Silva (2000), we have

$$\mathbf{R} + \mathbf{C} + \mathbf{E} = \frac{\rho c_p (T_s - T_o)}{r_o} + \frac{\rho c_p [P_s(T_s) - P_v]}{\gamma^*} \tag{7.7}$$

where $r_o = r_R r_H/(r_R + r_H)$ and $\gamma^* = P_a \gamma r_v / r_o$ and r_V is the resistance to heat loss by evaporation at the skin surface (s m^{-1}).

Considering the concept of equivalent temperature (McArthur 1987), after some manipulation, we have

$$\begin{aligned} \mathbf{R} + \mathbf{C} + \mathbf{E} &= \frac{\rho c_p}{r_o}\left[\left(T_s + \frac{P_s(T_s)}{\gamma^*}\right) - \left(T_o + \frac{P_v}{\gamma^*}\right)\right] \\ &= \frac{\rho c_p (T_s^* - T_o^*)}{r_o} \end{aligned}$$

where

$$T_o^* = T_o + \frac{P_v}{\gamma^*} \ ^\circ\mathrm{C} \tag{7.8}$$

is the *equivalent operative temperature* of the environment.

Equation 7.8 can be particularly useful for the evaluation of the thermal comfort within animal housings and shelters, as it takes into consideration the three main avenues of heat exchange between animals and environment.

7.1.2.2 Temperature-Humidity Index (THI)

This index was originally conceived (Thom 1959) as a thermal comfort index for humans, and it has been used with that purpose by the U.S. Weather Bureau since 1959. However, it has been also widely used as an indicator of heat stress in animals. THI can be assessed by different equations, and among the most known of them, these are the following:

$$\mathrm{THI} = T_{\mathrm{a}} + 0.36\,T_{\mathrm{dp}} + 41.5 \tag{7.9}$$

$$\mathrm{THI} = 0.72(T_{\mathrm{a}} + T_{\mathrm{w}}) + 40.6 \tag{7.9a}$$

where T_{a} is the dry bulb, T_{w} the wet bulb and T_{dp} the dew-point temperature, all of them in °C.

According to Hahn (1985), a value such as THI $\leq$ 70 is an indication of a non-stressing environment; a value between 71 and 78 is critical; from 79 to 83 is an indication of danger; and above 83 is an emergency. Such a range of values would be valid for any livestock, not only for cattle.

This index has been used to assess the thermal comfort given out to animals by specified environments, since Johnson et al. (1962, 1963) and Cargill and Stewart (1966) reported significant decreases of the milk yield of cows in association with increased THI values. Even many other authors have used it to evaluate the thermal stress in several animal species under different conditions, by assuming an association of the index with the production ability of the animals.

Conceptually, it is difficult to ascertain whether THI is an appropriate measurement of heat stress in cattle (Dikmen and Hansen 2009). There has no explanation about the relative weighting of T_{a} and T_{u} *or* T_{dp} in the index. In addition, it has been observed (Ingraham et al. 1976; Buffington et al. 1981; West 2003) that correlations between THI and responses of cattle may underestimate the effects of humidity.

Based on meteorological data from the states of Arizona and Georgia (USA), Bohmanova et al. (2007) evaluated seven variations of the temperature-humidity index with respect to the milk production performance of dairy cows. Their results showed that those indices with higher weights on humidity were the best for humid climates, whereas indices with larger weights on air temperature were the best indicators of heat stress in the semiarid climate.

Silva et al. (2007) evaluated a number of 1,359 data on body temperature and respiratory rate of Holstein and Jersey cows in three locations of an equatorial semiarid region; those observed values were correlated with the variations of the environmental conditions, as measured by six different indexes. Some results are given in Table 7.1. This study showed very low correlations of THI with both body temperature (39.6°C for Holsteins, 39.5°C for the Jerseys) and respiratory rate (59.7 for Holsteins and 79.6 for Jerseys); the cows were observed in open field conditions under the following average conditions: air temperature 30.07 $\pm$ 0.08°C, partial vapour pressure 2.99 $\pm$ 0.01 kPa and wind speed 2.30 $\pm$ 0.04 m s^{-1}. It was concluded that THI is not an adequate index for cattle in equatorial regions.

Table 7.1 Correlation coefficients of six indexes with responses of Holstein and Jersey cows to the conditions of an equatorial semiarid environment

Index	T_R	F_R
THI (Thom 1959)	−0.053 *n.s.*	0.099**
BGHI (Buffington et al. 1981)	0.050 *n.s.*	0.155**
ETI (Baeta et al. 1987)	0.293**	0.520**
ESI (Moran et al. 2003)	0.209**	0.464**
HLI (Gaughan et al. 2002)	0.286**	0.842**
PRR (Eigenberg et al. 2002, 2003)	0.114**	0.344**

From Silva et al. (2007)
T_r = rectal temperature (°C), F_r = respiratory rate (breaths min^{-1}), *n.s.* = non-significant, ** = Significant, $P < 0.01$

7.1.2.3 Black-Globe Humidity Index (BGHI)

As it was mentioned earlier, thermal radiation is among the most significant environmental factors, being of critical importance for animals on pasture especially in tropical regions. Indexes as the THI which do not take radiation into account are quite inadequate to compare the thermal environment under a shelter, for example, with that under the sun.

Therefore, an adaptation of the THI for dairy cattle, by simply substituting the globe temperature for the air temperature in the original THI equation, was proposed by Buffington et al. (1981). The new formula was named as the *black-globe humidity index* (BGHI):

$$\text{BGHI} = T_g + 0.36\,T_{dp} + 41.5 \tag{7.10}$$

where T_g and T_{dp} are the black globe and the dew-point temperatures (°C), respectively. The authors mentioned that BGHI values lower than 70 have small effect on the performance of dairy cows, while the values BGHI > 75 leave to significant decrease in the food ingestion by the animals.

As it is shown in Table 7.1, this index was considered also as of very low efficiency for evaluating the environment for cattle in tropical regions.

7.1.2.4 Equivalent Temperature Index (ETI)

The effects of air temperature (T_a, °C), relative humidity (U_R, %) and wind speed (U, m s^{-1}) on the thermal balance of dairy cows were combined by Baeta et al. (1987) in their *equivalent temperature index*, given in degrees °C as

$$\begin{aligned}\text{ETI} = {} & 27.88 - 0.456\,T_a + 0.010754\,T_a^2 - 0.4905\,U_R + 0.00088\,U_R^2 \\ & + 1.1507\,U - 0.126447\,U^2 + 0.019876\,T_aU_R - 0.046313\,T_aU\end{aligned} \tag{7.11}$$

The above equation was applied to five high-producing Holstein cows with their summer hair coat, exposed to variable environmental conditions within a climatic chamber: 16–41°C air temperature, 40–90% relative humidity and wind from 0.5 to 6.5 m s^{-1}. The results showed that an increased ETI value within a given temperature caused a 38.3% decrease in the milk yield; at the same time, the rectal temperature increased up to 40.8°C. The following scale of ETI was considered as valid for the cows:

No problems	18–27°C
Caution	27–32°C
Extreme caution	32–38°C
Danger	38–44°C
Extreme danger	>44°C

Because the little number of animals tested and the short (3 days) treatment period, Berman (2005) considered as intriguing the results of the ETI on lactating dairy cattle, as compared with other studies. According to Hahn et al. (2003), ETI may provide representative results for short-term heat exposures that often occur in the summer season (in temperate regions). Nevertheless, the study by Silva et al. (2007) with Holstein and Jersey cows in an equatorial region showed ETI as one of the two best indexes for tropical conditions; it presented significant correlations with body temperature and respiratory rate (0.293 and 0.520, respectively). See Table 7.1.

7.1.2.5 Selection Index for Adaptation of Beef Cattle

Considering that livestock performance is directly related to animals' adaptation to their environment, Silva (1973, 1975) proposed the simultaneous selection for production and adaptation of the existing cattle in tropical regions, based on the genetic and phenotypic relationships among production traits and those associated to the adaptation. Steers and heifers of the Canchim breed aged up to 28 months were exposed for to direct sun from 09:00 to 15:00 h in a pen, during the summer; rectal temperature and respiratory rate were recorded before and after each treatment, together with average daily weight gain from weaning to 18 months of age. The estimated selection index was

$$\begin{aligned}\mathrm{SIA} = 100 &- 0.026\,T_{\mathrm{ri}} - 0.064(T_{\mathrm{rf}} - T_{\mathrm{ri}}) - 0.009\,F_{\mathrm{ri}} \\ &- 0.133\,\log(F_{\mathrm{rf}} - F_{\mathrm{ri}}) + 0.281\,W \end{aligned} \qquad (7.12)$$

where T_{ri} and F_{ri} are the rectal temperature (°C) and respiratory rate (breaths min^{-1}) taken before exposure to sun, while T_{rf} and F_{rf} are the respective measurements done after exposure, and W is the average daily weight gain from weaning to 18 months of age (kg). The SIA value is an estimate of the additive genetic merit of an animal to be used in the selection of bulls for breeding.

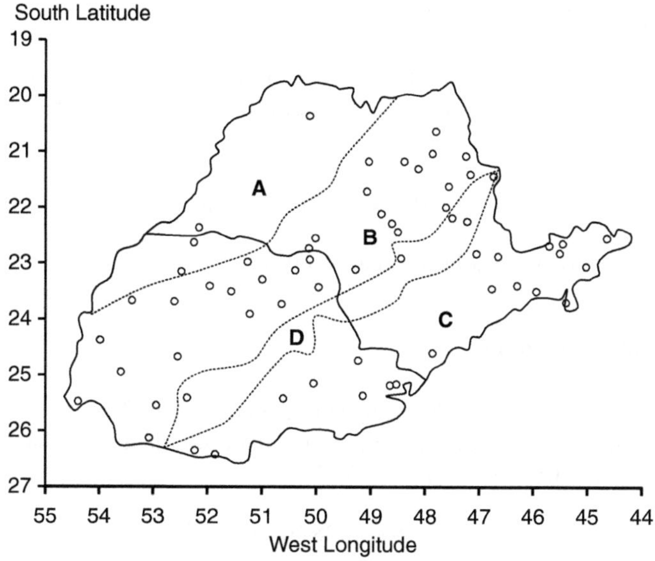

Fig. 7.1 Bioclimatic zoning of the states of São Paulo (*right up*) and Paraná (*left down*), Brazil, for breeding of Polwarth (*A*), Corriedale (*B*) and Suffolk (*C*) sheep. The *D* zone is an intermediary region where either Corriedale or Suffolk breeds can be bred. Points on the map are locations with meteorological stations that were used in the study (Modified from Barbosa et al. 1995)

7.1.2.6 Thermal Comfort Index for Sheep

This index was developed by Silva and Barbosa (1993) to assess environments for sheep breeding in the subtropical south-eastern region of Brazil. Its formula is

$$\mathrm{TCI} = 0.659\,T_\mathrm{a} + 0.511\,P_\mathrm{v} + 0.550\,T_\mathrm{g} - 0.042\,U \tag{7.13}$$

The above index was extensively tested in different regions in the states of São Paulo and Paraná by using three sheep breeds: Corriedale, Polwarth and Suffolk. The results showed that the body temperature of Polwarths remained steady at normal levels under TCI values from 20 to 37, increasing rapidly up to 40°C under ICT = 50. The temperature of Suffolk animals began to increase at ICT = 20, attaining to 40°C under ICT = 38. This last temperature level was attained by Corriedale animals under ICT = 43. Respiratory rate was also significantly affected according to the index variation: Under ICT < 25, the respiratory rate of the Polwarths remained below 90 breaths min^{-1} and increasing rapidly from ICT = 35 (124 breaths min^{-1}) to ICT = 48 (280 breaths min^{-1}). Equation 6.10 was compared with those of THI and BGHI, and the conclusion was that ICT presented best results for sheep than the two ones.

The index was later used by Barbosa et al. (1995) to establish a bioclimatic zoning of the states of São Paulo and Paraná as for sheep breeding. See Fig. 7.1.

7.1.2.7 Equivalent Thermal Stress Index

This index was proposed by Moran et al. (2003) for humans; notwithstanding it was not yet used for animals, it is potentially useful for this purpose. Its formula is the following:

$$\mathrm{ESI} = 0.63\,T_\mathrm{a} - 0.03\,U_\mathrm{R} + 0.002\,\mathbf{S} + 0.0054\,T_\mathrm{a}U_\mathrm{R} - 0.073\,(0.1 + \mathbf{S})^{-1} \quad (7.14)$$

where **S** is the short-wave solar irradiance (W m^{-2}).

7.1.2.8 Heat Load Index

This index was originally developed by Gaughan et al. (2002) to assess the thermal stress on feedlot beef cattle in Australia. It was later modified (Gaughan et al. 2008) into two parts: the first one for $T_\mathrm{g} < 25°\mathrm{C}$:

$$\mathrm{HLI} = 10.66 + 0.28\,U_\mathrm{R} + 1.3\,T_\mathrm{g} - U \quad (7.15)$$

and the second for $T_\mathrm{g} > 25°\mathrm{C}$:

$$\mathrm{HLI} = 8.62 + 0.38\,U_\mathrm{R} + 1.55\,T_\mathrm{g} - 0.5\,U + e^{2.4-U} \quad (7.16)$$

According to the above-cited authors, the environment evaluated by HLI can be classified as follows:

Thermo neutral conditions	≤70.0
Warm	70.1–77.0
Hot	77.1–86.0
Very hot	>86.0

The study by Silva et al. (2007) showed that the 2002 version of HLI was one of the two best thermal stress indexes for dairy cows in open tropical pastures, with significant correlations with the rectal temperature ($r = 0.286$) and the respiratory rate ($r = 0.542$). The new version of this index (Eq. 7.16) was tested by us on 1,000 data of Holstein cattle in the semiarid conditions of north-eastern Brazil, giving values from 59.2 to 117.2; the correlations with rectal temperature and respiratory rate were 0.303 and 0.620, respectively. It must be pointed that in this equatorial semiarid region, the wind is almost constant, at speeds from 0.1 to 3 m s^{-1} and even more; at the same time, air temperatures can reach 35–40°C and globe temperatures of 60°C or more are frequent.

7.1.2.9 Respiratory Rate Index

The respiratory rate has been long considered as an indicator of thermal discomfort in animals. However, its use as an estimator of the degree of thermal stress in cattle was suggested Gaughan et al. (2002), with the possible intention of limiting the individual measurements of the animals and avoid disturbance of their behaviour during field observations.[1]

Eigenberg et al. (2005) extended that concept by eliminating at all the direct observations of the animals. They proposed an indirect estimator of the respiratory rate based on environmental data only by using the formula:

$$\mathrm{RR} = 5.4\,T_{\mathrm{a}} + 0.58\,U_{\mathrm{R}} - 0.63\,U + 0.024\,\mathbf{S} - 110.9 \tag{7.17}$$

The following ranges of RR values (breaths min^{-1}) were suggested by those authors:

Normal	≤85
Warning	85–110
Danger	110–133
Emergency	>133

This index was used by Eigenberg et al. (2010) together with the HLI (Eq. 7.15) to evaluate shelter-roofing materials, obtaining results considered as encouraging ones. However, the results of the evaluation carried out by Silva et al. (2007) with dairy cows in a tropical region showed a low correlation coefficient of the RR index with the measured respiratory rate, $r = 0.344$ ($P < 0.01$). Such low correlation was probably due to the data used for building Eq. 7.16, which were produced by beef cattle under temperate climate conditions. A new version specific for dairy cows under tropical conditions would be of interest.

7.1.2.10 Index of Thermal Stress for Cattle

Silva et al. (2011) proposed a new index to evaluate the thermal environment of dairy cows in intertropical regions, with emphasis on equatorial conditions. Its formula is

$$\begin{aligned}\mathrm{ITSC} = {} & 7.9505 + 0.0667\,T_{\mathrm{g}} + 0.0673\,U + 0.0214\,U^2 - 1.9005\,P_{\mathrm{v}} \\ & + 0.1749\,P_{\mathrm{v}}^2 + 0.045\,T_{\mathrm{a}}P_{\mathrm{v}} - 0.0095\,T_{\mathrm{g}}U\end{aligned} \tag{7.18}$$

with determination coefficient $R^2 = 0.734$. This index was tested on 1,321 data from Holstein herds in the north-eastern region (Ceará and Rio Grande do Norte) and in São Paulo; there were found significant ($P < 0.01$) correlations with rectal temperature ($r = 0.472$), respiratory rate ($r = 0.793$), skin surface temperature ($r = 0.755$) and sweating rate ($r = 0.570$).

[1] However, it is possible to count the respiratory movements of the ribs at some distance from the animals, by using binoculars.

It was suggested the following scale of results for ITSC:

Comfort	≤8.0
Mild stress	8.1–10.0
Moderate stress	10.1–11.0
Very distressing	>11.0

7.2 Development of Indexes

7.2.1 Why New Indexes?

There has a real need of new, more sophisticated thermal stress indexes.

In their paper about interactions between climate and animal production, Hahn et al. (2003) asked what where the most useful approaches for further development of thermal stress indexes. They suggested that efforts should be done for improving on the basic THI concept, by considering thermal radiation and airflow. These authors recognised that, in certain cases, air temperature alone has been enough to represent the effect of hot thermal environments.

However, with respect to the evaluation of outdoor environments, as is the case of animals observed in the open range, we must consider as many variables as possible, both environmental and physiological ones. In fact, animal performance is generally a result of combinations of those variables, whose combinations are sometimes very complex and are not fixed, changing with time and the circumstances.

Because animal organisms evolve together with the environments in which they live, their physiological mechanisms – and eventually their morphology – change with time. For example, Holstein cows are bred throughout the world, but the several populations of this breed in the tropical regions are somewhat different with respect to the Holsteins bred in Europe, Canada and other temperate regions. Some of the differences that have been recognised even by the breeders are related to the sweating ability and to the body surface pigmentation. Such differences should be associated to internal changes that are less evident, though effective and that lead with time to changes in the population characteristics.

Therefore, care must be needed as for the generalised use of an index under tropical conditions, if it was calculated from observations done in temperate climates, even if the tests carried out in climatic chambers are used to justify such practice. Climatic chambers are artificial environments in which the animals present reactions (both behavioural and physiological) often much different from those they have in open field conditions. On the other hand, what was considered as a truth several decades ago does not meet the needs of present times; progress is constant, and everyone tries to learn the lessons, but nature always manages to get its own evolutionary way.

Then, the question about the need of new and better indexes must be answered with an emphatic YES.

7.2.2 *Nature of the Indexes*

An index can be of a nature different from that of another index, depending on the assumptions it is based on and the method of its determination. Several indexes have been built as multiple regression equations, while other types were established by different methods.

For example, Beckett (1965) proposed the use of the enthalpy concept in order to establish thermal comfort indexes for swine. This method was used by Moura et al. (1997) to evaluate swine and poultry housing in a tropical environment. They studied two housing types for swine in comparison with the external environment; by considering the thermo neutral zone of the animals at the end of the growth phase, the calculated enthalpy was 68.62 kJ kg^{-1} of dry air. In the housing type 1, the average enthalpy was 10% higher than that of the external environment, reaching 55% of the critical values; in the housing type 2, the enthalpy was 64%. The conclusion was that the enthalpy values can be an important tool for the evaluation of animal environment. Silva et al. (2008) used also the enthalpy to evaluate swine housings. However, a problem to be considered is that the use of enthalpy per se as an environmental index does not takes into account the thermal radiant energy, which is one of the most important stress factors for the animals, especially in tropical regions.

Other approaches have been considered, and Mitchell et al. (2001) proposed an index, the apparent equivalent temperature, for birds under transportation conditions.

It is now clear that in areas where environmental stress can be of significance, variation in animals' performance is dependent on interactions of gene effects governing production, reproduction and resistance to stress factors. The magnitude of those genetic and genotype-environment interactions depends on the species, breed and even the population considered. However, the use of genetic information for the improvement of livestock breeding in stressing environments has not been generally considered yet.

Silva (1973) estimated the genetic variation of rectal temperature, respiratory rate and blood haemoglobin level in Canchim cattle (5/8 Charollais × 3/8 Nelore). The increase in body temperature after exposure to sun in the hottest hours of the day presented moderate heritability coefficient (0.443) and high negative genetic correlation (−0.895) with the average daily weight gain.

Silva et al. (1988) determined the heritabilities of the sweating rate (0.222), skin pigmentation (0.112), hair coat pigmentation (0.303), hair coat thickness (0.233) and hair length (0.081) of Jersey cattle bred in a tropical environment; considering the importance of the hair length for the thermal comfort and the performance of cattle in hot climates, the respective heritability value was surprisingly low. For Holstein cattle in the same environment, it was found a coefficient of 0.20 (Pinheiro 1996).

On the other hand, evidence has been found that supports the existence of a major gene (*slick hair* gene) which would be dominant and responsible for producing a very short, sleek hair coat in cattle (Olson et al. 2003). This hair coat type is a trait of great value for cattle under tropical conditions, and it is favoured by the natural (and sometimes by the artificial) selection.

Present programmes of genetic improvement of livestock in tropical countries must take into account not production traits only (milk yield, weight gain, egg or wool production) but also those traits related to the interaction of organisms with environmental factors as the solar radiation, wind, air temperature and humidity. Given the increasing importance of livestock production in tropical regions, it is clear that effects of heat stress on animal biology and production are likely to be more important in the near future. Accordingly, there are needed more efforts in the development of indexes that can be used for the genetic improvement of populations on the basis of selection for adaptation traits.

7.2.3 Methods of Index Calculation

7.2.3.1 Multiple Regression 1

An interesting example is the procedure followed by Baeta et al. (1987) to establish their equivalent temperature index (ETI) for dairy cows.

Starting from the calculation of the thermal storage by the animals during the day, the authors calculated multiple correlations of milk yield, rectal temperature, metabolic rate, respiratory evaporation, total evaporation, feed intake, water consumption and body weight on the environmental variables (T_a, U_r *e* U). Thermal storage values were combined later with the regression coefficients for T_r, T_s and body weight. Assuming that $U_R = 40\%$ and $U = 0.5 \text{ m s}^{-1}$, a preliminary thermal storage value was estimated and then used to calculate a multiple regression on the environmental variables in order to obtain the first estimate of equivalent temperature, ET(*thermal storage*). A second value was calculated on the basis of the milk yield, ET(*milk*).

The means of the constants in the two equations, ET(*thermal* storage) and ET (milk), resulted into the final ETI equation.

7.2.3.2 Multiple Regression 2

There is calculated the multiple regression of a trait measured on the animals on a set of environmental variables. This method was used by Eigenberg et al. (2005) to establish their RR index. Gaughan et al. (2008) developed their HLI index by regression of the panting score of animals on the environmental variables.

However, the problem becomes more complex when several physiological traits must be considered. In such a case, it is possible to reduce the set of physiological traits into just one synthetic trait by means of a *principal components analysis*[2]; then, a multiple regression is carried out relating this synthetic trait to the set of environmental variables. This procedure was used by Silva et al. (2011) to obtain a thermal stress index for dairy cows in tropical environments and a summarised description of it will follow:

First, there were obtained the correlations among p traits measured in the animals (rectal temperature, respiratory rate, sweating rate and so on). The respective correlation coefficients constitute a $p \times p$ matrix $\mathbf{R}$, from which there are extracted the p latent roots or eigenvalues and the respective eigenvectors, $\mathbf{e}$, assuming as zero the determinant of the matrix $(\mathbf{R} - \lambda\mathbf{I})$, where λ is some eigenvalue and $\mathbf{I}$ is an identity matrix.[3]

Second, the eigenvector $\mathbf{e}$ corresponding to the greatest eigenvalue is applied to the n records of the p physiological traits, thus obtaining a new trait which is a synthesis of the original p variables.

Three, a series of multiple regressions of the new trait on some combination of the environmental traits is calculated, and the equation presenting the greater coefficient of determination is chosen.

Example. Let us have n observations on $p = 3$ traits (Y_1, Y_2 and Y_3) of animals that were exposed to $q = 3$ environmental measurements (X_1, X_2 and X_3). The following correlation coefficients were calculated among the p animal traits:

$$\mathbf{R} = \begin{bmatrix} 1 & 0.53 & -0.25 \\ 0.53 & 1 & 0.16 \\ -0.25 & 0.16 & 1 \end{bmatrix}$$

The eigenvalues and respective eigenvectors of the $\mathbf{R}$ matrix are

$$\begin{bmatrix} \lambda_1 \\ \lambda_2 \\ \lambda_3 \end{bmatrix} = \begin{bmatrix} 1.538803 \\ 1.119533 \\ 0.341665 \end{bmatrix}$$

$$\mathbf{e}_1 = \begin{bmatrix} 0.726213 \\ 0.673703 \\ -0.136892 \end{bmatrix} \quad \mathbf{e}_2 = \begin{bmatrix} -0.186245 \\ 0.384474 \\ 0.904153 \end{bmatrix} \quad \mathbf{e}_1 = \begin{bmatrix} 0.661762 \\ -0.631113 \\ 0.404682 \end{bmatrix}$$

---

[2] For details about principal component analysis and multivariate regression, see the books of Rencher (1995) and Johnson and Wichern (1988).

[3] An introduction to the concept of eigenvalues and eigenvectors is found in the book of Searle (1966), while practical algorithms for the calculations are given in Faddeeva (1959).

Note that

$$\sum_{i=1}^{p} \lambda_i = 1.538803 + 1.119533 + 0.341665 = \mathrm{tr}(\mathbf{R}) = 3$$

And

$$\mathbf{e'}_1\mathbf{e}_1 = \begin{bmatrix} 0.726213 & 0.673703 & -0.136892 \end{bmatrix} \begin{bmatrix} 0.726213 \\ 0.673703 \\ -0.136892 \end{bmatrix} = 1$$

The vector $\mathbf{e}_1$ is that eigenvector corresponding to the greatest eigenvalue λ_1, and it is then applied to the n records of the Y_i animal traits, thus obtaining n values of the new synthetic variable β:

$$\beta_k = 0.726213\, y_{1k} + 0.673703\, y_{2k} - 0.136892\, y_{3k}$$

where $k = 1,\ldots, n$. Finally, a multiple regression of β_k on the environmental variables is calculated in order to estimate the desired index:

$$I = \alpha + b_1 x_{1k} + b_2 x_{2k} + b_3 x_{3k}$$

where b_1, b_2 and b_3 are the partial regression coefficients of regression of β_k on the variables X_1, X_2 and X_3, respectively.

7.2.3.3 Enthalpy

The use of the enthalpy concept was proposed by several authors, as it was previously mentioned, aiming to evaluate animal housing. The following formula was modified from that of Villa Nova et al. (1972) and can be used to estimate enthalpy in terms of kJ kg^{-1} of air:

$$H = 6.7 + 0.243\, T_\mathrm{a} + 2.216\left[\frac{P_\mathrm{v}}{P_\mathrm{s}(T_\mathrm{a})} 10^{7.5T_\mathrm{a}/(T_\mathrm{a}+237.3)} - 1\right] \ \mathrm{kJ/kg} \qquad (7.19)$$

where T_a is air temperature (°C), $P_\mathrm{s}(T_\mathrm{a})$ the saturation air vapour pressure (kPa) at this temperature and P_v is the partial air vapour pressure (kPa).

7.3 Problems

Problem 7.1. Jersey cows bred in a location at 6°04′05″ south, 35°20′ west and 52 m altitude presented rectal temperature of 39.2°C and skin temperature 35.5°C. The following measurements of environmental variables were obtained: air temperature 31°C, partial air vapour pressure 2.7 kPa, mean radiant temperature 50°C and wind speed 1.6 m s^{-1}; the wind blows parallel to the animals' body axis. The body of the animals was assumed as a horizontal cylinder with hemispherical ends, 1.5 m length and 0.75 m diameter, and the hair coat is settled. Calculate the equivalent operative temperature.

Data

L_t = latitude = $-6°04'05'' = -6.068056°$
L_g = longitude = $35°20' = 35.333333°$
z = altitude = 52 m
T_a = air temperature = 31°C
P_v = partial vapour pressure = 2.7 kPa
T_{rm} = mean radiant temperature = 50°C = 323.15 K
U = wind velocity = 1.6 m s^{-1}
T_r = rectal temperature = 39.2°C
T_s = body surface temperature = 35.5°C = 308.15 K
L = body length = 1.5 m
D = body diameter = 0.75 m
V = body volume = $\pi L\,(0.5\,D)^2 = 0.66268$ m^3

Solution
Thermal properties of the atmosphere (Eqs. 1.38, 1.39, 1.40, 1.41, 1.42, 1.43, 1.44, 1.45, 1.46, 1.47, 1.48, 1.49, 1.50, 1.51, 1.52, and 1.53), for the temperature $T_m = 0.5(T_a + T_s) = 0.5(31 + 35.5) = 33.25$°C:

$$\begin{aligned} g &= 9.78013 + 8.18 \times 10^{-5} L_t + 1.168 \times 10^{-5} L_t^2 - 3.1 \times 10^{-6} z \\ &= 9.78013 + 8.18 \times 10^{-5}(6.068056) + 1.168 \times 10^{-5}(6.068056)^2 3.1 \times 10^{-6}(52) \\ &= 9.780895 \ \ \mathrm{m\,s^{-2}} \end{aligned}$$

$$\begin{aligned} P_a &= 101.325 \exp\left\{-\frac{zg}{287.04\,T_a}\right\} \\ &= 101.325 \exp\left\{-\frac{52\,(9.780895)}{287.04\,(33.25 + 273.15)}\right\} = 100.741 \text{ Pa} \end{aligned}$$

$$\begin{aligned} c_p &= 1.0052 + 4.577 \times 10^{-4} exp\left\{\frac{T_m}{32.07733}\right\} \\ &= 1.0052 + 4.577 \times 10^{-4} exp\left\{\frac{33.25}{32.07733}\right\} = 1.00649 \ \ \mathrm{J\,g^{-1}\,°C^{-1}} \end{aligned}$$

$$\lambda = 2500.788 - 2.37374\,T_m = 2421.861 \text{ J g}^{-1}$$

$$\gamma = \frac{c_p}{0.6223\,\lambda} = \frac{1.00649}{0.6223\,(2{,}421.861)} = 0.000668 \ °\mathrm{C}^{-1}$$

$$\rho = \frac{3{,}484.358274\,P_a}{T_m} = \frac{3{,}484.358274\,(100.741)}{33.25 + 273.15} = 1{,}145.619 \ \ \mathrm{g\,m^{-3}}$$

$$k = \rho c_p (1.888 \times 10^{-5} + 1.324 \times 10^{-7} T_m)$$
$$= 1{,}145.619\,(1.006512)\left[1.888 \times 10^{-5} + 1.324 \times 10^{-7}(33.25)\right]$$
$$= 0.02685 \;\; \mathrm{W\,m^{-1}\,^{\circ}C^{-1}}$$

$$v = 1.32743 \times 10^{-5} + 9.22286 \times 10^{-8} T_m$$
$$= 1.32743 \times 10^{-5} + 9.22286 \times 10^{-8}(33.25) = 1.6341 \times 10^{-5} \;\; \mathrm{m^2\,s^{-1}}$$

$$D_v = 2.12138 \times 10^{-5} + 1.4955 \times 10^{-7} T_m = 2.6186 \times 10^{-5} \;\mathrm{m^2\,s^{-1}}$$

Adimensional Numbers

As the wind speed is >0.5 m s^{-1}, then there is forced convection.

The characteristic dimension of the body is estimated by the cubic root of the body volume:

$$d = \left[\pi L (0.5\,D)^2\right]^{1/3} = 0.66268^{1/3} = 0.871836 \;\mathrm{m}$$

$$P_r = \rho\, c_p\, v\, k^{-1} = 1145.619 \frac{(1.006512)(1.6341 \times 10^{-5})}{0.02685} = 0.701768$$

$$R_e = U d v^{-1} = \frac{1.6\;(0.871836)}{(1.6341 \times 10^{-5})} = 85{,}364.2263$$

$$N_u = 0.332 R_e^{1/2} P_r^{1/3}$$
$$= 0.332(85{,}364.2263)^{1/2}(0.701768)^{1/3} = 86.2 \qquad (2.43)$$

$$S_c = \frac{v}{D_v} = \frac{1.6341 \times 10^{-5}}{2.6186 \times 10^{-5}} = 0.624036$$

$$S_h = 0.332\, R_e^{1/2} S_c^{1/3} = 0.332\,(85{,}364.2263)^{1/2}\,(0.624036)^{1/3} = 82.892$$

Thermal Resistances

$$r_H = \frac{\rho\, c_p\, d}{k N_u} = \frac{1{,}145.619\,(1.006512)(0.871836)}{0.02685\,(86.2)} = 434.3526 \;\; \mathrm{s\,m^{-1}} \qquad (2.32)$$

$$r_T = 225 - 5.44\, T_s = 225 - 5.44\,(35.5) = 31.88 \;\mathrm{s\,m^{-1}} \text{ (from Table 4.5)}$$

$$r_R = \frac{\rho c_p}{4\varepsilon_s \sigma T_s^3} = \frac{1145.619(1.006512)}{4(0.98)(5.67\times 10^{-8})(35.5+273.15)^3} = 176.4376\ \text{s m}^{-1} \quad (4.21a)$$

$$r_O = \frac{r_R r_H}{r_R + r_H} = \frac{(176.4376)(434.3526)}{176.4376 + 434.3526} = 125.47\,\text{s m}^{-1} \quad (7.7b)$$

$$r_V = \frac{d}{D_v S_h} = \frac{0.871836}{2.6186\times 10^{-5}\,(82.892)} = 401.655\ \text{s m}^{-1} \quad (2.53a)$$

Equivalent Operative Temperature

$$T_o = \frac{r_R T_a + r_H T_{rm}}{r_R + r_H} = \frac{176.4376\,(31) + 434.3526\,(50)}{176.4376 + 434.3526} = 44.512\ °\text{C}$$

$$\gamma^* = \frac{P_a \gamma r_V}{r_o} = \frac{100.741\,(0.000668)(401.655)}{125.47} = 0.215425$$

$$T_o^* = T_o + \frac{P_v}{\gamma^*} = 44.512 + \frac{2.7}{0.215425} = 57.05\ °\text{C}$$

This is a very high value, indicating thermal stress on the animals. If animals were protected against the high thermal radiation, under a lower mean radiant temperature, say $T_{rm} = 35°\text{C}$, then the equivalent operative temperature would be about 46.3°C.

Problem 7.2. Considering the environmental records given in Problem 7.1, calculate the enthalpy of the environment.

Data

T_a = air temperature = 31°C
P_v = partial vapour pressure = 2.7 kPa

Solution
Saturation vapour pressure at temperature T_a:

$$P_v = 0.61078 \times 10^{7.5\,T_a/(T_a+237.5)}$$
$$= 0.61078 \times 10^{7.5\,(31)/(31+237.5)} = 4.4855\,\text{kPa}$$

Enthalpy

$$H = 6.7 + 0.243\,T_a + 2.216\left[\frac{P_v}{P_s(T_a)}10^{7.5\,T_a/(T_a+237.3)} - 1\right]$$
$$= 6.7 + 0.243\,(31) + 2.216\left[\frac{2.7}{4.4855}10^{7.5\,(31)/(31+237.5)} - 1\right] = 21.8\ \text{kJ/kg}$$

Problem 7.3. Two locations in an equatorial region (5° and 6° south latitude, respectively) presented the environmental averages shown in the following table, together with rectal temperature and respiratory rate averages obtained in the same period of time from Holstein dairy cows bred in the open field. Calculate the respective indexes HLI, RR and ITSC (Eqs. 7.16, 7.17, and 7.18).

Variable	Location A	Location B
Air temperature, °C	31.2	29.4
Globe temperature, °C	40.7	41.1
Wind speed, m s^{-1}	2.17	1.03
Partial vapour pressure, kPa	3.00	3.21
Relative humidity, %	66.9	78.2
Solar radiation, W m^{-2}	912.7	787.4
Rectal temperature, °C	39.3	39.8
Respiratory rate, breaths min^{-1}	75.8	66.4

Solution

Location A

$$\begin{aligned}\mathrm{RR} &= 5.4\,T_a + 0.58\,U_R - 0.63\,U + 0.024\,\mathbf{S} - 110.9\\ &= 5.4\,(31.2) + 0.58\,(66.9) - 0.63\,(2.17) + 0.024\,(912.7) - 110.9\\ &= 116.9\end{aligned}$$

$$\begin{aligned}\mathrm{ITSC} = {} & 7.9505 + 0.0667\,T_g + 0.0673\,U + 0.0214\,U^2 - 1.9005\,P_v\\ & + 0.1749\,P_v^2 + 0.045\,T_a P_v - 0.0095\,T_g U\\ = {} & 7.9505 + 0.0667\,(40.7) + 0.0673\,(2.17) + 0.0214\,(2.17^2)\\ & - 1.9005\,(3.0) + 0.1749\,(3.0^2) + 0.045\,(31.2)(3.0)\\ & - 0.0095\,(40.7)(2.17) = 10.2\end{aligned}$$

$$\begin{aligned}\mathrm{HLI} &= 8.62 + 0.38\,U_R + 1.55\,T_g - 0.5\,U + e^{2.4-U}\\ &= 8.62 + 0.38\,(66.9) + 1.55\,(40.7) - 0.5\,(2.17) + e^{2.4-2.17} = 97.3\end{aligned}$$

Location B

$$\begin{aligned}\mathrm{RR} &= 5.4\,T_a + 0.58\,U_R - 0.63\,U + 0.024\,\mathbf{S} - 110.9\\ &= 5.4\,(29.4) + 0.58\,(78.2) - 0.63\,(1.03) + 0.024\,(787.4) - 110.9\\ &= 111.5\end{aligned}$$

$$\begin{aligned}\text{ITSC} =&7.9505 + 0.0667\,T_g + 0.0673\,U + 0.0214\,U^2 - 1.9005\,P_v\\ &+ 0.1749\,P_v^2 + 0.045\,T_a P_v - 0.0095\,T_g U\\ =&7.9505 + 0.0667\,(41.1) + 0.0673\,(1.03) + 0.0214\,(1.03)^2\\ &- 1.9005\,(3.21) + 0.1749\,(3.21)^2 + 0.045\,(29.4)(3.21)\\ &- 0.0095\,(41.1)(1.03) = 10.3\end{aligned}$$

$$\begin{aligned}\text{HLI} &= 8.62 + 0.38\,U_R + 1.55\,T_g - 0.5\,U + e^{2.4-U}\\ &= 8.62 + 0.38\,(78.2) + 1.55\,(41.1) - 0.5\,(1.03) + e^{2.4-1.03}\\ &= 105.5\end{aligned}$$

Summary of the Results

	RR	ITSC	HLI
Location A	116.9	10.2	97.3
Location B	111.5	10.3	105.5

The RR results were high for both locations, and by the respective scale, the cows in those places are at risk of danger (RR = 110–133). At the same time, those RR values are (by definition) estimated values of the respiratory rates of the animals under the given environmental conditions; however, the really observed F_r values were much lower (75.8 and 66.4 breaths min^{-1}, respectively); at the same time, body temperatures were 39.3 and 39.8°C, within the normal limits for cattle in tropical regions. Therefore, the RR index seems to be not able to evaluate the effects of the equatorial environment on the cows.

The HLI showed values higher than that referred as "very hot" (>86) in the respective scale. However, the observed air temperatures were 31.2 and 29.4°C, which were not high enough to increase the respiratory rate above the normal limits. Perhaps a new scale would be established for cows in tropical environments. As for the ITSC index, it showed similar results for both locations, whose environmental conditions were classified as "mild stress" (ITSC from 10 to 11). It must be remembered that this index was established by using data from animals bred in an equatorial region.

References

Baeta, F.C., Meador, N.F., Shanklin, M.D., Johnson, H.D.: Equivalent temperature index at temperatures above the thermoneutral for lactating dairy cows. In: Summer Meeting of American Society of Agricultural Engineers, 1987, Baltimore, MD, June 28–July 1, 21 p. ASAE, St. Joseph, MI (1987)

Barbosa, O.R., Silva, R.G., Scolar, J., Guedes, J.M.F.: Utilização de um índice de conforto térmico no zoneamento bioclimático da ovinocultura [Use of a Thermal Comfort Index in the Bioclimatic Zoning of Sheep Breeding]. Braz. J. Anim. Sci. **24**, 661–671 (1995)

Beckett, F.E.: Effective temperature for evaluating or designing hog environments. Trans. ASAE **8**, 163–166 (1965)

Berman, A.: Estimates of heat stress relief needs for Holstein dairy cows. J. Anim. Sci. **83**, 1377–1384 (2005)

Bianca, W.: Relative importance of dry- and wet-bulb temperature in causing heat stress in cattle. Nature (London) **195**, 251–252 (1962)

Bohmanova, J., Misztal, I., Cole, J.B.: Temperature-humidity indices as indicators of milk production losses due to heat stress. J. Dairy Sci. **90**, 1947–1956 (2007)

Buffington, D.E., Collazo-Arocho, A., Canton, G.H., Pitt, D.: Black globe-humidity index (BGHI) as comfort equation for dairy cows. Trans. ASAE **24**, 711–714 (1981)

Cargill, B.F., Stewart, R.E.: Effect of humidity on total heat and total vapor dissipation of Holstein cows. Trans. ASAE **9**, 702–706 (1966)

Dikmen, S., Hansen, P.J.: Is the temperature-humidity index the best indicator of heat stress in lactating dairy cows in a subtropical environment? J. Dairy Sci. **92**, 109–116 (2009)

Eigenberg, R.A., Brown-Brandl, T.M., Nienaber, J.A., et al.: Dynamic response of feedlot cattle to shade and no-shade. In: Meeting of the ASAE, St. Joseph (2002) [mimeo]

Eigenberg, R.A., Brown-Brandl, T.M., Nienaber, J.A., Hahn, G.L.: Dynamic indicators of heat stress in shaded and non-shaded feedlot cattle. Part 2. Predictive relationships. Biosyst. Eng. **91**, 111–118 (2005)

Eigenberg, R.A., Brown-Brandl, T.M., Nienaber, J.A.: Shade material evaluation using a cattle response model and meteorological instrumentation. Int. J. Biometeorol **34**, 601–607 (2010)

Eigenberg, R.A., Nienaber, J.A., Brown-Brandl, T.M.: Development of a livestock safety monitor for cattle. In: Meeting of the ASAE, St. Joseph (2003) [mimeo]

Epstein, Y., Moran, D.S.: Thermal comfort and the heat stress indices. Ind. Heal. **44**, 388–398 (2006)

Faddeeva, V.N.: Computational Methods of Linear Algebra, 252 p. Dover Publications, New York (1959)

Gaughan, J.G., Goopy, J., Spark, J.: Excessive heat load index for feedlot cattle. In: Report of the Meat and Livestock-Australia Project, FLOT.316. MLA Ltd., Sydney (2002)

Gaughan, J.G., Mader, T.L., Holt, S.M., Lisle, A.: A new heat load index for feedlot cattle. J. Anim. Sci. **86**, 226–234 (2008)

Hahn, G.L.: Management and housing of farm animals in hot environments. In: Yousef, M.K. (ed.) Stress Physiology of Livestock, vol. 2. CRC Press, Boca Raton (1985)

Hahn, G.L., Mader, T.L., Eigenberg, R.A.: Perspective on development of thermal indices for animal studies and management. In: Lacetera, N., et al. (eds.) Interactions Between Climate and Animal Production. European Association of Animal Production Technical Series Nº 7, pp. 31–44. Wageningen Academic Publishers, Wageningen (2003)

Hazel, L.N.: The genetic basis for constructing selection indexes. Genetics **28**, 476–490 (1943)

Herrington, L.P., Winslow, C.E.A., Gagge, A.P.: The relative influence of radiation and convection upon vasomotor temperature regulation. Am. J. Physiol. **120**, 133–143 (1937)

Hillman, P.E., Lee, C.N., Parkhurst, A.: Impact of hair color on thermoregulation of dairy cows to direct sunlight. In: Annual International Meeting of the American Society of Agricultural Engineers, 94, Sacramento, CA, ASAE Paper no. 014031, 29 July–1 Aug 2001

Ingraham, R.H., Stanley, R.W., Wagner, W.C.: Relationship of temperature and humidity to conception rate of Holstein cows in Hawaii. J. Dairy Sci. **59**, 2086–2090 (1976)

Johnson, H.D., Ragsdale, A.C., Berry, I.L., Shanklin, M.D.: Effects of various temperature-humidity combinations on milk production of Holstein cattle. Mo. Agric. Exp. Sta. Res. Bull. nr. 791 (1962)

Johnson, H.D., Ragsdale, A.C., Berry, I.L., Shanklin, M.D.: Temperature-humidity effects including influence of acclimation in feed and water consumption of Holstein cattle. Mo. Agric. Exp. Sta. Res. Bull. nr. 846 (1963)

Johnson, R.A., Wichern, D.W.: Applied Multivariate Statistical Analysis, 2nd edn., 607 p. Prentice-Hall, Englewood Cliffs (1988)

Kerslake, D.M.K.: The Stress of Hot Environments, p. 316 p. Cambridge University Press, Cambridge (1972)

Matzarakis, A., Rutz, F., Mayer, H.: Modelling radiation fluxes in simple and complex environments – application of the RayMan model. Int. J. Biometeorol. **51**, 323–334 (2007)
McArthur, A.J.: Thermal interaction between animal and microclimate: a comprehensive model. J. Theor. Biol. **126**, 203–238 (1987)
Mitchell, M.A., Kettlewell, P.J., Hunter, R.R., Carlisle, A.J.: Physiological stress response modeling – application to the broiler transport thermal environment. In: Proceedings of the 6th International Livestock Environment Symposium, Louisville, 21–23 May 2001
Moran, D.S., Pandolf, K.B., Shapiro, Y., Laor, A., Heled, Y., Gonzalez, R.R.: Evaluation of the environmental stress index for physiological variables. J. Therm. Biol. **28**, 43–49 (2003)
Moura, D.J., Nääs, I.A., Silva, I.J.O., Sevegnani, K.B., Corria, M.E.: The use of enthalpy as a thermal comfort index. In: Proceedings of the 5th Livestock Environment Symposium, vol. 1, pp. 577–583, Bloomington, MN (1997)
Olson, T.A., Lucena, C., Chase Jr., C.C., Hammond, A.C.: Evidence of a major gene influencing hair length and heat tolerance in *Bos taurus* cattle. J. Anim. Sci. **81**, 80–90 (2003)
Pinheiro, M.G.: Variação genetic de caracteristicas da capa externa de vacas da raça Holandesa em ambiente tropical [Genetic Variation of Hair Coat Characteristics of Holstein Cows in Tropical Environment]. Doctoral thesis, University of São Paulo, Ribeirão Preto, Brazil (1996)
Rencher, A.C.: Methods of Multivariate Analysis, 627 p. Wiley, New York (1995)
Rhoad, A.O.: The Iberia heat tolerance test for cattle. Trop. Agric. **21**, 162–164 (1944)
Riemerschmidt, G.: The amount of solar radiation and its absorption on the hair coat of cattle under South African and European conditions. J. S Afr. Vet. Med. Assoc. **14**, 121–141 (1943)
Searle, S.R.: Matrix Algebra for the Biological Sciences, 296 p. Wiley, New York (1966)
Shioya, S., Terada, F., Iwama, Y.: Physiological responses of lactating dairy cows under hot environments. Eiyoseiri-kenkyukaiho **4**, 61–68 [English abstract] (1997)
Silva, R.G.: Improving tropical beef cattle by simultaneous selection for weight and heat tolerance. Heritabilities and correlations of the traits. J. Anim. Sci. **37**, 637–642 (1973)
Silva, R.G.: índices de seleção para ganho de peso e tolerância ao calor no gado Canchim e progresso genético esperado em população simulada [Selection Indexes for Weight Gain and Heat Tolerance of Canchim Cattle and Genetic Progress in a Simulated Population]. Doctoral thesis, 98 p, University of São Paulo, Ribeirão Preto (1975)
Silva, R.G.: Um modelo para a determinação do equilíbrio térmico de bovinos em ambientes tropicais [A model for determination of thermal balance of cattle in tropical environments]. Braz. J. Anim. Sci. **29**, 1244–1252 (2000)
Silva, R.G., Barbosa, O.R.: Thermal comfort index for sheep. In: Proceedings of the 13th International Congress of Biometeorology, vol. 2, pp. 383–392, Calgary, AB, Canada (1993)
Silva, R.G., Arantes-Neto, J.G., Holtz-Filho, S.V.: Genetic aspects of the sweating rate and coat characteristics of Jersey cattle. Brazil. J. Genet. **11**, 335–347 (1988)
Silva, R.G., Guilhermino, M.M., Morais, D.A.E.F.: Thermal radiation absorbed by dairy cows in pasture. Int. J. Biometeorol. **54**, 5–11 (2010). doi:10.1007/s00484-009-0244-lSilva, R.G., LaScala Jr., N., Tonhati, H.: Radiative properties of the skin and haircoat of cattle and other animals. Trans. ASAE **46**, 913–918 (2003)
Silva, R.G., Morais, D.A.E.F., Guilhermino, M.M.: Evaluation of thermal stress indexes for dairy cows in tropical regions. Braz. J. Anim. Sci. **36**, 1192–1198 (2007)
Silva, I.J.O., Pandolfi, H., Piedade, S.M.S.: Influência do sistema de alojamento no comportamento e bem-estar de matrizes suínas em gestação [Influence of housing system on the behavior and well-being of pregnant sows]. Braz. J. Anim. Sci. **37**, 1319–1329 (2008)
Silva, R.G., Maia, A.S.C., Costa, L.L.M.: A thermal stress índex for dairy cows in intertropical regions. In: Proceedings of the 19th International Congress of Biometeorology, 4–8 December 2011, Auckland, New Zealand. Submission #016
Thom, E.C.: The discomfort index. Weatherwise **12**, 57–60 (1959)
Villa Nova, N.A., Ometto, J.C., Salati, E.: Aspectos termodinâmicos da atmosfera, 24 p. CENA/ESALQ, Piracicaba (Boletim Didático no 6) (1972)
Yamamoto, S.: The assessment of thermal environment for farm animals. In: Proceedings of the 5th World Congress on Animal Production, vol. 1, pp. 197–204, Tokyo (1983)
West, J.W.: Effects of heat stress on production of cattle. J. Dairy Sci. **86**, 2131–2144 (2003)

Chapter 8
Special Methods

Abstract The methods of metabolism determination by indirect calorimetry are detailed. Some structural and functional aspects of ventilated capsules and facial masks are discussed. Practical methods to evaluate cutaneous and respiratory evaporation in mammals and birds are described. Techniques for the sampling of hairs and the evaluation of hair coat characteristics.

Keywords Carbon dioxide • Cutaneous evaporation • Facial masks • Hair coat • Metabolic rate • Methane • Nitrogen • Ventilated capsules • Oxygen • Respiratory evaporation

8.1 Metabolic Rate Evaluation

As it was shown in Chap. 3, the measurement of metabolic rate is based on direct and indirect procedures. While the direct methods involve determination of the total amount of heat produced, in the indirect ones, there was evaluated the heat production by aerobic processes in the tissues. Such heat production is determined through the oxygen consumption, together with carbon dioxide and methane production; sometimes, nitrogen content of urine is also considered. This is the most widely used procedure. In Table 3.3, there are presented the equations for the evaluation of the metabolic rate in ruminants, nonruminants and birds.

There is generally considered that O_2 and CO_2 together contribute with 99% of the heat produced, while 1% only can be attributed to the CH_4 and the N (Brouwer 1965; Kleiber 1975). Therefore, the metabolic rate can be estimated without significant error through the amount of O_2 consumed and CO_2 produced, as follows:

Ruminants:

$$M = 16.18\,O_2 + 5.016\,CO_2 \quad \text{kJ} \tag{8.1}$$

R. Gomes da Silva and A.S. Campos Maia, *Principles of Animal Biometeorology*, Biometeorology 2, DOI 10.1007/978-94-007-5733-2_8,

Nonruminants:

$$M = 16.57\,O_2 + 4.5\,CO_2 \quad kJ \tag{8.2}$$

Birds:

$$M = 16.2\,O_2 + 5.0\,CO_2 \quad kJ \tag{8.3}$$

However, McLean (1972) showed that the metabolic heat production can be evaluated by the amount of consumed O_2 only with a precision of about 2%, except in cases of abnormal feeding conditions; this author suggested the following formula:

$$M = 20.44\,O_2\ kJ \tag{8.4}$$

By using the coefficients presented in Table 3.3 and based on the results of Brouwer (1965), Lighton (2008) and Nienaber et al. (2009), the metabolic rate of goats exposed to sun and under shade in an semiarid region was determined by Maia et al. (2011) as determined the metabolic rate of goats exposed to sun and under shade in an equatorial semiarid region, by using the following formula:

$$\mathbf{M} = F_R V_T A^{-1}\left(0.7633\,Q_O\,O_{cons} + 0.25\,Q_{CO}CO_{prod}\right)\ W.m^{-2} \tag{8.5}$$

where F_R is the respiratory rate (breaths s^{-1}), V_T is the tidal volume (L breath^{-1}) and Q_O and Q_{CO} are heat coefficients of O_2 and CO_2, respectively (J·L^{-1}). The body surface area was estimated from the body weight (kg) as $A = 0.13\,W^{0.556}$ (m^2). The values O_{cons} and CO_{prod} are the proportions of O_2 consumed and CO_2 produced, respectively:

$$O_{cons} = O_{atm} - O_{exp}$$
$$CO_{prod} = CO_{exp} - CO_{atm}$$

For birds, Nascimento et al. (2011) used the same Eq. 8.5 but changing the coefficients 0.7633 and 0.25 by 0.7642 and 0.2358, respectively.

The amount of heat generated, as evaluated by the O_2 consumed and the CO_2 produced, depends on the substrate considered in the metabolic reactions, as it is illustrated in Table 8.1.

A major difficulty in the estimation of the metabolic heat by using the above equations is the determination of V_T and F_R values; however, these values can be obtained from the facial mask method used by Maia et al. (2011), as illustrated by Fig. 8.1. Such a mask can be made from a plastic bottle (5 L for cattle, 0.5 L for sheep and goats, 10 mL for chicken). See also Figs. 8.2, 8.3, and 8.4.

The mask for cattle must have a saliva collector placed just bellow it, and the space between animal's head and the mask is closed by silicone or rubber foam. Two PVC tubes (3 cm internal diameter) are placed just in front of the animal's

Table 8.1 Heat produced by metabolic reactions (resting metabolism) according to the type of substrate

Substrate	O_2 consumed (kJ L^{-1})	CO_2 produced (kJ L^{-1})	CO_2/O_2
Lipids	19.7	27.8	0.80
Proteins	19.2	23.8	0.70
Carbohydrates	21.2	21.2	1.00
Mean	20.03	24.27	

Adapted from Kleiber (1975), Blaxter (1989), Schmidt-Nielsen (2002) and Randall et al. (2001)

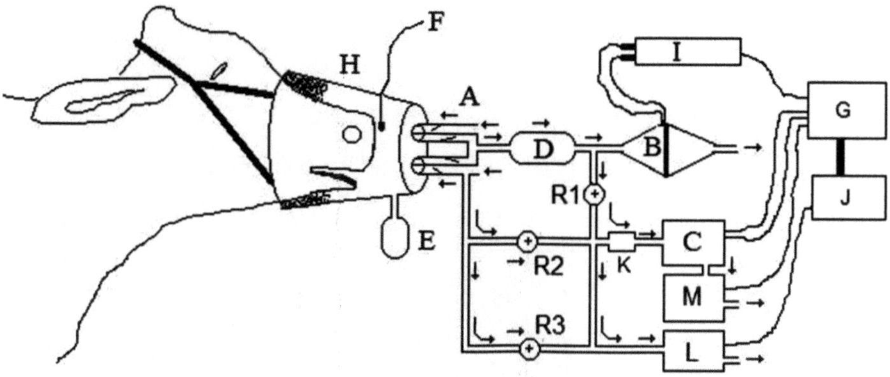

Fig. 8.1 Schematic representation of the facial mask method used for indirect calorimetry in ruminants. (*A*) Air input and output, (*B*) air flux metre, (*C*) O_2/CO_2 analyzer, (*D*) air mixer, (*E*) saliva collector, (*F*) temperature sensor, (*G*) data centre, (*H*) facial mask, (*I*) spirometer, (*J*) computer, (*K*) air dryer, (*L*) H_2O analyzer, (*M*) methane analyzer, (R1, R2, R3) sluices (Maia et al. 2011)

nostrils, both provided of valves (one for the inspired and the other for the expired air). The expired air flows into the gas mixer and then to the flux metre; when R1 sluice is open and R2 and R3 are closed, the expired air goes into the O_2/CO_2, methane and H_2O analyzers, which give the proportion of those expired gases. When R1 is closed and R2 and R3 open, samples of the inspired air flow into the gas analyzers, thus giving the proportion of the gases in the inspired air. The tidal volume and the respiratory rate are given by the flux metre B, while the temperature of the expired air is taken by a temperature sensor (*F*) placed in front of the animal's nostrils.

8.2 Evaluation of Latent Heat Loss

8.2.1 Respiratory Surfaces

In Chap. 4, it was explained that the latent heat loss from respiratory surfaces can be described by dividing Eq. 4.33:

$$\mathrm{E_R} = \lambda \dot{m} \rho^{-1} (\Psi_{\mathrm{EXP}} - \Psi_{\mathrm{A}})\ \mathrm{W}$$

Fig. 8.2 Facial mask used to evaluate the metabolic heat production of cattle (Maia et al. 2011)

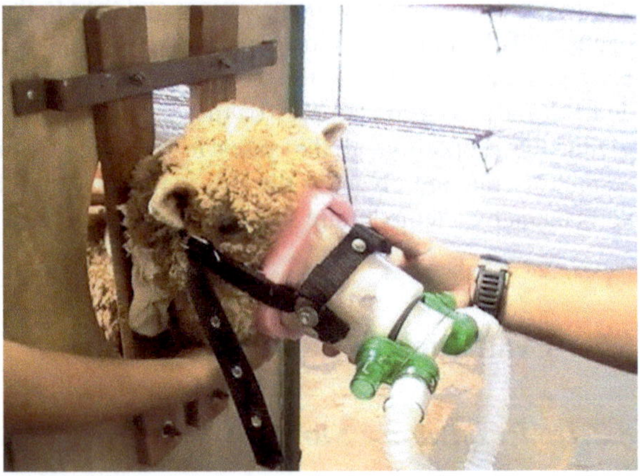

Fig. 8.3 Facial mask used to evaluate the metabolic heat production of a sheep

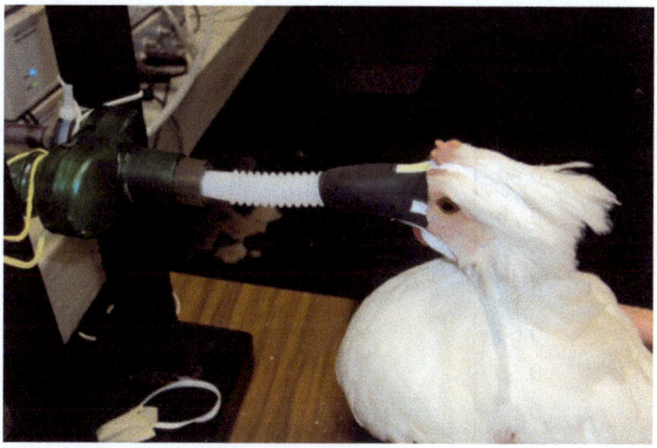

Fig. 8.4 Facial mask used to evaluate the metabolic heat production of chickens (Nascimento et al. 2011)

where $\dot{m}$ is the mass flux rate (kg·s^{-1}), λ is the latent heat of vapourisation (J·g^{-1}) and ρ the air density (g·m^{-3}) at the temperature of the expired air (T_{EXP}); Ψ_A and Ψ_{EXP} are the absolute humidities of the atmosphere and the expired air (g·m^{-3}).

The main problem with the above equation is that it is given in terms of W, because in general we need an expression in terms of W m^{-2}. In that case, instead of dividing the above equation by the body, an adequate option would be the use of Eq. 4.37a, which is

$$\mathbf{E}_R = \frac{\rho c_p \left(P_s(T_{exp}) - P_v\right)}{P_a \gamma r_{Vr}} \ \mathrm{W\,m^{-2}}$$

where $P_S(T_{EXP})$ is the saturation vapour pressure at the expired air temperature (kPa) and r_{Vr} is the resistance for the vapour transfer at the respiratory surfaces, which is a function of the respiratory rate F_R (Eq. 4.38). This equation allows us to use the facial mask previously described in order to evaluate the latent heat loss from respiratory surfaces.

If an animal inhales the atmospheric air at an absolute humidity Ψ_A through a facial mask and exhales it at the absolute humidity Ψ_{EXP}, then

$$\Psi_{EXP} - \Psi_A = \beta(\Psi_{mask} - \Psi_A) \quad (8.6)$$

where β is the fraction of the difference between the absolute humidity of the air leaving the mask (Ψ_{mask}) and that of the expired air (Ψ_{EXP}), given by

$$\beta = 1 - \frac{e^K - 1}{K(e^K + 1)} \quad (8.7)$$

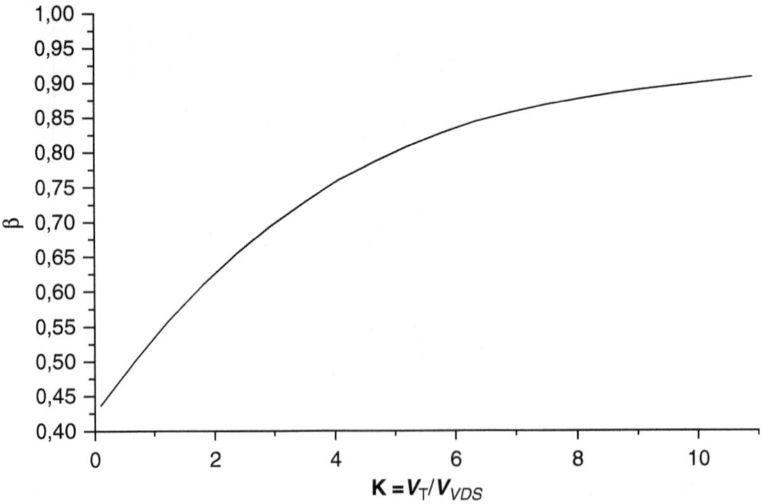

Fig. 8.5 Fraction of the difference between the absolute humidity of the air leaving the mask and that of the air exhaled by the animal (β), as a function of the relation of the tidal volume to the volume of dead space within the mask

$$K = V_{\mathrm{T}}/V_{\mathrm{VDS}}$$

where V_{T} is the tidal volume and V_{VDS} is the volume of the ventilated dead space within the mask. Then, Ψ_{mask} will be the very absolute humidity of the expired air only if the dead space within the mask be zero ($V_{\mathrm{VDS}} = 0$); in such a case, $\beta = 1$ and $K \to \infty$, according to McLean (1963b).

If the value of V_{VDS} is great enough, the humidity content of the air into the mask can increase progressively until it reached a level between the atmospheric humidity and that of the air exhaled within the mask ($\beta = 0.5$, $K \to 0$). Therefore, the animal exhales within the mask at a humidity level higher than that leaving out the mask; however, it does not mean that the evaporative loss is incorrectly measured, for it varies according to the humidity of the air inhaled by the animal within the mask, which increases in equal proportion relatively to that of the atmosphere (depending on V_{VDS}). It is very difficult to develop a mask for animals with $V_{\mathrm{VDS}} \to 0$. In Fig. 8.5, it is apparent that as the K value increases Ψ_{mask} tends to Ψ_{EXP}.

Therefore, when a mask with input and output sluices is applied to an animal, the loss of humidity is reduced, unless the dead space is much reduced in relation to the tidal volume. In such a case, the measured evaporative loss will be the real value of the respiratory humidity lost by the animal.

The mask used by Maia et al. (2005a) showed $V_{\mathrm{VDS}} = \pm 1$ L on the average, while the tidal volume varied from 2.5 to 9 L. Consequently, its β was 0.66–0.89 (average 0.83).

8.2.2 Cutaneous Surfaces

8.2.2.1 Capsule Description

Cutaneous evaporation can be evaluated by several methods, as the colorimetric one (Schleger and Turner 1965; Silva 2000; Pereira et al. 2010), the ventilated capsule method (McLean 1963a; Gatenby 1980), the gravimetric method (Holmes 1985; Finch et al. 1982; Silva and Starling 2003) and the spot calorimeter (Nienaber et al. 2009).

When using the ventilated capsule method, the flux of latent heat from the body surface can be estimated as the product of the evaporation rate at the surface ($\mathbf{\Phi}_{\mathrm{Ev}}$) by the latent heat of vapourisation at the surface temperature (λ, J·g^{-1}), considering that

$$\mathbf{\Phi}_{\mathrm{Ev}} = \frac{\mathbf{\Phi}_{\mathrm{cap}}\left(\Psi_{\mathrm{cap}} - \Psi_{\mathrm{A}}\right)}{A_{\mathrm{cap}}} \ \mathrm{g.m^{-2}.s^{-1}} \tag{8.8}$$

where Ψ_{cap} is the absolute humidity of the air going away from the capsule (g m^{-3}) and A_{cap} is the section area of the capsule (m^2); Φ_{cap} is the air flux throughout the capsule, which is given by

$$\Phi_{\mathrm{cap}} = \pi r^2 U_{\mathrm{cap}} \ \mathrm{m^3\ s^{-1}} \tag{8.9}$$

where r is the radius (m) of the capsule output tube and U_{cap} the speed of the air (m s^{-1}) passing through the capsule. Both U_{cap} and the air temperature within capsule can be evaluated by means of a hot-wire anemometer placed in the capsule output.

Figure 8.6 depicts the schematic description of the ventilated capsule used by Maia et al. (2005b) and based on that described by McLean (1963a). It was made from a 6.3-cm-long section of a plastic PVC tubing of 7 cm diameter. The atmospheric air is sucked into the tube A by a pump placed in the point D; this tube has a projection within the capsule that is placed 1 cm above the hair coat surface and has several holes in order to spread the entering air over the test area. Tubes B and D are connected to a H_2O analyzer. If the sluice I_2 is open while I_1 is closed, the partial vapour pressure of air entering the capsule, $P_{\mathrm{v}}(T_{\mathrm{A}})$, is measured by the H_2O analyzer; if I_2 is closed and I_1 open, it evaluated the vapour pressure of the air leaving the capsule, $P_{\mathrm{S}}(T_{\mathrm{cap}})$.

Cutaneous evaporation must be evaluated in the different body regions, as, for example, the trunk (about 18 cm below the backbone), the neck and the hindquarter. The different body regions have different sweating rates, and therefore, the greater the number of measured regions, the more accurate will be the estimate of animal's cutaneous evaporation.

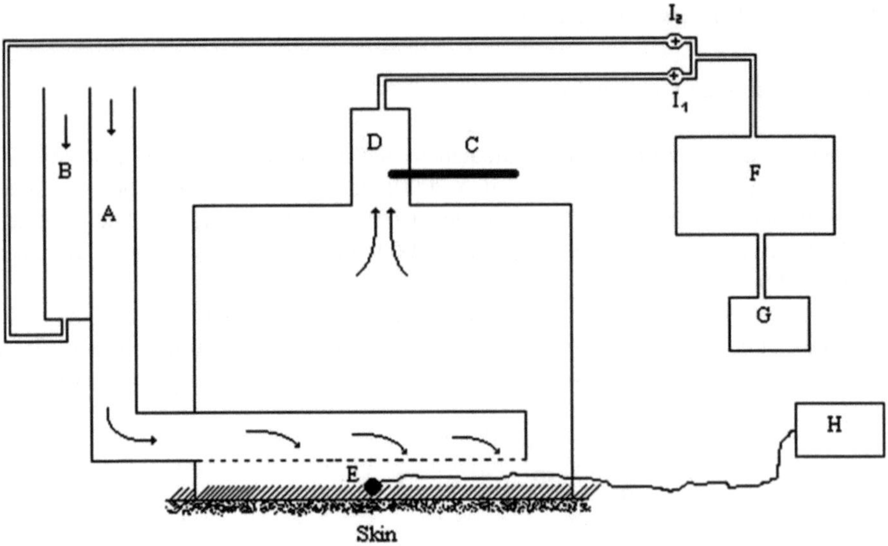

Fig. 8.6 Ventilated capsule used by Maia et al. (2005b). A = capsule air input, B = atmospheric air input, C = hot-wire anemometer, D = air output and pump, E and H = infrared thermo anemometer, F = H_2O/CO_2 analyzer, G = gas compressor, I_1 and I_2 = sluices

8.2.2.2 Theoretical Aspects of the Ventilated Capsule

According to McLean (1963a), when a ventilated capsule is applied to the cutaneous surface, the physical conditions over that surface can be modified as follows:

1. The rate of air flow over the skin changes, thus affecting the heat losses by both convection and evaporation from the area covered by the capsule.
2. Any change of the temperature and humidity of the air within the capsule affects also both convection and evaporation.
3. The mean radiant temperature of the body area covered by the capsule is changed, affecting the local heat exchange by radiation.
4. The pressure applied to the capsule in order to secure it on the skin surface leaves to changes in the local blood flow, which can affect all mechanisms of local heat exchange.
5. The speed of the air through the capsule must be regulated in order to maintain the air flux, Φ_{cap}, in such a way that the gradient temperature within the capsule be null or very close to zero. It is possible that variations in U_{cap} can affect the skin surface temperature under the capsule and consequently the sensible heat exchange, so making the estimate of evaporative heat loss questionable.

As it was observed by McLean (1963a), an increase of Φ_{cap} results in the fall of the air temperature under the capsule and an increased heat loss by evaporation, $\mathbf{E}_S$, considering that $\log(\mathbf{E}_S)$ as a function of $\log(\Phi_{cap})$. However, there had determined cases in which Φ_{cap} did not affect the temperature and the gradient temperature was zero; such Φ_{cap} value was called as the critical flux, Φ_{crit}. Taneja (1959) observed an increase of the evaporative rate from 75 to 200 g m^{-2} h^{-1} when Φ_{cap} varied from 0.5 to 2 L m^{-1} under the same air temperature conditions.

The cutaneous evaporative rate under a ventilated capsule can be assumed as equivalent to that occurring on a non-covered skin area, if the critical air flux is that used in the ventilation. However, if other air flux is used, the evaporation to be measured is changed by a factor:

$$(\Phi\,\mathrm{cap}/\Phi\,\mathrm{crit})^{\omega} \tag{8.10}$$

where ω measures the degree by which evaporation is affected by the ventilation rate; it will be small and $\omega \to 0$ for a capsule with a great internal volume that is also able to promote a homogeneous spreading of the air over the test area.

McLean (1963a) strongly recommends the use of capsules with an air flux $\Phi_{cap} = 1.2$ L min^{-1}, similar to that used in his experiments. On the contrary, the evaporation rate $\mathbf{E}_S$ should be weighted by the correction factor 8.10.

The main disadvantage of the above-described capsule is that leakage of air can occur through the hair coat below the capsule. It is possible to detect such a leakage by putting a hot-wire anemometer at the air output of the capsule; the measured air flux must be constant, so that there has no leakage; on the other hand, the measured $\mathbf{E}_S$ should be discarded.

8.3 Enteric Methane

There has some amount of CH_4 in the air expired by animals, which is produced by fermentations within the digestive system. It can be evaluated by means of a portable methane analyzer coupled to a facial mask, as it can be seen in Fig. 8.1.

The enteric methane can be evaluated from the expired air by using the following equation:

$$\mathbf{E}_{CH4} = F_R V_T \left(CH4_{exp} - CH4_{atm}\right)\ \mathrm{L\,s^{-1}} \tag{8.11}$$

where $CH4_{exp}$ and $CH4_{atm}$ are the CH_4 proportions in the expired air and in the atmosphere, respectively.

8.4 Evaluation of Hair Coat Traits

8.4.1 Coat Thickness

When the coat layer (wool, hairs or plumage) is not much thick, its depth can be measured by means of a thin metallic rule provided with a slide. The rule is

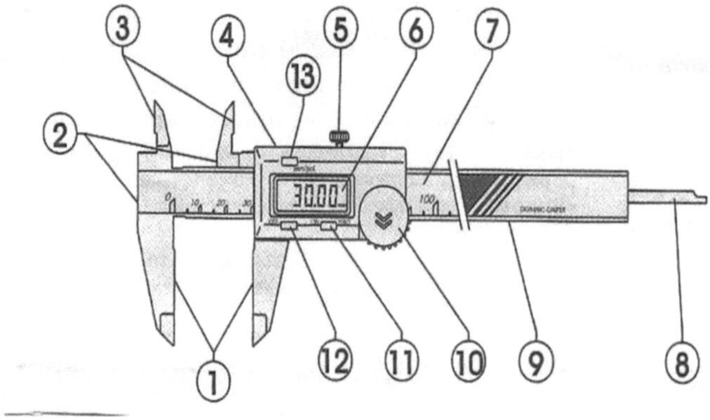

Fig. 8.7 Digital calliper or pachymeter can be used to evaluate animals' coat layer. Coat thickness is measured by using projection 8, while the jaws (*1*) are used to evaluate both the length and the thickness of the individual hairs. The result is given on display (6), which can be zeroed or changed by using buttons 11, 12, or 13. The other numbered parts are not used.

introduced into the coat layer until touching the skin surface; the slide is then displaced downwards until the outer surface of the coat.

The better device one can use even when the coat is too thin is a digital calliper or pachymeter, as illustrated in Fig. 8.7.

It is important to remember that coat thickness must be measured before any hair sample is done and always in the same place. If the coat thickness is measured in places other than those used for hair sampling, there would be no compatibility among the measures of these and of other coat traits.

8.4.2 *Hair Sampling*

The best method to take samples of a hair coat is yet that proposed by R.R. Shrode and described by Lee (1953). It used a common electrician's pliers whose jaws are ground down to a fairly small thickness, in order to allow an easy penetration of the hair coat. A nail is placed between the handles just at the pliers' pivot, keeping the jaws slightly apart so that the rectangular space formed by the jaw ends is about 0.2 cm^2 (measured with a calliper).

The sampling device is pressed on to the skin at a right angle to the surface; the nail is removed and then the jaws are closed tightly, catching the hairs and allowing them to be plucked out. Care must be taken about the following points: (a) The hairs must be brushed parallel to the line of the pliers' jaws before the sample is taken, in order to avoid pulling the hairs which merely overlap the sample area; (b) the handlers must be strongly pressed (otherwise, some hairs of the sampling area can escape to be plucked off); and (c) several samples must be taken from each body region.

The hair sample is then stored into a paper or plastic envelope or even in a glass vial.

8.4.3 Hair Coat Numeric Density

The hairs contained in a sample are spread out on a sheet of white paper and individually counted under some reading lens with the aid of a needle. This count must be converted into number of hairs per cm^2 of skin, as follows.

If the pliers' jaws were 8 mm wide and the space between them were 2.5 mm, then any sample taken with that device will refer to a real skin area of 0.8×0.25 cm $= 0.2$ cm^2. If that sample was given as 250 hairs, then the respective hair coat numeric density will be 1,250 hairs cm^{-2}.

When there have great differences among the hairs of a sample, it would be necessary to count two or more hair types separately; however, there has a wide variation in general, making it very difficult to separate hairs into one or two classes only (Udo 1978).

8.4.4 Hair Length

Because the wide variation generally found in the length of the hairs within a sample, it was suggested by Kassab (1964) that the longest hairs only were measured. This procedure can be justified by the fact that the coat thickness is affected in a great extent by the longest hairs. Udo (1978) used this method to evaluate the hair length in Holstein cows by measuring the 50 longest hairs of each sample, while Maia et al. (2003) considered the ten longest hairs only. On the other hand, Silva et al. (1988) measured all the hairs of each sample from Jersey cows.

The hairs of the sample can be spread on a graph paper sheet, and their lengths measured directly with the aid of a needle. Otherwise, the measurements can be done with a calliper or pachymeter.

8.4.5 Hair Diameter

Medullated hairs present greater diameters than the non-medullated ones; therefore, the hair sample must be divided into a class for the medullated hairs and another class for the non-medullated ones. The measurements are done by using a microscope provided with a millimetre scale.

Before measurements, the hair sample must be treated for 10 min in ether and then in hydrogen dioxide (H_2O_2) at a ten-volume concentration for 18 h, in order to make its medullation visible. The hairs are then washed in distilled water and mounted on a microscope glass slide with glycerol and a glass slide. The observations are done in the microscope under 400$\times$ magnification, considering 50 medullated hairs and 50 non-medullated ones (Udo 1978).

The hair diameter can also be measured by using a digital micrometre, as it was done by Maia et al. (2003).

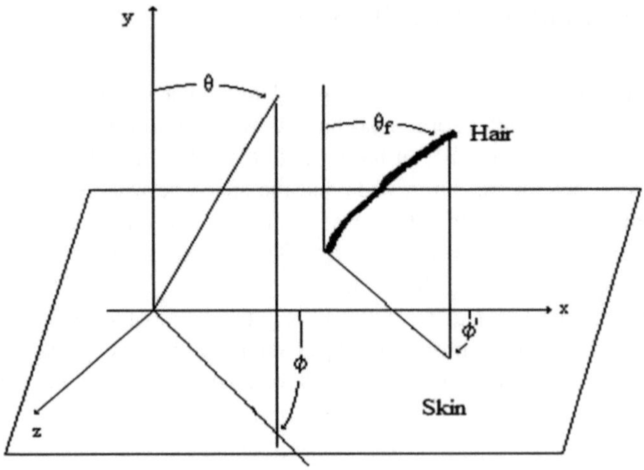

Fig. 8.8 Hair geometry: θ_f = angle of the hair in relation to the normal to the skin; ϕ' = angle between the hair projection on the sin and the hair dominant direction (Adapted from Davis and Birkebak 1974)

8.4.6 Thermal Conductivity of the Coat

The effective thermal conductivity of the hair coat (k_{ef}, mW·m^{-1}·K^{-1}) can be evaluated by assuming absence of air displacements within it and considering the system of Cartesian coordinates illustrated in Fig. 8.8 (Davis and Birkebak 1974).

According to Davis and Birkebak (1974), one needs only to calculate the energy flux in the y direction, in order to obtain k_{ef} by means of the following equation:

$$k_{ef} = k_1 \cos^2\theta_f + k_p \text{sen}^2\theta_f \tag{8.12}$$

For convenience,

$$k_1 = \left[\frac{\rho_{ef}}{\rho_p}\right]k_f + \left[1 - \left(\frac{\rho_{ef}}{\rho_f}\right)k_a\right] \tag{8.13}$$

where k_f, and k_a were thermal conductivities of the hairs and the air, respectively. According to Davis and Birkebak (1974), $k_f = 0.26$ W·m^{-1}·K^{-1} and $k_a = 0.025$ W·m^{-1}·K^{-1}.

Equation 8.12 demands knowledge of the fraction area parallel to the skin and the fraction volume of the coat that is effectively occupied by hair mass, at a given distance from the skin. Such a relation is

$$\frac{\rho_{ef}}{\rho_f} = \left(\frac{0.25N\pi D^2}{\cos\theta_f}\right) \tag{8.14}$$

where θ_f (decimal degrees) is the angle of the hairs in relation to the normal to the skin:

$$\theta_f = \arccos\left(\frac{E}{C}\right) \tag{8.15}$$

Thus,

$$\frac{\rho_{ef}}{\rho_f} = \frac{0.25CN\pi D^2}{E} \tag{8.16}$$

Thermal conductivity in a direction perpendicular to the hair axis (k_p mW·m^{-1}·K^{-1}) can be estimated after Kowalski (1978) in the following form:

$$k_p = \left(\frac{k_a(l_c - D)}{l_c}\right) + \left(\frac{Dk_a k_f}{Dk_a + (l_c - D)k_f}\right) \tag{8.17}$$

where $l_c = \frac{1}{\sqrt{N}}$.

Equation 8.12 gives the conductivity in the direction y within a hair coat layer formed by hairs that are oriented to the θ direction, but as the hairs of an actual hair coat were randomly oriented (Kowalski 1978), an average k_{ef} value (from values obtained with Eq. 8.12 for all directions) can be given by

$$k_{ef} = 0.5(k_1 + k_p) \tag{8.18}$$

However, radiative transfer through the coat (k_r, mW·m^{-1}·K^{-1}) is not considered in Eq. 8.18, because k_r can be estimated by theoretical calculations (Cena and Monteith 1975a, b) in which the thermal conductivity of the coat due to the radiation is defined as

$$k_r = \frac{4b}{3P} \tag{8.19}$$

where $b = 4\sigma\bar{T}_c^3$, σ is the Stefan-Boltzmann constant (5.67 × 10^{-8} W·m^{-2}·K^{-1}) and $\bar{T}_c$ is the temperature within the hair layer. Cena and Monteith (1975b) demonstrated that the b value increases from 5.7 W·m^{-2}·K^{-1} for $\bar{T}_c = 293$ K to 7.0 W·m^{-2}·K^{-1} for $\bar{T}_c = 313$ K. Of course, if the maximum temperature difference through the hair coat is less than 20 K, then b can have a value defined by a mean temperature $\bar{T}_c$. Then, it assumed a value $\bar{T}_c = 303$ K and consequently $b = 6.3097$ W·m^{-2}·K^{-1}.

Cena and Monteith (1975a) defined P as the fraction of the radiant energy that is intercepted by a unit hair thickness given by

$$P = ND(\tan\theta_f)$$

Consequently, k_r can be determined by

$$k_r = \frac{8.4129}{ND(\tan \theta_f)} \tag{8.20}$$

Finally, by combining Eqs. 8.17 and 8.20, the thermal conductivity of the coat due to the conduction and the radiation through the hair layer can be estimated as

$$k_{ef} = 0.5(k_1 + k_p) + k_r \tag{8.21}$$

References

Blaxter, K.L.: Energy metabolism in animals and man, 336 p. Cambridge University Press, Cambridge (1989)

Brouwer, E.: Report of subcommittee on constants and factors. In: Blaxter, K.L. (ed.) Energy Metabolism, pp. 441–443. London Academic press (1965)

Cena, K., Monteith, J.L.: Transfer processes in animal coats. I. Radiative transfer. Proc. R. Soc. Lond. B **188**, 377–393 (1975a)

Cena, K., Monteith, J.L.: Transfer processes in animal coats. II. Conduction and convection. Proc. R. Soc. Lond. B **188**, 395–411 (1975b)

Davis, L.B., Birkebak, R.C.: On the transfer of energy in layers of fur. Biophys. J. **14**, 249–268 (1974)

Finch, V.A., Bennett, I.L., Holmes, C.R.: Sweating responses in cattle and its relation to rectal temperature, tolerance of sun and metabolic rate. J. Agric. Sci. **99**, 479–487 (1982)

Gatenby, R.M.: Evaporimeter measuring sweat rate of cattle outdoors. J. Therm. Biol. **5**, 21–27 (1980)

Holmes, C.R.: Application of a weighing system for measuring total evaporative water losses in large ruminants. J. Thermal. Biol. **10**, 5–7 (1985)

Kassab, S.A.: On the maternal and some other influences on birth weight, growth and hair coat in two Dutch cattle breeds. Meded. Landbouwhogeschool Wageningen 64–1. Veenman & Zonen, Wageningen (1964)

Kleiber, M.: The Fire of Life: An Introduction to Animal Energetic, 2nd edn. Krieger Publications, New York (1975)

Kowalski, G.J.: An analytical and experimental investigation of the heat loss through animal fur. Ph.D. thesis, Mechanical Engineering Department, University of Wisconsin, Madison, 306 p (1978)

Lee, D.H.K.: Manual of Field Studies on Heat Tolerance of Domestic Animals, 161 p. Food and Agriculture Organization of the United Nations, Rome (1953)

Lighton, J.R.B.: Measuring Metabolic Rates: A Manual for Scientists, 201 p. Oxford University Press, New York (2008)

Maia, A.S.C., Silva, R.G., Bertipaglia, E.C.A.: Características do pelame de vacas Holandesas em um ambiente tropical: um estudo genético e adaptativo [Characteristics of the hair coat of Holstein cows in a tropical environment: a study on genetic and adaptative aspects]. Brazil. J. Anim. Sci. **32**, 843–853 (2003)

Maia, A.S.C., Silva, R.G., Loureiro, C.M.B.: Respiratory heat loss of Holstein cows in a tropical environment. Int. J. Biometeorol. **49**, 332–336 (2005a)

Maia, A.S.C., Silva, R.G., Loureiro, C.M.B.: Sensible and latent heat loss from the body surface of Holstein cows in a tropical environment. Int. J. Biometeorol. **50**, 17–22 (2005b)

Maia, A.S.C., Domingos, H.G.T., Araújo, F.Q.A., Chiquitelli-Neto, M., Silva, R.G.: Thermoregulation in goats managed in semiarid region: a Study of production, gain and loss of heat. In: Proceedings of the 19th International Congress of Biometeorology, Auckland, NZ (2011)

McLean, J.A.: Measurement of cutaneous moisture vaporization from cattle by ventilated capsules. J. Physiol. **167**, 417–426 (1963a)

McLean, J.A.: The partition of insensible losses of body weight and heat from cattle under various climatic conditions. J. Physiol. **167**, 427–447 (1963b)

McLean, J.A.: On the calculation of heat production from open-circuit calorimetric measurements. Brit. J. Nutr. **27**, 597–600 (1972)

Nascimento, S.T., Maia, A.S.C., Leite, L.G.: Respiratory flow and metabolism settlement in poultry. In: Proceedings of the 19th International Congress of Biometeorology, Auckland, NZ (2011)

Nienaber, J.A., DeShazer, J.A., Xin, H., Hillman, P.E., Yen, J.T., Ferrell, C.F.: Chapter 4. In: DeShazer, J.A. (ed.) Measuring Energetics of Biological Processes, pp. 73–109. American Society of Agricultural and Biological Engineers, Moscow (2009)

Pereira, A.M.F., Infante, A.A., Tito, E.L., Baccari Jr., F., Almeida, J.A.: A device to improve the Schleger and Turner method for sweating rate measurements. Int. J. Biometeorol. **54**, 37–43 (2010)

Randall, D., Burggren, W., French, K.: Eckert Animal Physiology: Mechanisms and Adaptation, 5th edn. W.H. Freeman, New York (2001)

Schleger, A.V., Turner, H.G.: Sweating rates of cattle in the field and their reaction to diurnal and seasonal changes. Aust. J. Agric. Res. **16**, 92–106 (1965)

Schmidt-Nielsen, K.: Animal Physiology: Adaptation and Environment, 5th edn. Cambridge University Press, Cambridge (2002)

Silva, R.G.: Introdução à Bioclimatologia Animal, 286 p. Nobel/FAPESP, São Paulo (2000)

Silva, R.G., Starling, J.M.C.: Evaporação cutânea e respiratória em ovinos sob altas temperaturas ambientes [Cutaneous and respiratory evaporation in sheep at high environmental temperatures]. Brazil. J. Anim. Sci. **32**, 1–6 (2003)

Silva, R.G., Arantes Neto, J.G., Holtz Filho, S.V.: Genetic aspects of the variation of the sweating rate and coat characteristics of Jersey cattle. Brazil. J. Genet. **11**, 335–347 (1988)

Taneja, G.C.: Cutaneous evaporative loss measured from limited areas and its relationship with skin, rectal, and air temperatures. J. Agric. Sci. **52**, 50–61 (1959)

Udo, H.M.: Hair coat characteristics in Friesian heifers in the Netherlands and Kenya. Meded. Landbouwhogeschool Wageningen 78–6. Veenman & Zonen, Wageningen, (1978) 135 p

Appendix

Modelling Tissue Temperature Distribution

Mathematical models are more and more used as tools in the study of biological systems, because of their advantages in relation to the experimental methods in terms of costs; besides, they can be used to predict situations in which a direct experiment would not possible. Realistic thermal control models had been developed for humans (Albuquerque Neto 2010; Ferreira and Yanagihara 2009; Xu and Lu 2011). In the case of animals, most of the proposed models (Cena and Monteith 1975a, b, c; Cena and Clark 1978; McArthur 1987; Arkin et al. 1991; Ehrlemark and Sallvik 1996; Gebremedhin and Wu 2001, 2003; Silva 2000; Turnpenny et al. 2000; McGovern and Bruce 2000) were focused on ruminants and present several questions not solved yet. In fact, energy transfer through live tissues is a process more complex than those involving engineering materials and includes mechanisms as blood circulation, evaporation of sweat and other fluids, metabolic heat production, heat loss through the skin, hairs and plumage (Xu and Lu 2011). The first attempt to use the equation of heat transfer in biological tissues (human forearm) was that of Pennes (1948), considered as valid yet (Wissler 1998). That equation is now known as *bio-heat equation* and involves the production of metabolic heat and the blood transportation; despite its limitations, it is yet the most used model for biological tissues in general. Other models had been proposed for human tissues, notwithstanding (Xu and Lu 2011).

Let us have a homogeneous medium in which there has a macroscopic displacement (advection) and the distribution of temperatures is represented by Cartesian coordinates $T(x, y, z)$. By considering the principle of conservation of energy and an infinitely small volume differential (dx, dy, dz), the first law of thermodynamics can be stated for a given time interval, as it is illustrated in Fig. A.1.

Conduction heat flux occur in directions perpendicular to the surfaces q_x, q_y and q_z, respectively; in the opposite directions, they are $q_{x+\mathrm{d}x}$, $q_{y+\mathrm{d}y}$ and $q_{z+\mathrm{d}z}$, respectively. Mass transfer by convection is given by the terms i_x, i_y and i_z, respectively; in the opposite directions, it is given by $i_{x+\mathrm{d}x}$, $i_{y+\mathrm{d}y}$ and $i_{z+\mathrm{d}z}$, respectively. Thermal

R. Gomes da Silva and A.S. Campos Maia, *Principles of Animal Biometeorology*, Biometeorology 2, DOI 10.1007/978-94-007-5733-2,

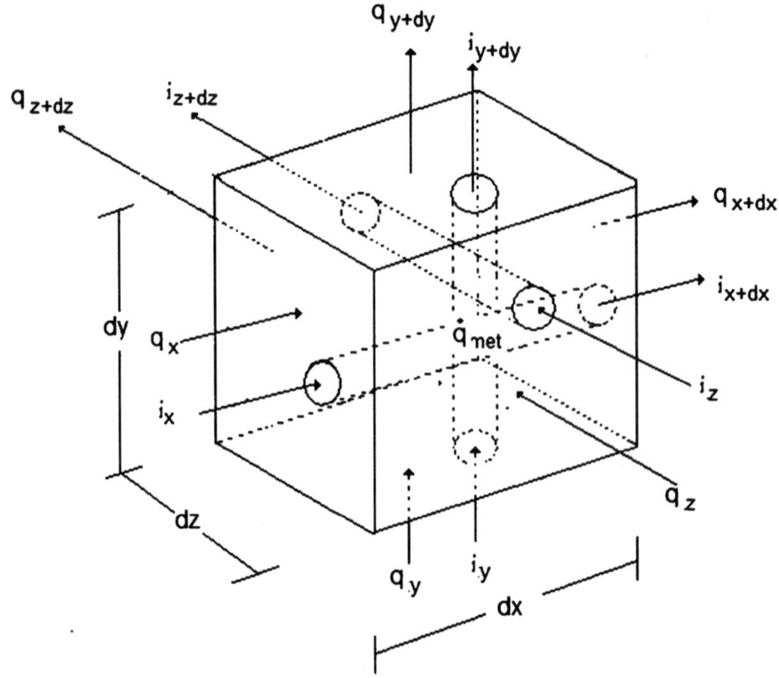

Fig. A.1 Heat transfer in an infinitely small volume characterised by the dimensions dx, dy and dz. There are considered both heat transfer and mass transfer. See text for explanation

energy is produced within the medium at a rate $\dot{q}_{\text{met}}$ per unit volume (W m^{-3}), while $\rho c_{\text{p}} \partial T / \partial t$ is the rate of temperature variation with time per unit volume.

Thermal energy balance in the volume shown in Fig. A.1 can be evaluated as

$$\dot{E}_{\text{E}} - \dot{E}_{\text{S}} + \dot{E}_{\text{G}} = \dot{E}_{\text{A}} \tag{A.1}$$

where

$\dot{E}_{\text{G}} = \dot{q}_{\text{met}} \text{d}x\text{d}y\text{d}z$ = rate of heat generation within the volume

$\dot{E}_{\text{A}} = \rho c_{\text{p}} \dfrac{\text{d}T}{\text{d}t} \text{d}x\text{d}y\text{d}z$ = rate of variation of the energy content

Considering that heat transfer is one-dimensional in the x direction, the conduction and mass transfer rates that constitute the energy input $(\dot{E}_{\text{E}})$ and output $(\dot{E}_{\text{S}})$ can be given by

$$\dot{E}_{\text{E}} = q_x + i_x$$
$$\dot{E}_{\text{S}} = q_{x+\text{d}x} + i_{x+\text{d}x}$$

Substituting into Eq. 8.11, we have

$$q_x + i_x - q_{x+\mathrm{d}x} - i_{x+\mathrm{d}x} + \dot{q}_{\mathrm{met}}\mathrm{d}x\mathrm{d}y\mathrm{d}z = \rho c_{\mathrm{p}} \frac{\mathrm{d}T}{\mathrm{d}t} \mathrm{d}x\mathrm{d}y\mathrm{d}z \tag{A.2}$$

Then,

$$\begin{aligned} q_{x+\mathrm{d}x} &= q_x + \frac{\partial q_x}{\partial x}\mathrm{d}x \\ i_{x+\mathrm{d}x} &= i_x + \frac{\partial i_x}{\partial x}\mathrm{d}x \end{aligned}$$

and substituting again in Eq. A.2,

$$q_x - q_x - \frac{\partial q_x}{\partial x}\mathrm{d}x + i_x - i_x - \frac{\partial i_x}{\partial x}\mathrm{d}x + \dot{q}_{\mathrm{met}}\mathrm{d}x\mathrm{d}y\mathrm{d}z = \rho c_{\mathrm{p}} \frac{\mathrm{d}T}{\mathrm{d}t} \mathrm{d}x\mathrm{d}y\mathrm{d}z \tag{A.3}$$

In a steady condition, Eq. A.3 reduces to

$$-\frac{\partial q_x}{\partial x}\mathrm{d}x - \frac{\partial i_x}{\partial x}\mathrm{d}x + \dot{q}_{\mathrm{met}}\mathrm{d}x\mathrm{d}y\mathrm{d}z = 0 \tag{A.4}$$

Heat flux by conduction can be determined by Fourier's law:

$$q_x = -k\mathrm{d}y\mathrm{d}z\frac{\partial T}{\partial x} \tag{A.5}$$

For situations in which the mass i_x is transferred to a point beyond the volume limits (cases involving liquid displacements), the input and output terms include also the energy (thermal or mechanical) which is transported by advection by the mass in transfer. If the fluid is an ideal gas having a constant c_{p} value, the flux of mass will involve heat transfer at a rate given by

$$i_x = \rho c_{\mathrm{p}} V \mathrm{d}y\mathrm{d}z (T_{\mathrm{in}} - T_{\mathrm{out}}) \tag{A.6}$$

Substituting Eqs. A.5 and A.6 into formula A.4 and dividing by the differential volume dxdydz, we obtain

$$k\left(\frac{\partial^2 T}{\partial x^2}\right) - \rho c_{\mathrm{p}} \frac{\partial}{\partial x}(\mathrm{V}(T_{\mathrm{in}} - T_{\mathrm{out}})) + \dot{q}_{\mathrm{met}} = 0 \tag{A.7}$$

If T_{sai} and k are constants and $(\partial V/\partial x) = 0,$ then

$$\frac{\mathrm{d}^2 T}{\mathrm{d}x^2} - \frac{\rho c_{\mathrm{p}}}{k} V \frac{\mathrm{d}T}{\mathrm{d}x} + \frac{\dot{q}_{\mathrm{met}}}{k} = 0 \tag{A.8}$$

Equation A.8 is a general one to describe heat transfer in one dimension within a system of Cartesian coordinates having constant properties. Notwithstanding this equation is widely applied to most physical systems, its use for live systems presents some problems. Among such problems, there are the variations in blood vase dimensions and in the flux of blood through them. Besides, some properties of live tissues (as thermal conductivity, density and specific heat) can vary significantly between short distances; therefore, we must rewrite Eq. A.8 as follows:

$$\frac{\mathrm{d}^2T}{\mathrm{d}x^2} + \frac{\dot{q}_{\mathrm{met}} + \dot{q}_{\mathrm{perf}}}{k} = 0 \tag{A.9}$$

where $\dot{q}_{\mathrm{perf}}$ is a heat source due to a perfusion process. Pennes proposed the term perfusion by supposing that blood flowing in the capillaries within any small tissue volume enters at arterial temperature (T_{art}, K) and comes out at local tissue temperature (T, K). The rate at which heat is gained by the tissue equals the rate of blood heat loss. If the perfusion rate (ω, m^3 s^{-1}) is the flow of a volume of blood through the tissue, heat loss from the blood can be evaluated by using the following equation:

$$\dot{q}_{\mathrm{pef}} = \omega\rho_{\mathrm{b}}c_{\mathrm{pb}}(T_{\mathrm{art}} - T) \tag{A.10}$$

where ρ_{b} and c_{pb} are the density and the specific heat of blood, respectively. By substituting Eq. A.10 into Eq. A.9, we have

$$\frac{\mathrm{d}^2T}{\mathrm{d}x^2} + \frac{\dot{q}_{\mathrm{met}} + \omega\rho_{\mathrm{b}}c_{\mathrm{pb}}(T_{\mathrm{art}} - T)}{k} = 0 \tag{A.11}$$

In order to simplify this equation, we can define the excess temperature θ as

$$\theta = T - T_{\mathrm{a}} - \frac{\dot{q}_{\mathrm{met}}}{\omega\rho_{\mathrm{b}}c_{\mathrm{pb}}} \tag{A.12}$$

where $\tilde{m}^2 = (\omega\rho_{\mathrm{b}}c_{\mathrm{pb}})/k$. If T_{art}, $\dot{q}_{\mathrm{met}}$ and ω are constant values, Eq. A.11 can be rewritten as

$$\frac{\mathrm{d}^2\theta}{\mathrm{d}x^2} - \tilde{m}^2\theta = 0 \tag{A.13}$$

Equation A.13 is a second-order ordinary differential equation, linear and homogeneous with constant coefficients and the following general solution:

$$\theta(x) = C_1\mathrm{e}^{\tilde{m}x} + C_2\mathrm{e}^{-\tilde{m}x} \tag{A.14}$$

In order to determine constants C_1 and C_2 in Eq. A.14 and then finding a solution, two appropriate boundary conditions must be established. For the present case and considering an excess temperature θ, the boundary conditions are as follows:

$$\theta(x=0)$$
$$\theta_c = T_c - T_{\text{art}} - \frac{\dot{q}_{\text{met}}}{\omega \rho_s c_s}$$
$$\theta(x=L)$$
$$\theta_L = T_i - T_{\text{art}} - \frac{\dot{q}_{\text{met}}}{\omega \rho_s c_s}$$

Substituting into Eq. A.14, we have respectively

$$\theta_c = C_1 + C_2$$
$$\theta_L = C_1 \mathrm{e}^{\tilde{m}L} + C_2 \mathrm{e}^{-\tilde{m}L}.$$

Thus, we can obtain C_1 and C_2 values by means of the following matrix system:

$$\begin{bmatrix} C_1 \\ C_2 \end{bmatrix} = \begin{bmatrix} 1 & 1 \\ \mathrm{e}^{\tilde{m}L} & \mathrm{e}^{-\tilde{m}L} \end{bmatrix}^{-1} \begin{bmatrix} \theta_c \\ \theta_L \end{bmatrix}$$
$$\begin{bmatrix} C_1 \\ C_2 \end{bmatrix} = \left(\mathrm{e}^{-\tilde{m}L} - \mathrm{e}^{\tilde{m}L}\right)^{-1} \begin{bmatrix} \mathrm{e}^{-\tilde{m}L} & -1 \\ -\mathrm{e}^{\tilde{m}L} & 1 \end{bmatrix} \begin{bmatrix} \theta_c \\ \theta_L \end{bmatrix}$$

whose solution gives

$$C_1 = \frac{\theta_c \mathrm{e}^{-\tilde{m}L} - \theta_L}{\mathrm{e}^{-\tilde{m}L} - \mathrm{e}^{\tilde{m}L}} \quad \text{and} \quad C_2 = \frac{-\theta_c \mathrm{e}^{\tilde{m}L} + \theta_L}{\mathrm{e}^{-\tilde{m}L} - \mathrm{e}^{\tilde{m}L}},$$

Those solutions are substituted into Eq. A.14 in order to obtain the temperature distribution within a live tissue as

$$\theta(x) = \frac{\left(\theta_c \mathrm{e}^{-\tilde{m}L} - \theta_L\right)\mathrm{e}^{\tilde{m}x}}{\mathrm{e}^{-\tilde{m}L} - \mathrm{e}^{\tilde{m}L}} + \frac{\left(-\theta_c \mathrm{e}^{\tilde{m}L} + \theta_L\right)\mathrm{e}^{-\tilde{m}x}}{\mathrm{e}^{-\tilde{m}L} - \mathrm{e}^{\tilde{m}L}}$$

After some manipulation, we have finally

$$\theta(x) = \frac{-\theta_c\left(\mathrm{e}^{\tilde{m}(L-x)} - \mathrm{e}^{-\tilde{m}(L-x)}\right) - \theta_L\left(\mathrm{e}^{\tilde{m}x} - \mathrm{e}^{-\tilde{m}x}\right)}{-\left(\mathrm{e}^{\tilde{m}L} - \mathrm{e}^{-\tilde{m}L}\right)} \tag{A.15}$$

As senh $x = 0.5(\mathrm{e}^x - \mathrm{e}^{-x})$, Eq. A.15 can be modified to

$$\theta(x) = \theta_c \frac{(\theta_L/\theta_c)\,\mathrm{senh}\,\tilde{m}x + \mathrm{senh}\,\tilde{m}(L-x)}{\mathrm{senh}\,\tilde{m}L} \tag{A.16}$$

The rate of heat loss from the tissue to the skin or the subcutaneous fat layer can be evaluated by applying Fourier's law under $x = L$; then,

$$q_T\Big|_{x=L} = -k_T A \frac{\mathrm{d}T}{\mathrm{d}x}\Big|_{x=L} = -k_T A \frac{\mathrm{d}\theta}{\mathrm{d}x}\Big|_{x=L}$$

$$q_T\Big|_{x=L} = -k_T A \theta_c \frac{\mathrm{d}}{\mathrm{d}x}\left[\frac{(\theta_L/\theta_c)\mathrm{senh}\ \tilde{m}x + \mathrm{senh}\ \tilde{m}(L-x)}{\mathrm{senh}\ \tilde{m}L}\right]$$

The respective derivative with respect to x leaves to

$$q_T\Big|_{x=L} = -k_T A \tilde{m} \theta_c \left[\frac{(\theta_L/\theta_c)\cosh\ \tilde{m}x + \cosh \tilde{m}(L-x)}{\mathrm{senh}\ \tilde{m}L}\right]$$

By considering that $x = L$, the rate of heat loss will be

$$q_T = -k_T A \tilde{m} \theta_c \left[\frac{(\theta_L/\theta_c)\cosh\ \tilde{m}L - 1}{\mathrm{senh}\ \tilde{m}L}\right] \ \mathrm{W} \tag{A.17}$$

while the heat flux from the tissue will be

$$q''_{\mathrm{T}} = -k_T \tilde{m} \theta_c \left[\frac{(\theta_L/\theta_c)\cosh\ \tilde{m}L - 1}{\mathrm{senh}\ \tilde{m}L}\right] \ \mathrm{W\ m^{-2}} \tag{A.18}$$

Example. Consider a 15-day-old piglet with a hairless body surface, standing in a place at 21° south latitude and 595 m high where it is exposed to a 0.5 m s^{-1} wind blowing perpendicularly at the body axis. Its body can be taken as a horizontal cylinder with 0.3 m length and 0.15 m diameter; it is assumed as a mass of muscle covered by a 30-mm-thick fat/skin layer. At a point $L_{\mathrm{m}} = 30$ mm within the muscle, the temperature of that mass is equal to the body temperature, $T_c = 38°\mathrm{C}$. Thermal conductivity of the muscle is $k_{\mathrm{m}} = 0.5$ W m^{-1} K^{-1}. The metabolic heat production within the muscle is $\dot{q} = 700$ W m^{-3}; perfusion rate is $\omega = 0.0005$ s^{-1}; blood density and specific heat are $\rho_{\mathrm{b}} = 1{,}000$ kg m^{-3} and $c_{\mathrm{pb}} = 3{,}600$ J kg^{-1} K^{-1}, respectively; and arterial blood temperature, T_{art}, is equal to the body temperature. Other data are given in the following figure. It is assumed that the heat generated by metabolism and in function of the perfusion into the fat/skin layer can be disregarded. It is desired to evaluate (a) the temperature distribution through those layers and (b) the rate of heat loss through the body surface of the animal.

Assumptions:

1. Steady state conditions.
2. One-dimensional heat transfer through muscle and fat/skin layers.
3. Uniform rates of metabolic heat generation, perfusion rate, arterial temperature, blood properties and thermal conductivities.
4. Solar irradiance is negligible.

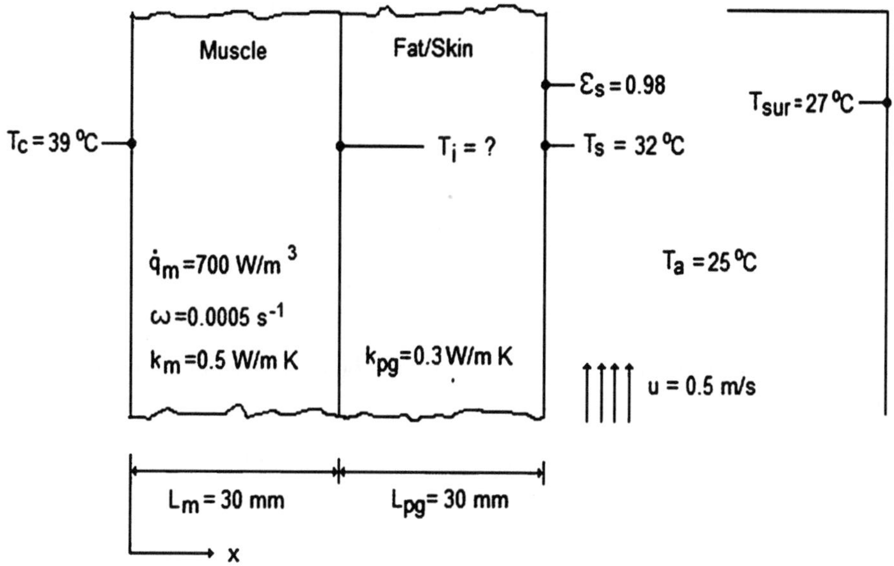

Fig. A.2 Schematic model of the heat transfer through tissues

Fig. A.3 Tissue thermal resistances

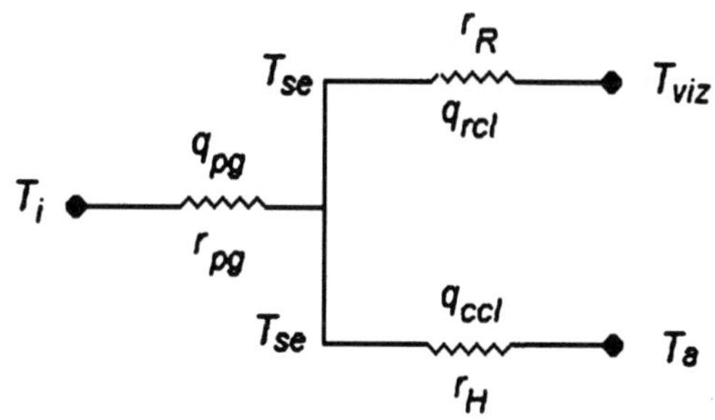

Solution. Sensible heat transfer between skin surface and atmosphere depends on the atmospheric temperature T_a and the temperature of the surroundings (T_{sur}), as it is shown in Figs. A.2 and A.3. Temperature T_{sur} can be assumed to be equal to the mean radiant temperature (T_{mr}) which was defined in Chap. 1 (Eq. 1.34). Those temperatures can be combined into only one, the operative temperature T_o (Eq. 7.6). Now, considering the series resistances, we have

$$q_t = \frac{1}{r_{pgo}}(T_i - T_o)$$

where r_{pgo} (W^{-1} K) is the resistance to conduction through the fat/skin layer, which is in series to the parallel resistances for convection and radiation transfer in the boundary layer. By definition,

$$r_{pgo} = r_{pg} + r_o$$

where

$$r_{pg} = \frac{L_{pg}}{k_{pg}A} = \frac{0.03}{(0.3)(0.176714)} = 0.5659$$

In order to calculate r_o, it is necessary to obtain the values of r_R and r_H as follows:

$$r_R = \left[A\varepsilon_s\sigma(T_{se} + T_{rm})\left(T_{se}^2 + T_{rm}^2\right)\right]^{-1}$$
$$r_H = \frac{d_a}{A\ kN_u}$$

For these calculations, the thermal properties of the atmosphere, the black globe convective coefficient and the nondimensional numbers N_u, R_e, G_r and P_r (Eqs. 2.35, 2.34 and 2.42) are needed, considering that the characteristic dimension of the body is $d = 0.15$ m because wind blows its axis perpendicularly. For the convection coefficient of the black globe (h_g), the formulas proposed by Silva (2002) were used. Finally, the mean radiant temperature was obtained as

$$T_{rm} = \left[\varepsilon_g^{-1}h_g(T_g - T_a) + \sigma T_g^4\right]^{0.25} = 303.18 \text{ K} = 30.03\ °\text{C}$$

and the resistance to radiant heat exchange will be

$$r_R = \left[(0.17671)(0.98)\left(5.67 \times 10^{-8}\right)(305.23 + 303.18)\left(305.23^2 + 303.18^2\right)\right]^{-1}$$
$$= 0.91308 \ \ W^{-1} \text{ K}$$

As for the nondimensional numbers, N_u was calculated by using Eqs. 2.39 (free convection) and 2.44 (forced convection). Once determined that convection is a mixed one, the final value for the Nusselt number was given as

$$N_u = \left[\left(N_u^N\right)^{3.5} + \left(N_u^F\right)^{3.5}\right]^{\frac{1}{3.5}}$$
$$= \left[(18.0483)^{3.5} + (35.5606)^{3.5}\right]^{\frac{1}{3.5}} = 36.5606$$

where N_u^N stands for the free convection and N_u^F for the forced one. On the other hand, the following thermal properties of the atmosphere were calculated for the temperature $T_m = 0.5(T_a + T_{se}) = 28.5°C$.

Therefore, the resistance to heat transfer by convection was calculated as follows:

$$r_H = \frac{d_a}{A\ kN_u} = \frac{0.15}{(0.1767)(0.024909)(36.5606)} = 0.93967\ \mathrm{W\ K^{-1}}$$

The operative temperature was

$$T_o = \frac{r_H T_{rm} + r_R T_a}{r_R + r_H} = 300.7\ \mathrm{K} = 27.6\ °\mathrm{C}$$

where

$$r_o = \frac{r_R r_H}{r_R + r_H} = 0.4596\ \mathrm{W^{-1}\ K}$$

Heat transfer in the muscle layer is governed by Eq. A.13. Boundary conditions are specified in terms of temperatures T_c and T_b, but T_i is yet unknown. In terms of a temperature excess θ, the rate of heat loss from the muscle can be assumed as equal to the rate of heat passing through the fat/skin layer to the environment. Therefore, by combining Eqs. A.17 and A.2, we have

$$-k_m A\,\tilde{m}\theta_c\left[\frac{(\theta_L/\theta_c)\cosh\tilde{m}L_m - 1}{\operatorname{senh}\tilde{m}L_m}\right] = \frac{T_i - T_o}{r_{pgo}}$$

Now we can define

$$T_i = \frac{T_o \operatorname{senh}\tilde{m}L_m + k_m A\,\tilde{m}r_{pgo}\left[\theta_c + \left(T_{art} + \frac{\dot{q}_m}{\omega\rho_s c_{ps}}\right)\cosh\tilde{m}L_m\right]}{\operatorname{senh}\tilde{m}L_m + k_m A\,\tilde{m}r_{pgo}\cosh\tilde{m}L_m}$$

where

$$\tilde{m} = \sqrt{(\omega\rho_s c_{ps})/k_m} = \sqrt{(0.0005)(1{,}000)(3{,}600)/0.5} = 60\ \mathrm{m^{-1}}$$

$$\operatorname{senh}\tilde{m}L_m = \operatorname{senh}[(60)(0.03)] = 2.9421$$

$$\cosh\tilde{m}L_m = \cosh[(60)(0.03)] = 3.1075$$

$$\theta_c = T_c - T_{art} - \frac{\dot{q}_{met}}{\omega \rho_s c_{ps}} = 39 - 39 - \frac{700}{[(0.0005)(1000)(3600)]}$$
$$= -0.3889 \text{ K}$$

$$T_i = \frac{(27.5539)(2.9421) + (0.5)(0.1767)(60)(1.02557)[-0.3889 + (39 + 0.3889)(3.10747)]}{2.9421 + (0.5)(0.1767)(60)(1.02557)(3.1075)(3.1075)}$$

$$T_i = 37.53°\text{C}$$

Finally, considering that

$$r_{pgo} = r_{pg} + r_o = 1.0255 \text{ W}^{-1} \text{ K}$$

the rate of heat loss was

$$q_t = \frac{T_i - T_o}{r_{pgo}} = \frac{37.5269 - 27.5539}{1.0255} = 9.7252 \text{ W}$$

and the heat flux was

$$q_t'' = \frac{q_t}{A} = \frac{9.7252}{0.1767} = 55.0336 \text{ W m}^{-2}$$

The distribution of temperatures in the muscle layer can be estimated by means of the following equation:

$$T_{(x_m = L_m)} = T_{art} + \frac{\dot{q}_{met}}{\omega \rho_s c_{ps}} + \theta_c \frac{(\theta_L/\theta_c)\text{senh}\, \tilde{m}L_m + \text{senh}\, \tilde{m}(L_m - x)}{\text{senh}\, \tilde{m}L_m}$$

while for the fat/skin layer the equation is

$$T_{(x_{pg} = L_{pg})} = x\frac{(T_{se} - T_i)}{L_{pg}} + T_i$$

In the case the surface temperature of the fat/skin layer is unknown, it would be estimated by making

$$-k_m A\, \tilde{m}\theta_c \left[\frac{(\theta_L/\theta_c)\cosh \tilde{m}L_m - 1}{\text{senh}\, \tilde{m}L_m}\right] = \frac{T_i - T_{se}}{r_{pg}}$$

and then solving for T_{se}

$$T_{se} = \frac{T_i\left(\text{senh}\, \tilde{m}L_m + k_m A\, \tilde{m} r_{pg} \cosh \tilde{m}L_m\right) - k_m A\, \tilde{m} r_{pg}\left[\theta_c + \left(T_a + \frac{\dot{q}_m}{\omega \rho_s c_{ps}}\right)\cosh \tilde{m}L_m\right]}{\text{senh}\, \tilde{m}L_m}$$

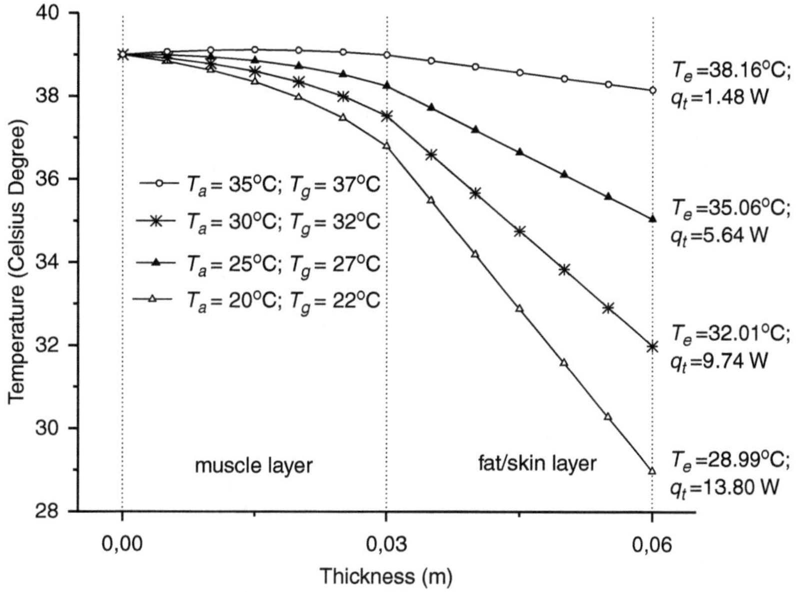

Fig. A.4 Distribution of temperatures in muscle and fat/skin layers according to the variation of air and black globe temperatures

An iterative process can be used to estimate T_{se}, as follows:

1. *Give an initial value for* T_{se}. *For example, if the air temperature is 30°C and the globe temperature 32°C, an initial* T_{se} *value of 34°C can be used.*
2. *Solve all equations by using this initial value, obtaining* $q_t = 5.6206$ *W and then a new value* $T_{se} = 35.08$*°C.*
3. *The new value is compared to the initial one. If there has a difference, the calculations are carried out again with the new* T_{se} *and so on, until the results converge. In the present example, the convergence was attained after four iterations, giving the final values* $T_{se} = 35.078$ *and* $q_t = 5.6213$ *W.*

The above described procedures can be used to perform several simulations. It is possible, for example, to study the distribution of temperature in tissues and the subsequent heat loss in animals under different conditions. Figure A.4 shows the dependence of temperature distribution, T_{se} and q_t in relation to variations in T_o and T_g.

The results presented by the model illustrated in Fig. A.4 show that, under air temperature of 20°C and globe temperature of 22°C, the estimated surface temperature was 28.99°C and the rate of sensible heat transfer was 13.80 W. When the air temperature changed to 35°C and the globe temperature to 37°C, the estimated surface temperature was 38.16°C, and the rate of sensible heat transfer reduced to 1.48 W.

References

Albuquerque Neto, C.: Modelo integrado dos sistemas térmico e respiratório do corpo humano, 186 p. Tese (Doutorado em Engenharia Mecânica) – Escola Politécnica da Universidade de São Paulo, São Paulo (2010)

Arkin, H., Kimmel, E., Berman, A., Broday, D.: Heat transfer properties of dry and wet furs of dairy cows. Trans. ASAE **34**(6), 2550–2558 (1991)

Cena, K., Clark, J.A.: Thermal insulation of animal coats and human clothing. Phys. Med. Biol. **23**(4), 565–591 (1978)

Cena, K., Monteith, J.L.: Transfer processes in animal coats. I. Radiative transfer. Proc. R. Soc. Lond. B **188**, 377–393 (1975a)

Cena, K., Monteith, J.L.: Transfer processes in animal coats. II. Conduction and convection. Proc. R. Soc. Lond. B **188**, 395–411 (1975b)

Cena, K., Monteith, J.L.: Transfer processes in animal coats. III. Water vapour diffusion. Proc. R. Soc. Lond. B **188**, 413–423 (1975c)

Ehrlemark, A.G., Sallvik, K.G.: A model of heat and moisture dissipation from cattle based on thermal properties. Trans. ASAE **39**(1), 187–194 (1996)

Ferreira, M.S., Yanagihara, J.Y.: A transient three-dimensional heat transfer modelo f the human body. Int. Commun. Heat Mass Transf. **36**(7), 718–724 (2009)

Gebremedhin, K.G., Wu, B.: A model of evaporation cooling of wet skin surface and fur layer. J. Therm. Biol. **26**, 537–545 (2001)

Gebremedhin, K.G., Wu, B.W.: Characterization of flow field in a ventilated space and simulation of heat exchange between cows and their environment. J. Therm. Biol. **28**, 301–319 (2003)

McArthur, A.J.: Thermal interaction between animal and microclimate: a comprehensive model. J. Theor. Biol. **126**, 203–238 (1987)

McGovern, R.E., Bruce, J.M.: A model of the thermal balance for cattle in hot conditions. J. Agric. Eng. Res. **77**(1), 81–92 (2000)

Pennes, H.H.: Analysis of tissue and arterial blood temperature in the resting human forearm. J. Appl. Physiol. **1**, 93–122 (1948)

Silva, R.G.: Um modelo para a determinação do equilíbrio térmico de bovinos em ambiente tropical. Braz. J. Anim. Sci. **29**, 1244–1252 (2000)

Silva, R.G.: Carga térmica radiante: revisando a técnica de avaliação (Radiant heat load: revisiting its evaluation method). Revista Engenharia Rural (Piracicaba) **13**, 77–84 (2002)

Turnpenny, J.R., McArthur, A.J., Clark, J.A., Wathes, C.M.: Thermal balance of livestock. 1. A parsimonious model. Agric. For. Meteorol. **101**, 15–27 (2000)

Wissler, E.H.: Pennes'1948 paper revisited. J. Appl. Physiol. **85**(1), 35–41 (1998)

Xu, F., Lu, T.: Introduction to skin biothermomechanics and thermal pain, 414 p. Springer, Beijing (2011)

Index

R. Gomes da Silva and A.S. Campos Maia, *Principles of Animal Biometeorology*,
Biometeorology 2, DOI 10.1007/978-94-007-5733-2,

V

W

MIX
Papier aus verantwortungsvollen Quellen
Paper from responsible sources
FSC® C105338

If you have any concerns about our products,
you can contact us on
ProductSafety@springernature.com

In case Publisher is established outside the EU,
the EU authorized representative is:
Springer Nature Customer Service Center GmbH
Europaplatz 3, 69115 Heidelberg, Germany

Printed by Libri Plureos GmbH
in Hamburg, Germany